NORMAL TABLE OF XENOPUS LAEVIS (DAUDIN)

NORMAL TABLE OF
XENOPUS LAEVIS (DAUDIN)

A SYSTEMATICAL AND CHRONOLOGICAL SURVEY
OF THE DEVELOPMENT FROM THE FERTILIZED EGG
TILL THE END OF METAMORPHOSIS

Edited by
P.D. Nieuwkoop and J. Faber

WITH A NEW FOREWORD BY
JOHN GERHART *and* MARC KIRSCHNER

Routledge
Taylor & Francis Group

NEW YORK AND LONDON

Published by Routledge

270 Madison Ave, New York NY 10016
2 Park Square, Milton Park, Abingdon, Oxon, OX14 4RN

Transferred to Digital Printing 2010

Copyright © 1994 Pieter D. Nieuwkoop and J. Faber
All rights reserved

Library of Congress-in-Publication Data

Normal table of Xenopus laevis (Daudin) : a systematical and chronological
    survey of the development from the fertilized egg till the end of metamor-
    phosis / edited by P. D. Nieuwkoop and J. Faber : with a foreword by John
    Gerhart and Marc Kirschner.
        p. cm.
    Originally published: 2. ed. Amsterdam : North-Holland Pub. Co.,
1967.
    "Issued by the Hubrecht Laboratory, Utrecht."
    Includes bibliographical references (p. 193) and Index.
    ISBN 0–8153–1896–0
    1. Xenopus laevis — Development. I. Nieuwkoop, Pieter D. II. Faber,
J., 1926– , III. Hubrecht-Laboratorium (Embryologisch Institut)
QL668. E265N67   1994
597.8'4—dc20   94-11518

Cover and front matter design by Marc Shifflett

Publisher's Note
The publisher has gone to great lengths to ensure the quality of this reprint
but points out that some imperfections in the original may be apparent.

FOREWORD

It has been almost forty years since the first edition of Nieuwkoop and Faber's Normal Table of Xenopus. This book was the culmination of an international collaboration of zoologists to describe the embryology of Xenopus, including the origins of the organ systems, with the purpose of promoting the use of Xenopus as a model amphibian for embryological study. The few books that are reprinted from such an ancient vintage are generally historical curiosities, of interest to collectors and historians of science. Yet this book is now reprinted for the same purpose as was there forty years ago, to stimulate, and make comparable research on a particular experimental system for the understanding of fundamental principles of embryonic development. When this book was written, the period of classical embryology was almost at an end. During the better part of these past forty years, many of us who devoutly used this book, wandered in the desert looking for new approaches to development, guided by a Normal Table that always kept the deep problems of embryonic organization and developmental anatomy in front of us. Today the climate is very different. The problems of organogenesis and early development are enticing many young scientists to enter the field armed with new molecular approaches. They as much as anyone are in need of an accurate readable map of the temporal and spatial changes in the anatomy of the developing organism. They will find in this book a meticulous description of Xenopus development that can serve as a guide for the study of most of the important problems in vertebrate embryology. Throughout this book the editors and authors have shown a high concern for the usefulness of the material to readers engaged in experimentation. Pieter Nieuwkoop and Job Faber and their contributors should certainly feel satisfied that the first edition of the book effectively contributed to the founding of an international community of Xenopus embryologists, so much so that their students now demand the same access to this extraordinary document.

John Gerhart
Marc Kirschner
April 14, 1994

NORMAL TABLE OF XENOPUS LAEVIS

CONTENTS

CHAPTER I

GENERAL INTRODUCTION

The project of the elaboration of a Normal Table of the development of the South African toad, *Xenopus laevis*, has been set up by the staff of the Hubrecht Laboratory for two main reasons.

Xenopus laevis has become a common laboratory animal since the detection of its usefulness for pregnancy tests and other hormonal reactions. It has turned out to be a very good laboratory animal, which, being an aquatic form, can easily be kept and reproduces readily in captivity after hormonal stimulation. The eggs develop quite satisfactorily and the larvae can be reared very smoothly up to metamorphosis, which period is also passed without great difficulties. At present *Xenopus laevis* is therefore found in many biological, medical and veterinary institutes throughout the world. Although some approaches have already been made towards dividing its development into convenient stages (cf. PETER, 1931, WEISZ, 1945), a great want was generally felt for an extensive and well elaborated Normal Table of this species.

Although only very fragmentarily known, the development of *Xenopus laevis* is quite interesting from a descriptive and comparative as well as from an experimental embryological point of view. Preliminary experiments demonstrated that the egg of *Xenopus laevis*, notwithstanding its very rapid development, is quite suitable for experimental work. Since the number of Anuran species for which this is true, is very restricted, a good basic knowledge of the normal development of this species seems to open new perspectives for experimental analysis in this group of Vertebrates. The rather aberrant development of this systematically somewhat isolated species suggests, moreover, interesting possibilities for descriptive and comparative embryological studies.

The second reason for setting up the project falls in line with the international character of the Hubrecht Laboratory. This institution tries to promote embryological research and tries to stimulate international cooperation. Since the study of the normal development of a species can adequately be subdivided into a large number of well circumscribed separate subjects, a Normal Table can be elaborated from data provided by an international team, stimulating international cooperation in a very practical form.

For these two reasons in 1949 the first editor took the initiative for the organization of this project.

The division of the development of *Xenopus* into stages for the purposes of the table has been carefully considered. Up till now stages in the various Normal Tables have not been directly comparable with each other, since the tables were made independently. This very unsatisfactory situation has led already to an enormous confusion in the literature, the more so since the Normal Tables cited are usually not available and partially even entirely unknown to the majority of the readers. We have therefore decided that the most adequate Normal Table should be used as a basis of comparison. As such we have chosen the Normal Table of *Amblystoma punctatum* established by R. G. HARRISON. Unfortunately this Normal Table has never been published; only the external appearance of the successive stages has been reproduced (cf. HAMBURGER, 1947; RUGH, 1948). In the experimental literature it has, however, been referred to very extensively. The stages of the present Normal Table have been made as nearly as possible analogous with those of HARRISON's table. (See also Chapter III.)

In order to resolve the present confusion of stages in the various tables of Anuran development, a comparison between them has been compiled. This is presented in Chapter VIII on page 189.

A further difficult decision had to be taken with regard to the form in which this Normal Table should appear. The individual contributions varied so much in size and form—some were much more extensive and detailed than others—that most of the contributions needed considerable adaptation for the sake of uniformity. In several cases the text had to be shortened greatly. The editors tried nevertheless to preserve as far as possible the personal character of the individual contributions. The study of this material by a team of about twenty scientists has led evidently to a more detailed description of the development of the various organ systems than that given in other Normal Tables. This fact, together with serious typographical objections, made publication in the usual tabular form almost impossible. The editors therefore decided to give up the tabular form. The alternative was the writing of a systematically arranged continuous text. (See further in Chapter VI.) The first editor has therefore changed the text of the various contributions into the present form, for which he is entirely responsible. The individual collaborators remain, however, responsible for the data of their personal contributions.

In order to make the Normal Table more adequate for stage determinations of the entire embryo as well as of the individual organ systems, a separate chapter containing well defined stage criteria has been added. This chapter contains the external and internal stage criteria, while also the corresponding age and size measurements are given.

It seemed desirable that a short chapter on the systematic position and geographical distribution of *Xenopus laevis* should be included, and that also some data should be given about the natural ecological conditions of this species and about methods of rearing it under laboratory conditions. (See Chapter iv and v on page 9 and page 13).

We hope that the new form in which this Normal Table appears, will meet most of the requirements to be fulfilled for a comprehensive but nevertheless easily accessible Normal Table. Suggestions for improvements in form and organization will be highly appreciated in order that the most adequate form of publication may gradually be achieved, and a standard might be set up for the publication of other Normal Tables.

Any national, but particularly any international project set up with a large number of collaborators, faces the difficulty of special time requirements, which difficulty in this project was greatly increased by the technical troubles met with in working out the older stages, the animals being of considerable dimensions. The personal responsibilities of the individual collaborators often interfered with an investigation of the material within the time-limits proposed, so that dead-lines had to be changed several times. In some cases more serious difficulties arose, so that new collaborators had to be found, a problem to be coped with up till the last moment. Notwithstanding these unavoidable complications the work has gradually become completed. It is a very great pleasure for the editors to present this work to their colleagues, and to express here their thanks and appreciation to all the collaborators for their personal interest and the accuracy with which they have carried out their part of the work.

This project has been made possible by generous financial support from Dutch and from International Organizations. For the collecting of the material in South Africa financial support has been received from the following organizations: "Jan Dekker Stichting"; "Utrechts Universiteits Fonds"; "Hollandsche Maatschappij van Wetenschappen"; "Provinciaal Utrechts Genootschap"; "Nederlands Natuur- en Geneeskundig Congres"; "Nederlandse Organisatie voor Zuiver Wetenschappelijk Onderzoek" and "Biologische Raad van Nederland". The publication of the work in book-form has been made possible by the generous support received from the "International Council of Scientific Unions". The editors want to express their most sincere thanks to all these organizations for the interest and support received.

They are highly indebted also to the Professors C. G. S. de Villiers and C. A. du Toit of the Zoological Institute, University of Stellenbosch, for the hospitality and help given in their institute to the second editor during his stay in South Africa, and to Dr. D. Hey of the Jonkershoek Inland Fisheries Department in Stellenbosch for his permission and help during the collecting of the material from the ponds of the Jonkershoek Fish Hatchery.

The plates, which are based on pencil drawings by the second editor, were made by the artist Mr. J. J. PRIJS of Utrecht, whom we should like to thank very cordially for his valuable work.

We should like to express our sincere gratitude for the personal cooperation received from the publishers, which have contributed very much to the ultimate form of this work.

Our sincere thanks are also due to Miss D. THOMASON, Ph. D., of the Royal Free Hospital School of Medicine, University of London, who has made some final corrections of the English text.

* *
*

CHAPTER II

THE ORGANIZATION OF THE NORMAL TABLE PROJECT

As mentioned in the general introduction, this Normal Table project was completed from data provided by an international team. The opportunity of practising international team work was met with great interest and appreciation, and led to the collaboration of about twenty scientists from three continents. Among them South Africa and the Netherlands are rather well represented. The South African zoologists were very much interested in this project, which concerned one of their native amphibien species. The number of Dutch collaborators increased particularly during the last one or two years when time became gradually too restricted to interest a foreign scientist in a subject from which the original collaborator had to withdraw. The development of the ear, by Miss M. TH. C. VAN EGMOND, the development of the skin, by G. A. VAN ERKEL and the development of the skull and visceral muscles up till metamorphosis, by P. H. VAN DOESBURG, have thus recently been studied at the Hubrecht Laboratory, while both editors filled in some gaps which became apparent when the Normal Table was being written.

The following collaborators have taken part in the elaboration of the Normal Table:

The division of the material into stages was first planned by the editors in collaboration with CHR. P. RAVEN (Utrecht, Holland). The definitive division into stages and the fixation of the material was carried out by J. FABER during his stay in South Africa.

The external development, including the pigmentation pattern and the external appearance of the lateral line system, was studied on living material in South Africa by J. FABER.

The systematic position and geographical distribution of *Xenopus laevis* have been discussed by H. W. PARKER, (London, England).

The ecological data and rearing methods have been given by J. FABER.

The development of the separate organ systems has been studied by:

The early development from the fertilized egg up to a late neurula stage (stage 20) by P. D. NIEUWKOOP (Utrecht, Holland).

The early development from a late neurula stage up to an advanced tail bud stage (stage 29/30) by J. PASTEELS (Brussels, Belgium).

The development of the skin and lateral line system by G. A. van Erkel (Utrecht, Holland).

The development of the brain by J. Ariëns Kappers (Groningen Holland).

The development of the cephalic nerves and ganglia by Chr. P. Raven (Utrecht, Holland).

The development of the spinal cord, ganglia and nerves [1]) by J. J. Kollros (Iowa City, U.S.A.).

The development of the olfactory organ by S. Toivonen and L. Saxén (Helsinki, Finland).

The development of the eye by Miss J. H. Bijtel (Groningen, Holland).

The development of the ear vesicle by Miss M. Th. C. van Egmond (Utrecht, Holland).

The development of skeleton and muscles of the head up to stage 54 by P. H. van Doesburg (Utrecht, Holland).

Idem, from stage 55 till the end of metamorphosis by S. N. Sedra and M. I. Michael (Alexandria, Egypt).

The development of the muscles of the trunk by P. A. J. Ryke (Potchefstroom, South Africa).

The development of the skeleton of the trunk by A. L. Smit (Stellenbosch, South Africa).

The development of skeleton and muscles of the tail by D. R. Newth (London, England).

The development of skeleton and muscles of shoulder girdle and forelimbs by R. van Pletzen (Bloemfontein, South Africa).

The development of skeleton and muscles of the pelvic girdle and hindlimbs by D. R. Newth (London, England).

The development of heart and vascular system up to stage 48 by Mrs. N. A. H. Millard (Cape Town, South Africa).

Idem, from stage 48 till the end of metamorphosis by J. Faber (Utrecht, Holland).

The development of the nephric system by R. Cambar and R. Williaume (Bordeaux, France).

The development of the gonads and adrenals by E. Vannini (Modena, Italy).

The development of the oropharyngeal cavity, including thyroid gland, thymus and lungs by S. Toivonen and L. Saxén (Helsinki, Finland).

The development of the intestinal tract and glands up to stage 57 by B. I. Balinsky (Johannesburg, South Africa).

Idem, from stage 57 till the end of metamorphosis by J. Faber (Utrecht, Holland).

---

[1]) Some additional data on the development of the spinal nerves, the meninges and the sympathetic nervous system have been given by P. D. Nieuwkoop (Utrecht, Holland).

CHAPTER III

MATERIAL

After thorough consideration it was decided that the material should be collected in South Africa, in order to obtain embryos which had developed under the *most natural* conditions. Extensive material has therefore been collected by the second editor during the South African spring and summer season 1949/50 at Stellenbosch, C. P., Union of South Africa.

The material originates almost entirely from one large earth pond, which had been fertilized with manure, compost and superphosphate of lime. During the breeding season the water was green with algae, later copepods became more and more abundant, and when the larvae approached metamorphosis the plankton consisted mainly of blue thread-algae. Eggs were obtained by placing large bundles of long grass or reeds in the water along the perimeter of the pond. Larvae were collected by means of nets of different mesh. The last stages of metamorphosis were obtained by letting early metamorphic stages metamorphose in the laboratory, since animals in late stages of metamorphosis usually hide in the mud and are very hard to obtain. Since the animals do not feed during metamorphosis the period in the laboratory will hardly have interfered with normal development.

This material might differ in more than one respect from material reared under laboratory conditions. Since these may vary from place to place, such laboratory material was considered unsuitable for this study. A comparison with the natural material used for this Normal Table might moreover lead to a search for the most satisfactory laboratory conditions.

Absolute measurements are seldom given in this Normal Table since the material reared under laboratory conditions will usually differ too much in size from the material collected for this Normal Table.

The material collected has been grouped into ten complete series each containing about five specimens per stage. Each series was put at the disposal of two to three collaborators. The material was usually sectioned in transverse, frontal and sagittal directions, so that for each direction only one or two specimens were available. The variation in the material of one and the same stage could therefore not be studied. It has however become evident from several contributions compared with the series in Utrecht that a general variation of approximately half a stage to either side has to be taken into account.

The material was fixed in LENHOSSEK fixative. The collaborators have in general used their own staining techniques.

As already mentioned in the general introduction the division into stages of this Normal Table has been done according to HARRISON's Normal Table of *Amblystoma punctatum*. It is however quite obvious that the development of *Xenopus laevis* differs in many respects from that of *Amblystoma punctatum*. The differences become more and more pronounced as development proceeds, so that from a certain stage on a direct comparison is no longer justified. Only up to stage 19 the development is therefore exactly comparable with that of *Amblystoma punctatum*. During subsequent development up to stage 38 only approximately comparable stages could be distinguished, while the later stages up to the end of metamorphosis had to be made independently of HARRISON's table, which moreover ends with stage 46. For the developmental period during which the hindlimbs develop (stages 50 to 56) the subdivision proposed by BRETSCHER (1949) has been followed with slight modifications.

The stages were based on external morphological criteria only, age and size being too variable for such a purpose. This led to the difficulty that over certain periods of development suitable external criteria were lacking. In the period of development between stage 28 and stage 40 this problem has been solved by reducing the number of stages. Here a number of combined stages have been established. Between stage 46 and 47 the absence of suitable external criteria led to a small gap in the description of the internal development of some organ systems which at that period are developing very rapidly, as e.g. the cranium, ear vesicle, olfactory organ and intestinal tract. Finally a number of intermediate stages have been distinguished during the period of gastrulation in order to make the Normal Table more suitable for experimental work.

* * *

CHAPTER IV

THE TAXONOMIC POSITION AND GEOGRAPHICAL DISTRIBUTION OF THE GENUS XENOPUS

Clawed Toads did not come to the notice of European zoologists until the beginning of the nineteenth century when DAUDIN (1802, p. 85, Pl. 30, fig. 1) briefly described and figured one under the name *Bufo laevis*, remarking that "by its form it has some affinities with the *Pipa*". MERREM (1820, p. 180) accepted this close relationship with the Surinam Toad but dissociated the two from the true toads (*Bufo*) under the generic name *Pipa*. GRAY (1825, p. 214) isolated this genus from the remaining frogs and toads, proposing for it a monogeneric higher group, the Piprina, but WAGLER (1827, p. 726) dissociated the African from the South American species, introducing the name *Xenopus* for the former. Despite the discovery of more species of both genera and of additional, related, genera in both Africa and South America, the association of the two as a super-generic group has persisted ever since almost without dissent. TSCHUDI (1838, p.p. 26, 89) associated the two as the PIPAE. DUMÉRIL & BIBRON (1841, p.p. 53, 761) placed them together in the family "Pipaeformes" and at the same time isolated them as the "Phrynaglosses" distinct from all the other frogs and toads, the "Phaneroglosses". GÜNTHER (1858, p. VI) separated the Ethiopian and Neotropical genera as two families within the "Aglossa" and a similar arrangement was adopted by many subsequent authors, for instance DUMÉRIL (1859, p. 231), STEINDACHNER (1867), BOULENGER (1882). Significant proposals during this period were made by COPE (1865, p. 98) who added the extinct family Palaeobatrachidae to the Aglossa, thereby indicating a possible ancestry for *Xenopus* and *Pipa* and by MIVART (1869) who doubted the monophyly of the Aglossa, stating that the lingual and eustachian characters of the group were a secondary condition. LATASTE (1879) and BLANCHARD (1885), however, gave greater weight to the condition of the vertebrae than to these two characters and proposed two suborders, OPISTHOCOELA for the Discoglossidae and Aglossa, and PROCOELA for the remaining anura. This indicated the nearest living relatives of *Xenopus* and *Pipa* to be the Palaearctic genera *Discoglossus*, *Bombina* and *Alytes*. The arrangement met little acceptance at the time though COPE (1889, p. 248) placed the Aglossa as an offshoot from the Discoglossidae at the base of the anuran phyletic tree. In 1916, however, NICHOLLS re-emphasized the importance of the vertebral condition as a taxonomic character and his work was used and elaborated by NOBLE (1922). In the resultant scheme *Xenopus*, *Hymenochirus*, *Pseudhymenochirus* and *Pipa* were grouped as one family (Pipidae) in a sub-order OPISTHOCOELA which was regarded as a primitive offshoot from the main anuran stem; the Discoglossidae were also included in the sub-order. Subsequent investigations showed that two genera, *Liopelma* and *Ascaphus*, previously included in the Discoglossidae, had amphicoelous vertebrae and were also more primitive than other living anura in some other respects. Consequently NOBLE (1931, p. 485, fig. 153), in a revised classification, showed the OPISTHOCOELA as an offshoot from the AMPHICOELA which were placed at the base of the main anuran stem.

Recent investigations of the anatomy and morphogenesis of the OPISTHOCOELA and AMPHICOELA have tended to support this scheme. It has, however, come to be recognised that some of the apparently "primitive" features of these groups may be due to the secondarily aquatic mode of life and to neoteny so that, with no fossil evidence of their phylogeny, opinions differ on the relationships of the Liopelmids and Discoglossids to the Pipids as well as concerning the inter-relationships of the genera composing the last family. No brief summary of the arguments is possible but reference should be made to the views expressed by MIRANDA RIBEIRO (1926), DE VILLIERS (1924, 1929, 1934), PATERSON (1946, 1949, 1951), SLABBERT & MAREE (1945), MILLARD (1945, 1949), VAN PLETZEN (1953) and DUNN (1948). The latter concludes that there is only a single valid genus in South America: *Pipa*. The African genera, *Xenopus*, *Hymenochirus* and *Pseudhymenochirus*, differ from this in characters that are mostly primitive but which "seem to be scattered in a rather uncorrelated manner"; every African species has some of them and *Pseudhymenochirus* appears to possess "a complete melange of characters formerly thought to be confined to *Xenopus* or to *Hymenochirus*, with none clearly peculiar to itself". It would, however, be premature to unite the three and they, with the extinct genus *Eoxenopoides*, together form a definable subfamily, the Xenopodinae.

The systematic position of *Xenopus* is thus:

CLASS Amphibia
  SUB-CLASS Apsidospondyli
    ORDER Anura
      SUB-ORDER Opisthocoela
        FAMILY Pipidae
          SUBFAMILY Xenopodinae

Type genus *Xenopus* WAGLER 1827: type species *Xenopus laevis* (DAUDIN 1802).

The genus has a geographical range throughout Africa south of the Sahara and comprises six species as follows:

(1) *Xenopus laevis* (DAUDIN).

Bufo laevis DAUDIN, 1802, Hist. Nat. des Rainettes des Grenouilles et des Crapauds, p. 85, Pl. 30, fig. 1. Originally described without indication of locality.

Distribution: From the Cape northwards to Angola in the west of the continent and to Lake Rudolf on the east; thence westward to northern Camerun (MONARD 1951). Within this area there is geographical variation which is still inadequately studied. PARKER (1936) recognised four, or perhaps five, subspecies:

(a) *Xenopus laevis laevis* (DAUDIN).

Namaqualand; Cape Province; Natal; Zululand; Orange Free State; Transvaal; S. Rhodesia; Mozambique; Nyasaland.

(b) *Xenopus laevis petersi* BOCAGE.

Xenopus petersi BOCAGE, 1895, Herpét. d'Angola et du Congo, p. 187. Originally described on the basis of specimens from several widely distributed localities in Angola, the subspecies has since been recognised from Angola; N. Rhodesia; Tanganyika Territory.

(c) *Xenopus laevis victorianus* AHL.

Xenophus victorianus AHL, 1924, Zoolog. Anz., **60**, p. 270. Type-locality: Lake Victoria. Specimens have also been assigned to this form from various localities in Uganda, (Lake Victoria; Kyagwe; Lake Edward; Mabira Forest; Fort Portal) and from nearby areas of the Belgian Congo and on Lake Tanganyika.

(d) *Xenopus laevis bunyoniensis* LOVERIDGE.

Xenopus laevis bunyoniensis LOVERIDGE, 1932, Proc. biol. soc. Washington, **45**, p. 114. This name has been applied to specimens not only from the type-locality, Lake Bunyoni, but from other regions in Uganda (Lake Mutanda; Lake Murene; Kigezi; Nyakabandi) and adjacent Belgian Congo (Mamvu). LOVERIDGE regards this as a form of the cold upland waters but PARKER (1932; 215) has argued that its small size and dark colour are due to malnutrition and vision impaired by parasitic infection.

(e) *Xenopus laevis borealis* PARKER.

Xenopus laevis borealis PARKER, 1936, Ann. Mag. nat. Hist. (10), **18**, p. 596. Type-locality: Marsabit, Lake Rudolf. It has also been reported from Lake Nakuru, Lake Naivasha, Leikipia and Mt. Elgon.

(2) *Xenopus gilli* ROSE & HEWITT.

Xenopus gilli ROSE & HEWITT, 1927, Trans. R. Soc. South Africa, **14**, 4, p. 343, Pl. 16, figs. 1, 3, 4.

This species was described from the Cape Peninsula and has not been found elsewhere.

(3) *Xenopus mülleri* (PETERS).

Dactylethra mülleri, PETERS, 1844, Monatsber. Ak. Berlin, p. 37. Type-locality: Mozambique.

Distribution: N. Rhodesia; Nyasaland; Portuguese East Africa, Tanganyika Territory; Kenya Colony; eastern Belgian Congo; French equatorial Africa; northern Nigeria; Dahomey.

(4) *Xenopus clivii* PERACCA.

Xenopus clivii PERACCA, 1898, Boll. Mus. Torino, **13**, 321, p. 3. Type-localities: Saganeiti and Adi Caié, Eritrea.

Distribution: Eritrea; Ethiopia; Turkana Province, Kenya.

(5) *Xenopus tropicalis* GRAY

Silurana tropicalis GRAY, 1864, Ann. nat. Hist., (3), **14**, p. 315. Type-locality: Lagos.

Distribution: Forested regions from Portuguese Guinea eastwards to the Ubangi-Shari region of French Tropical Africa and southwards to the mouth of the River Congo and northern Angola (Dundo).

(6) *Xenopus fraseri* BOULENGER.

Xenopus fraseri BOULENGER, 1905, Proc. zool. Soc. London, 1905 (2), p. 248, Pl. 4, fig. 24. Type-locality: Nigeria or Fernando Po.

Distribution: Forested regions from Fernando Po, the Cameroons, French Congo and north Angola in the west to the north-eastern border of the Belgian Congo.

*　*　*

CHAPTER V

SOME ECOLOGICAL DATA AND METHODS OF REARING OF XENOPUS LAEVIS UNDER LABORATORY CONDITIONS

XENOPUS LAEVIS IN NATURE

In South Africa *Xenopus laevis* lives in practically any kind and amount of water. Its amazing abilities of overland migration are well known, and make it improbable that there should be differences between local races. The frogs live by the thousands in silty farm ponds totally devoid of any higher plant vegetation.

In the neighbourhood of Stellenbosch and Cape Town, *Xenopus laevis* breeds both in moor waters with a relatively low pH, and in lakes rich in lime with a rather high pH.

In the ponds of the Jonkershoek Inland Fish Hatchery the animals spawn abundantly at a pH as high as 9.0, caused by the presence of large quantities of lime in the fertilizing-mixture used in the ponds.

During the breeding season (September to December) the spawning of the frogs on a certain day mainly depends on the temperature of the preceding day. In the full breeding season abundant spawning takes place if, on the preceding day, the afternoon surface temperature (i.e. the maximum surface temperature of the day) has risen above a level of about 21° C. In the beginning of the breeding season this critical level lies lower, but spawning is far less abundant.

In nature oviposition starts about eight to ten hours after a temperature maximum surpassing a certain critical level has been attained, and takes place while the temperature is again decreasing and already approaching its minimum.

REARING METHODS

Spawning can be induced in *Xenopus laevis* by injection of gonadotropic hormones, e.g. human gonadotropic pregnancy urine extract (Ciba, Basel). Injection is made into the dorsal lymph sac, piercing the skin of the thigh and the septum between the lymph sacs of the thigh and the back (see OCHSÉ, 1948).

The male is injected two days before eggs are required with maximally

300 I.U. in 0.5 cc of aqua dest. On the next afternoon the male, whose forelimbs should now show black "nuptial pads", is once again injected with maximally 300 I.U. in 0.5 cc of aqua dest., and the female with maximally 600 I.U. in 1 cc of aqua dest [1]). The animals are put into a large covered container, sheltered from light and provided with twigs for oviposition. A metal gauze with edges bent downwards may be put on the bottom to prevent the animals from eating fallen eggs.

Spawning occurs in the early morning, after the temperature has been gradually raised artificially during the night and has passed the critical temperature, but can be postponed by keeping the water below that temperature. After spawning, which may last up to 24 hours, the adults are taken out, their excrements are removed and the temperature of the water is kept at 20—25° C. During the rearing of the larvae there should finally not be more than about three larvae per litre of water in the container. The water must be aerated until the larvae begin to breathe by their lungs (indicated by the appearance of air bubbles at the surface). If plant powder is used as food (see below) it is not necessary to change the water, at least not during the first weeks, when the larvae are very susceptible to damage by netting. Care should be taken that the water is absolutely free of copper and chlorine ions. Chlorine can be removed by allowing the water to stand in contact with the air for several days.

Feeding starts during the fifth day. Excellent results have been obtained with nettle powder (*Herba urticae*), or alfalfa leaf powder as food. Others have obtained good results with dried yeast (THOMASON, personal communication) and with liver powder (cf. DEANESLY and PARKES, 1945), but a suspension of ground fresh beef liver used for comparison was found to be unsatisfactory (FABER), as it produced growth inhibition and abnormalities.

Finest degree nettle powder is put into a small bag of fine cloth, which is then submerged and squeezed out into the water until this is slightly opaque. The residue of the powder, not passing through the cloth, is discarded. An excellent food suspension can also be made from fresh nettles by means of a liquidizer. Usually the elaborate purifying method recommended by OCHSÉ is not necessary. The larvae are fed once a day, and the amount of food should be such that the water is cleared up before the next feeding. Removal of the excrements was found to be unnecessary.

The larvae do not need much light; they develop just as satisfactorily in total darkness. Certainly too much light interferes with normal development.

As a narcotic for the larvae a few crystals of trichloro-isobutylic alcohol were used, in an amount just sufficient to cause immobility.

---

[1]) The hormone doses required may vary with the condition of the animals. OCHSÉ (1948) used 400–600 I.U. per animal, others have obtained good results with much lower doses.

During metamorphosis the larvae stop feeding, become very sluggish and keep to the bottom of the container. In order to save them from drowning they should then be transferred to shallow water. Already before the end of metamorphosis they start feeding again, this time on small worms such as *Enchytreae* and *Tubifex*.

Under favorable conditions the animals can reach sexual maturity in the laboratory in about half a year. In general it is necessary not to underfeed the rapidly growing postmetamorphic animals, since animals fed insufficiently will remain small, even when sexually mature. Juvenile as well as adult animals should regularly receive vitamin D and Calcium added to the food, in order to prevent skeletal abnormalities and brittleness of the bones.

Another very important factor is that of space. Not only should the growing juveniles have enough space per individual (not much less than three litres),but the dimensions of the containers should be such that at least in one direction the animals can swim freely over fairly long distances. Also very shallow water is unfavourable.

The growing animals are gradually switched over to *Lumbricus* and later on partly or entirely to *beef heart* cut into small blocks. The adults are fed only twice a week. Detritus is removed regularly.

It is recommended to keep the animals in running water, the contents of the container being renewed at least once in 24 hours. A heating element of constant capacity is placed in the container, and the proper water temperature, about 18° C. for the adults, is regulated by the flow of water. The containers should be covered with gauze, and an emergency outlet should be provided at some distance below the gauze, in order to prevent drowning in case the water level should rise owing to clogging of the normal outlet. No sand or plants are needed in the container, only some fragments of roof-tiles to provide suitable shelter for the animals. The animals can be kept in ordinary daylight, or in artificial light which is interrupted during the night and adapted to the day-length of the season.

* * *

CHAPTER VI

THE SYSTEMATIC DESCRIPTION OF THE INTERNAL DEVELOPMENT OF XENOPUS LAEVIS

INTRODUCTION

The description of the internal development has been written in a continuous form and has been arranged systematically.

In the early development no distinction between organ systems can be made, so that up to STAGE 10 the embryo will be described as a whole. From STAGE 10 to STAGE 15 the formation and subsequent development of the germ layers will be followed. Since most organ systems have not yet been individualized at STAGE 15, for the following period the organism will be divided into a number of organ complexes, the development of which will be described up to a stage at which nearly all the individual organ systems have been individualized. The later development will be treated systematically, viz. according to individual organ systems.

The development of those organ systems in which important changes occur during metamorphosis, has been subdivided into the development up to metamorphosis and the development during metamorphosis. In the development of several organ systems successive periods of development have moreover been distinguished. Finally, in some divisions an anatomical description has been given of a certain stage, either at the beginning (the fertilized egg), in the middle (just before metamorphosis) or at the end of development, as a link between larval and metamorphic development and when the complexity of the system concerned made it desirable. This is reflected in the subdivision and heading of the text.

The subdivision of the text into successive periods of development leads inevitably to a certain scattering of the data concerning the development of certain structures over successive subdivisions. The editors hope that the alphabetical index may for the greater part remove this difficulty.

The external development of *Xenopus laevis* has not been included in this chapter. The editors found it more suitable to give the data on the external development in a short form in the chapter "stage criteria". The pigmentation pattern and the externally visible development of the lateral line system were however too extensive subjects to be inserted into the stage criteria, and had moreover little value for stage characterization. They have therefore been inserted into the systematic description of the internal development, as a separate section of the development of the skin.

This chapter has been subdivided into three main divisions and twenty-one divisions. These divisions have been grouped successively as ectodermal, mesodermal and entodermal derivatives. Their sequence is as follows:

*   *   *

THE EARLY DEVELOPMENT UP TO STAGE 15

THE FERTILIZED EGG

The *fertilized egg* in which the germinal vesicle has disappeared (STAGE 1) shows a characteristic plasm distribution. A well defined *cortical layer* can be distinguished, free of yolk material and rich in pigment in the animal half of the egg, except for the animal pole region where the contents of the germinal vesicle have partially displaced the pigment. This cortical layer is thick at the animal and dorsal side, and decreases in thickness in dorso-ventral and animal-vegetative directions. Underneath the cortical layer, which is probably identical with HOLTFRETER's "surface coat" (1943), a *subcortical plasm* is located as a rather thick layer in the animal pole region, tapering out towards the equator of the egg. This decrease in thickness is less pronounced at the dorsal side of the egg, where the plasm also extends further in the vegetative direction. This plasm consists of a rather dense cytoplasm laden with very small yolk platelets and dispersed pigment granules. Near the vegetative pole of the egg a separate plasm with grey pigment and relatively small yolk platelets can be observed immediately underneath the very thin cortical layer. In this plasm some *cytoplasmic inclusions*, identical with the so-called "cytoplasme germinal" of BOUNOURE (1939), are situated near the vegetative pole of the egg.

In the interior of the egg some further plasm areas can be distinguished. The *inner animal plasm* consists of cytoplasm heavily laden with medium-size yolk platelets. The *inner vegetative plasm* (future nutritive yolk material) is heavily laden with large yolk platelets. In between both inner plasms a rather dense plasm, the *central plasm*, containing relatively small yolk platelets, fills up the centre of the egg.

Both polar axes, the *animal-vegetative* and the *dorso-ventral axis*, are clearly indicated respectively by the yolk distribution and by the unequal distribution of the cortical layer and the subcortical plasm.

THE PERIOD OF CELL CLEAVAGE

The fragmentation of the large egg cell starts with the first cleavage and extends over a considerable period of development. Until the beginning of gastrulation, cleavage represents the main visible developmental process. Cleavage is most easily studied at the two, four, eight, sixteen and thirty-two cell stages (represented respectively by the STAGES 2,3,4,5, and 6 [1])).

---

[1]) These stages represent respectively advanced two, four, eight, sixteen and thirty two cell stages since in each stage the next nuclear division is already in progress.

It is characterized by a penetration of subcortical plasm into the interior of the egg along the future cleavage plane, particularly in the animal half of the egg, and by a furrowing of the cortical layer. During the first cleavage the *furrow* penetrates into the egg over not more than half the egg radius at the animal pole (a meridional cleavage), not beyond one third of the egg radius at the level of the equator, while the penetration is reduced to about one tenth of the egg radius at the vegetative side. Separation of the blastomeres therefore occurs by the "de novo" formation of a very thin *partition wall* in continuity with the ring-shaped cleavage furrow. After the establishment of the partition wall the external cleavage furrows gradually become more and more shallow by the extension of the partition walls and during later cleavages by the formation of a *cleavage cavity*, the future *blastocoelic cavity*, which is first visible at the four cell stage (STAGE 3). After the formation of the cleavage cavity, the cleavage furrows cut less and less far into the interior of the egg, e.g. they extend only over one fifth of the egg radius during the third cleavage and are still more shallow at later stages.

The first two cleavages are meridional, dividing the egg respectively into two and four blastomeres, usually of unequal size. The first plane of cleavage coincides more or less with the dorso-ventral *plane of bilateral symmetry* (STAGE 2). The second cleavage divides the egg into two smaller dorsal and two bigger ventral blastomeres, which is particularly obvious when seen from the animal pole. As a result of a somewhat oblique position of this cleavage plane, the four blastomeres seem to be of almost equal size when seen from the vegetative pole (STAGE 3).

The third plane of cleavage is "equatorial", and is situated at a distance of about one-third of the diameter of the egg from the animal pole, dividing the egg into one micromere and one macromere quartet (STAGE 4). In the micromere quartet the dorsal blastomeres are markedly smaller and less pigmented than the ventral ones. This state of affairs is not changed by the fourth cleavage, which is again meridional (STAGE 5). The cleavage cavity increases gradually by a further dilatation of the blastomeres, all of which are arranged in one layer around the central hollow. This hollow is mainly located in the animal half of the egg, and lies slightly excentrically towards the dorsal side at the sixteen cell stage (STAGE 5). The fifth cleavage, which is again equatorial, divides the egg into four "rows" of eight blastomeres, viz. an animal rosette of micromeres, a ring of micromeres, a slightly irregular ring of macromeres and a vegetative rosette of macromeres. The cleavages are still nearly synchronous in the four rows of blastomeres. The individual blastomeres, being arranged in this rather orderly way, protrude less into the cleavage cavity, which thus becomes more regular in form (STAGE 6).

Except for the formation and gradual extension of the central cleavage cavity, the topography of the various plasms does not change very much in comparison with the fertilized egg. The central plasm seems to be displaced towards the equator of the egg already during the first division. Cleavage and blastocoel formation lead to the following plasmatic distribution in the still single-layered embryo at the *morula* stage (STAGE 6½): The *animal blastomeres*, which are more or less cubical, consist, starting at the periphery, of a well developed, strongly pigmented cortical layer, a thick layer of dense subcortical plasm, central hyaline plasm with mitotic figures or nuclei, and a layer of inner animal plasm. The *dorsal equatorial blastomeres*, which are irregular in form and markedly bigger than the animal blastomeres, contain, starting at the periphery, a well developed cortical layer, a distinct layer of subcortical plasm, hyaline plasm with mitotic figures (located more towards the apical side of the cells), and a thick basal layer of inner animal plasm heavily laden with yolk platelets increasing in size in centripetal and in vegetative directions. The *lateral* and *ventral equatorial blastomeres* differ from the dorsal ones by the thinner cortical layer and rather thin layer of subcortical plasm. The big *vegetative blastomeres*, irregularly cylindrical in form except for the four central ones, which are more funnel-shaped, consist, starting at the periphery, of a very thin cortical layer with a few gray pigment granules, hardly any subcortical plasm, and a very thick layer of inner vegetative plasm in which small mitotic figures are embedded, surrounded by very little hyaline plasm. The *cytoplasmic inclusions* are still located near the vegetative pole, but may partially be displaced inwards along the cleavage furrows of the central vegetative blastomeres. In general, the transition between the various cell types is more gradual along the dorsal than along the lateral and ventral sides. By the time the morula stage is reached the divisions are no longer so closely synchronised, the animal blastomeres are in advance with respect to the vegetative ones.

THE BLASTULA STAGES

At the early *blastula* stage (STAGE 7) the synchronism of the cleavages has disappeared, cleavages being faster at the animal and dorsal sides than in the other areas. At this stage a more or less tangential cleavage occurs, changing the hitherto single-layered embryo into a double-layered one. The *outer layer* is characterized by the presence of the cortical layer which penetrates into the cleavage furrows only over one fourth of the thickness of this layer. It moreover contains by far the greater part of the subcortical plasm. In this outer layer two areas can be distinguished, one extending over the animal and equatorial regions, the other, the cells of which are very rich in yolk material and much thicker, extending over the vegetative side of the egg. There is a rather sharp boundary between these two areas at the ventral and lateral sides, whereas a more gradual transition can be observed at the dorsal side. The *inner cell material*, surrounding the blasto-coelic cavity, can more adequately be subdivided into three areas, viz. an animal, an equatorial and a vegetative one. The inner animal blastomeres contain some subcortical plasm and are very rich in inner animal plasm; the inner equatorial blastomeres contain hardly any subcortical plasm but consist mainly of inner animal plasm. The inner vegetative blastomeres finally consist exclusively of inner vegetative yolk material (the future nutritive yolk material). Except for the lateral and ventral sides where the boundary between equatorial and vegetative blastomeres is rather sharp, gradual transitions exist between the successive areas. The cleavage cavity of the preceding stages is now called the *blastocoelic cavity*.

At STAGE 7 the so-called *pregastrulation cell movements* are indicated by the *ascendance of cell plasm* along the cleavage furrows near the vegetative pole, whereby the cytoplasmic inclusions are displaced to a more interior position and by an *epiboly* of the animal and equatorial areas of inner and outer layers. A sharp delimitation does not yet exist between the outer and the inner layer, and intercellular spaces have not yet been formed. In the following stages the pregastrulation movements continue, and the various areas become more sharply distinguishable. The outer and inner layers become better delimited. The continuous cell cleavage and the epiboly result in a thinning out of the roof and side walls of the blastocoelic cavity. At the middle blastula stage (STAGE 8) the still rather thick roof of the blastocoel consists of two to three irregular cell layers. As the cells multiply and decrease in size, the number of cell layers tends to increase, but the concomitant epibolic extension of the animal area opposes this tendency.

At the late blastula stage (STAGE 9) the *animal area* consists of a single outer layer and about two inner layers of cells, but at the beginning of gastrulation (STAGE 10) the roof of the blastocoel near the animal pole consists of only one outer and one inner layer. Towards the equator the number of inner layers of cells increases, the outer layer remaining single. The same phenomenon occurs in the equatorial region. At STAGE 8 the *equatorial area* consists of four to five layers of cells at the dorsal side and three to four layers at the lateral and ventral sides. The blastomeres at the dorsal side however are smaller, so that the wall itself is not thicker. At STAGE 9 this number decreases at the dorsal side bordering the animal area (to one outer and about three inner layers), but is higher at more vegetative levels within the equatorial area (up to one outer and about six inner layers). At STAGE 10 the number varies respectively from one outer and two to three inner layers to one outer and five to six inner layers. In the *vegetative area* the number of layers of cells increases rapidly by the successive cell cleavages and by the ascendance of cell material from the vegetative pole. At STAGE 8 one outer and three to four inner layers of cells can be distinguished. At STAGE 10 the vegetative cell material consists of a large number of layers of cells of which only the outer one, being a single layer, is rather sharply demarcated.

During the *blastula* stages the various areas become more clearly distinguishable in the inner cellular material. Whereas at STAGE 8 a distinction between the *inner animal area, marginal zone* and *vegetative area* is only indicated, these three areas are already much more clearly demarcated at STAGE 9 and become still more sharply distinguishable at STAGE 10. They represent respectively the *presumptive sensorial layer of the ectoderm,* the *presumptive mesoderm* and the *presumptive nutritive entoderm.* Simultaneously the outer cell layer becomes better delimited from the inner cell material. At STAGE 8 intercellular spaces appear for the first time. This delimitation process continues rapidly, so that at STAGE 9 the outer layer can easily be separated from the inner cell material in the animal and equatorial areas of the embryo. Over this entire territory the outer layer consists of regular cubical cells and represents the extensive *presumptive epithelial layer of the ectoderm.* It rather sharply borders the much thicker outer entodermal layer of the vegetative side of the egg, which represents the *presumptive entodermal epithelial layer of the archenteron.*

The epibolic extension of the animal area, as one of the aspects of the *pregastrulation movements,* leads to a displacement of the entire marginal zone towards the vegetative side. Whereas at STAGE 9 the inner marginal zone is still located above the equator, except for its dorsal portion which extends below the equator, the lower borderline of the marginal zone has markedly descended at STAGE 10. The inner marginal zone has moreover become thicker at the dorsal than at the lateral and ventral sides.— The epiboly of the animal area is accompanied by a gradual increase of the blastocoelic cavity, which acquires its full size at STAGE 9, where its animal-vegetative dimension is about one quarter and its dorso-ventral and medio-lateral dimensions are about two third of the diameter of the egg. Meanwhile its inner surface becomes smooth.—The ascendance of the vegetative cell material is also reflected in a further displacement of the blastomeres with *cytoplasmic inclusions.* At STAGE 9 they usually acquire a position at about one fourth to one third of the thickness of the floor of the blastocoel measured from the vegetative pole. At this stage the inclusions shift from the periphery of the cells where they have been located up till now, towards the nucleus, with which they become closely associated.

THE EARLY GASTRULA STAGES

The following period of development is characterized by the complicated process of *gastrulation.* In this process the internal formation of the *mesodermal mantle* and the formation of the *entodermal archenteron* can be distinguished. As described by NIEUWKOOP and FLORSCHÜTZ (1950), the two processes do not proceed synchronously. During the beginning of gastrulation (STAGE 10) the inner marginal zone starts rolling inwards at the dorsal side before any invagination of the archenteron has taken place. In the outer cell layer of the presumptive entoderm, close to the border-line with the presumptive epithelial layer of the ectoderm, and at 50 to 60° dorsally from the vegetative pole, the future *blastoporal groove* is indicated, but only by the formation of bottle-necked cells, and by a concentration of the pigmented cortical layer.

At the early gastrula stage (STAGE 10¼) the internal blastopore lip has extended from the dorsal side to the lateral and even to some extent to the ventral side of the inner marginal zone ring. In the outer layer a distinct *dorsal blastoporal groove* has been formed, demarcated by a sharp pigment line. This shallow groove begins to extend in a lateral direction, without becoming much deeper. At STAGE 10¼ a part of the *definitive mesodermal mantle* has already been formed, from which the *presumptive prechordal mesoderm* begins to delimit itself by a loosening of its cells. At the crescent-shaped blastopore stage (STAGE 10½) the formation of the definitive meso-dermal mantle has made marked progress. At the dorsal side it has already reached the equator, whereas simultaneously the internal blastopore lip has been displaced towards the vegetative side of the embryo (now 30 to 40° from the vegetative pole) and has extended all around the entoderm.— The formation of the definitive mesodermal mantle leads to delimitation of the mesoderm from the central entoderm mass, a process which is first visible at the dorsal side by a loosening of the cells (STAGE 10¼) and which extends towards the lateral and ventral sides at STAGE 10½. It is through this loose cell material that the invaginating archenteron pene-trates, definitively separating the mesoderm from the entoderm mass. This cylindrical mesodermal mantle pushes forward the upper margin of the entodermal mass, thus gradually embracing, and later enclosing the blastocoelic cavity.

At STAGE 10½ the blastoporal groove, formed inside the area of presump-tive entoderm—leaving a narrow strip of peripheral entoderm,—has ex-tended to the dorso-lateral side of the vegetative area. A short slit-shaped *archenteron* has formed at the dorsal side only, penetrating upwards into the interior of the egg over 10 to 15°.

The *epibolic extension of the animal area* has led to a shifting of the borderline between the presumptive epithelial layer of the ectoderm and the outer entodermal layer towards the vegetative pole, thus reducing the *future yolk plug* area to a zone extending 30 to 40° around the vegetative pole. The presumptive sensorial layer of the ectoderm extends now approxi-mately to the equator of the egg at the lateral and ventral sides, while this border-line lies still at 10 to 20° above the equator at the dorsal side. The border-line is indicated by a rather abrupt change in the number of layers of cells. The *reduction of the vegetative area* (presumptive entoderm) to about half of the diameter of the egg is accompanied by a further ascendance of the entodermal cells in the neighbourhood of the vegetative pole (*continuation of the pregastrulation movements*). During the following stages all these processes continue, especially the extension of the archenteron. At the horse-shoe-shaped blastopore stage (STAGE 11) the pigment line has extended all around the *yolk plug*, while the actual groove has spread to the lateral sides only. Not before the large yolk plug stage (STAGE 11½) does the groove extend all round the yolk plug. At the dorsal side the slit-shaped archenteron, entirely lined with cubical entoderm cells—except for its anterior extremity where the cells are bottle-necked—extends over 40 to 50° at STAGE 11 and over 80 to 90° at STAGE 11½. Meanwhile the archenteron extends in a medio-lateral direction, so that at STAGE 11½ its middle and caudal portions measure approximately 30°. The advancing anterior portion still remains narrow.

The formation of the *definitive mesodermal mantle* has progressed concurrently with the development of the archenteron. At STAGE 11 the mesodermal mantle extends to 30 to 40° above the equator at the dorsal side, to 20 to 30° at the lateral side, and approximately up to the equator at the ventral side. At STAGE 11½ the mesodermal mantle reaches a point 20 to 30° from the animal pole at the dorsal side, whereas the ventral portion does not yet pass the equator. Consequently it is particularly the dorsal mesoderm which moves forward. Its *prechordal portion* consists of three to five layers of cells, while its *chordo-mesodermal portion*, which is rather thick, shows even five and more layers.

The invagination of the *archenteron* and the formation of a deep groove around the diminishing yolk plug leads to a shifting of the border-line between the outer entodermal layer and the presumptive epithelial layer of the ecto-neuroderm towards the blastoporal groove. Simultaneously the entoderm mass shifts inwards. The border-line between both areas of the external cell layer coincides with the blastoporal groove at STAGE 11½. The epithelial layer has now been thinned out to an epithelium with flattened cells.

The *presumptive sensorial layer of the ectoderm* which borders the rapidly reducing inner marginal zone, extends further and further caudad (i.e. towards the vegetative pole) and gets thinned out to a single layer with cubical cells. This extending sensorial layer of the neurectoderm and the rapidly moving dorsal, prechordal portion of the mesodermal mantle make contact for the first time at STAGE 10½. The contact area has markedly extended at STAGE 11 and already covers an area of 80 to 90° in length at STAGE 11½. At this stage the formation of the future *neural plate* is indicated by the thickening of the sensorial layer of the neurectoderm.

THE ADVANCED GASTRULA STAGES

The formation of the *ecto-*, *meso-*, and *entodermal germ layers* is completed in the following period of development in which the STAGES 12, 12½ and 13 have been distinguished. The main processes are (1) the further extension of the mesodermal mantle and the disappearance of the inner marginal zone by the formation of the more caudal portion of the mesodermal mantle, and (2) the extension and widening of the archenteron, accompanied by the reduction and the final disappearance of the yolk plug, and (3) the initial development of the neural plate in the sensorial layer of the ectoderm.

At the middle yolk plug stage (STAGE 12) a displacement of the embryonic pigment inside the epithelial layer of the ectoderm from a peripheral to a basal position is initiated. At this stage the mesodermal mantle—dorsally now called the *mesodermal archenteron roof*—reaches a point at some distance from the animal pole at the dorsal side, to about 45° at the lateral sides and to about 60° from this pole at the ventral side. At the small yolk plug stage (STAGE 12½) the mesodermal mantle has almost attained its final position, viz. up to a point close to the animal pole at the dorsal side and up to 40 to 50° from the animal pole at lateral and ventral sides. At the slit-blastopore stage (STAGE 13) it is only the prechordal portion of the mesodermal mantle which has stretched further in anterior and lateral directions. At STAGE 12 the prechordal and chordo-mesodermal portions begin to segregate and the *prechordal plate* gets thinner (about two layers of cells) than the *chordo-mesodermal plate* (three to four layers of cells). This process has continued at STAGE 12½. At STAGE 13 the prechordal plate has already changed into a thin membrane, except for its median strip. At STAGE 12½ the *notochord* also begins to segregate as a median thickening of the chordo-mesodermal layer, a process which progresses very rapidly, so that almost the entire notochord has been clearly individualized at STAGE 13. Meanwhile the *presumptive somite mesoderm* thickens and becomes double-layered. During this period of development the rolling in of the marginal zone has continued, and the sensorial layer of the ectoderm has extended further towards the blastopore. Except for the cell material still present in the thick inner blastopore lips the "invagination" of the mesoderm has been completed at STAGE 13.

The *entodermal invagination* has also made marked progress. Whereas at STAGE 12 a roundish yolk plug is still protruding, at STAGE 12½ this has been reduced to a small ovoid protuberance of the central entoderm mass, surrounded by a very deep blastoporal groove, penetrating far into the interior of the embryo. At STAGE 13 the yolk plug is usually no longer externally visible. Internally an entodermal protuberance, however, still exists. Meanwhile the *entodermal archenteron* has extended further and has begun to widen, a process indicated at STAGE 12. At this stage the archenteron extends over more than 90°. At STAGE 12½ the tip of the archenteron reaches the animal pole. The anterior portion of the archenteron has widened markedly and has extended in a medio-lateral direction. This process still continues at STAGE 13 when the archenteron reaches a point about 10° ventral to the animal pole. The definitive extension is not achieved before STAGE 13½, when a point about 45° ventral to the animal pole is attained. At these stages the entodermal roof of the archenteron forms a thin and single layer of cells. The extension of the archenteron gives rise to a ventrad displacement of the entire entoderm and of the blastocoelic cavity. The blastomeres with *special cytoplasmic inclusions* have been displaced by the gastrulation movements to the centre of the yolk mass, slightly caudal to the middle of the embryo at STAGE 13. The widening of the archenteron leads to a gradual reduction of the *blastocoelic cavity*. The rest of the blastocoelic cavity becomes entirely enclosed by the margin of the entoderm from STAGE 12½ to 13 onwards. A residue of the blasto-coelic cavity can however still be found at much later stages of development (STAGE 15 and beyond).

The extension of the dorsal mesodermal mantle is accompanied by a thickening of the overlying sensorial layer of the ectoderm, indicating the *future neural plate formation*. At STAGE 12 its anterior portion is considerably thickened, whereas its posterior portion is already somewhat stretched. At STAGE 12½ the neural area, which becomes more clearly demarcated, reaches approximately up to the animal pole. Its lateral and anterior margins are very much thickened, whereas its median area is rather thin, a phenomenon which is still more pronounced at STAGE 13, particularly along the midline.

After the completion of the gastrulation movements, during which cell layers move either in opposite directions (mesodermal mantle and over-lying ectoderm) or in the same direction but at different rates (mesodermal mantle and entodermal archenteron), a mutual *attachment of the various cell layers* develops. First a more intimate contact is established between the mesodermal archenteron roof and the neural plate area of the over-lying sensorial layer of the neurectoderm and between the former and the underlying entodermal archenteron roof. This contact begins at STAGE 12½ and is gradually increased during the next stages, particularly along the dorsal midline. The epithelial layer of the ectoderm which is hitherto not concerned in the neural plate formation, shows the first symptoms of attachment to the underlying sensorial layer at STAGE 13, where this contact is however still restricted to the dorsal median line. Nevertheless all the layers are still easily separable.

THE EARLY NEURULA STAGES

The next stages are particularly characterized by the *neural plate formation*, while at the same time the segregation in the mesodermal archenteron roof makes further progress. At the initial neural plate stage (STAGE 13½) the *medio-dorsal groove* of the neural plate is already well developed and the segregation of the neural area from the surrounding epidermis is indicated in its anterior half. This segregation process spreads in a cranio-caudal direction during the next period of development. At the "neural plate" stage (STAGE 14) the anterior, prechordal portion of the neural plate is markedly thickened except for its median area; its middle portion is a rather thin plate with thickened lateral margins; whereas its caudal portion does not yet show a sharp demarcation against the surrounding epidermis, so that the *neural folds* which begin to elevate more anteriorly still fade out in a caudal direction. At the early neural fold stage (STAGE 15) the entire neural plate is very much thickened except for its dorso-median strip, and already somewhat narrowed in its middle and caudal portions (*beginning of dorsal convergence movements*). The latero-anterior edges of the plate begin to protrude as neural fold formation. The median groove deepens and the *presumptive eye anlagen* begin to sink in at the latero-anterior edges of the plate. The *neural crest* begins to segregate from the neural plate and the surrounding epidermis by a loosening of its cells.

The further segregation in the *mesodermal archenteron roof* finds its expression in 1) a sharp delimitation of the notochord, also at its tip where it borders the median thickening of the thin prechordal plate, and 2) a better demarcation of the presumptive somite mesoderm. A continuous *myocoelic slit* is indicated in the presumptive somite mesoderm at STAGE 13½. At STAGE 14 the somite mesoderm is markedly thickened by the beginning of dorsal convergence, and is clearly demarcated from the lateral mesoderm, which still forms a thin, uniform layer. At STAGE 15 the somite mesoderm becomes reduced in a medio-lateral direction. A continuous *myocoelic cavity* is distinguishable for the first time.

At STAGE 14 the ventro-lateral free margin of the mesodermal mantle is gradually displaced towards the median line, a process continued in the following period of development. This part of the mesodermal mantle represents the double origin of the future *heart anlage*.

The *archenteron* has widened considerably during this period of development and has attained its maximal cranio-caudal extension of STAGE 13½ and its full medio-lateral extension at STAGE 14. In front of the neural plate, in the mesoderm-free area, the rather thin wall of the archenteron has made an extensive contact with the overlying epidermis, indicating the area in which the *cement gland anlage* and later the *stomodeal-hypophyseal anlage* will appear. The former becomes visible at STAGE 15 as a thickening of the epithelial layer of the ectoderm and by a corresponding concentration of the embryonic pigment (see further under the development of the *skin*, page 43). In the caudal wall of the anterior portion of the archenteron, the *foregut*, the *liver diverticulum* becomes visible at STAGE 13½, and deepens gradually during the following stages, in which only minor changes occur in the rest of the archenteron.

The mutual *attachment of the various cell layers* increases only slowly during this period, in which, particularly in the medio-dorsal region, the entodermal archenteron roof, the notochord, the neural plate and the overlying epithelial layer become more firmly attached to one another. At STAGE 15 the attachment begins to extend over the entire dorsal area of the embryo, particularly in the prechordal region.

The development from the fertilized egg (STAGE 1) up to the early neural fold stage (STAGE 15) has been treated chronologically for the entire embryo. This period was only subdivided into a number of characteristic phases (STAGES 1 to 6½, 7 to 10, 10¼ to 11½, 12 to 13 and 13½ to 15). In the following divisions the development of a number of organ complexes viz., a) "the early development of the *central nervous system, sense organs, ganglia and nerves*" on page 29; b) "the early development of the *axial system*" on page 35; and c) "the early development of the *alimentary system* and the *presumptive visceral skeleton, etc.*" on page 39, will be treated until a stage is reached from which the development of the individual organ systems can be described separately. "The development of the *skin, lateral line system*, etc." on page 43, "the development of the *heart* and *vascular system*, etc." on page 121 and "the development of the *nephric system*" on page 130 will be treated separately from STAGE 15 onwards.

* *
*

THE EARLY DEVELOPMENT OF THE CENTRAL NERVOUS SYSTEM, SENSE ORGANS, GANGLIA AND NERVES up to approx. STAGE 28 (placodes up to STAGE 37/38)

The initial development has been described under "the early development up to STAGE 15" on pages 26 and 27.

The *neural plate* is formed as a thickening of the sensorial layer of the ectoderm—from which also the *neural crest material* develops—during the STAGES 13 to 15. It segregates from the surrounding epidermal area in a cranio-caudal direction. Anteriorly a sharp demarcation is already present at STAGE 13½, whereas caudally the delimitation is not yet clear before STAGE 15. The thickening of the neural plate is first restricted to the anterior and lateral areas leaving a thin strip along the dorsal midline at STAGE 14.

THE NEURAL FOLD STAGES

The *neural fold formation* initiates the folding process of the neural plate and the segregation of the neural crest material which is absent only at the medio-anterior border of the plate. The *neural plate* (*s.s.*) begins to narrow, particularly in its middle and caudal portions at STAGE 15 and simultaneously the *median groove*, where the plate is firmly attached to the underlying notochord, becomes deeper. The invagination of the *eye anlagen* forms the earliest local differentiation of the neural plate, the antero-lateral edges of which become markedly thickened at STAGE 16. With the folding of the anterior half of the neural plate, the eye anlagen show already the first signs of a lateral evagination from the future neural tube at STAGE 17, a process which becomes more and more pronounced during the following stages during which the plate narrows and the elevated folds gradually approach each other. This formation of the neural tube takes place in a direct manner over the entire length of the neural anlage, so that in *Xenopus* a clear and well defined *neurenteric canal* develops where the neural plate passes into the archenteron roof around the dorsal blastoporal lip. This neurenteric canal is indicated at STAGE 17 and becomes deeper during the following stages. It closes to a tube by STAGE 21. It is evident that in the *neural tube formation* the epithelial layer of the ectoderm takes part, a process indicated by a first interdigitation of the cells of epithelial and sensorial layers not starting before STAGE 16.

This attachement develops rapidly, so that separation of the two layers becomes impossible at STAGE 17 except for the region around the blastopore. The attachment to the underlying layers also increases rapidly at these stages, so that only in the lateral and ventral regions and around the blastopore can separation be performed experimentally. In the caudal region the attachment remains weak up to STAGE 19.

The segregation of the *neural crest material* from the neural plate (s.s.) and from the surrounding epidermis begins at STAGE 15, indicated by a loosening of the cells. At STAGE 16 the neural crest has already been delimited in the anterior and middle regions of the neural anlage, while at STAGE 17 this process is just in progress in the caudal region. Simultaneously the sensorial layer of the epidermis shifts mediad underneath the epithelial layer, closing the gap formed by the withdrawal of the neural crest material. At this stage the neural crest material has already been displaced to the lateral sides of the future neural tube in the anterior half of the embryo. The anterior border of the massive neural crest, developing from the lateral margins of the neural anlage, lies at the level of the posterior half of the eye anlage at STAGE 18, at which stage also a first segregation takes place of *mes-* and *rhombencephalic neural crest*.

THE EYE VESICLE STAGES

The *eye anlagen* forming antero-lateral evaginations of the closing neural tube are originally almost massive formations in which only a slit-shaped extension of the neural groove penetrates at STAGE 18. At STAGE 19 at which the neural tube is formed by the first contact of the neural folds, but at which a fusion of the folds has not yet occurred, the eye anlagen clearly segregate from the anterior portion of the neural tube. The ependymal layer of the neural tube is still continuous with the epithelial layer of the epidermis at STAGE 20, a connection which is interrupted at STAGE 21. The *primary eye vesicle* formation does not take place before STAGE 21 at which the central cavity of the brain lined with ependyma (derived from the epithelial layer) rapidly penetrates into the optic stalks. At STAGE 22 the outer surface of the primary eye vesicles, which have a wide open communication with the archencephalic ventricles, has achieved a broad contact with the overlying epidermis (see further under "the further development of the *eye*", on page 77).

The *regional segregation* of the *brain* is indicated by a first subdivision into an *arch-* and a *deuterencephalic region* at STAGE 20. Both regions have a narrow ventricular cavity and massive walls. The modelling of the neural tube is initiated by the development of the *cephalic flexure* in the form of the *retro-infundibular fold* formation at STAGE 21. This is followed by a further segregation of the neural tube in *pros-*, *mes-* and *rhombencephalon* (*fore-*, *mid-* and *hindbrain*) at STAGE 22. Simultaneously the *ventricular formation* makes progress. The originally slit-shaped cavities (STAGE 20) extend dorso-ventrally and widen. At STAGE 21 the *archencephalic ventricle* has a dorso-ventrally elongated form. At STAGE 22 the archencephalic ventricle widens markedly. The slit-shaped lumen in the rhombencephalic portion of the brain has begun to widen anteriorly at STAGE 21 and has become slightly triangular, so that a very first indication of a roof formation appears. The *rhombencephalic ventricle* is already clearly triangular at STAGE 22, when the ventricle is capped by a flat roof still several cell layers thick. The more caudal region, that of the *spinal cord*, still entirely confluent with the rhombencephalic portion, is expanding transversely and becomes triangular in cross section (with base upwards) at STAGE 21. The broad spinal cord contains a dorso-ventrally elongated central canal which anteriorly begins to become constricted in its middle by the thickening of the side walls.

During the primary segregation in the neural tube the *neural crest material* begins to migrate and becomes subdivided into individual cell masses. The massive *mesencephalic neural crest* shifts craniad. Its anterior border lies cranial to the posterior half of the eye anlage at STAGE 19 and at the level of the anterior border at STAGE 20. In the head region the neural crest material migrates laterad and ventrad from STAGE 19 on so that it is already located between brain and future branchial region at STAGE 20. It simultaneously segregates into separate cell masses for the formation of the visceral skeleton. At STAGE 21 the *rostral* (*maxillary*) *mesectoderm* is still located dorsal to the posterior portion of the eye anlage, whereas the *mandibular mesectoderm* (from mesencephalic origin) has already migrated to a position ventro-cranial to the archencephalon for the formation of the mandibular arch. Both portions are sharply delimited from each other at STAGE 22. The *hyal mesectoderm* from the rhombencephalic neural crest in front of the ear placodes (see below) has penetrated into the region of the future hyoid arch at STAGE 22. The more caudal neural crest destined to form the branchial arches has migrated as a massive cell mass into the future branchial region at STAGE 22. It becomes gradually segregated into branchial portions. The *first branchial portion* segregates at STAGE 23, the *second* at STAGE 31, while the *third* is separated from the *fourth* at STAGE 37/38 (see further under "the early development of the *alimentary system* and the *presumptive visceral skeleton* and *musculature*", on page 39).

The *ear placodes* develop as dorso-lateral thickenings of the sensorial layer of the ectoderm at the level of the middle of the rhombencephalon at STAGE 21 to 22. Shortly after their appearance, several other head placodes become visible. The *dorsolateral placodes* develop as local thickenings of the sensorial layer of the ectoderm on both sides of the brain at

STAGE 22, while the *olfactory placodes* arise fom the same layer latero-dorsal to the prosencephalon, first in continuity with the stomodeal-hypophyseal anlage at STAGE 23. Prosencephalon and overlying ectoderm develop a temporary intimate contact in the areas of the olfactory placodes from STAGE 23 till about STAGE 29/30.

THE INITIAL TAIL BUD STAGES

The formation of the *spinal cord* is much retarded with respect to the development of the brain. The *trunk neural crest material* still caps the unconstricted central canal of the spinal cord at STAGE 20. The withdrawal of the neural crest and the completion of the neural tube begins at its anterior end at STAGE 22. The neural crest now comes to lie above the neural tube. The completion of the spinal cord extends slowly in a cranio-caudal direction, while simultaneously the central canal becomes constricted in the middle. At STAGE 24 the anterior half of the spinal cord is completed, while the constriction of the central canal already extends over two thirds of its length. At STAGE 25 only the posterior one fourth of the spinal canal is still capped by neural crest cells. Only the posterior one fifth of the central canal is unconstricted and still capped by neural crest at STAGE 26. This has been reduced to one sixth at STAGE 28 and to the extreme posterior section of the spinal cord at STAGE 29/30. The withdrawal of the neural crest from the tube is not completed before STAGE 31. At STAGE 25 the lateral migration of the trunk neural crest has begun, so that the dorsal ridge of neural crest is thinner in the midline than laterally. Successive steps of lateral migration of the neural crest and of the formation of segmented masses lying against the lateral sides of the spinal cord, a process started at STAGE 26 in the anterior region, can be found at following stages. At STAGE 28 some neural crest cells have already migrated as far as the ventral border of the spinal cord in the region of its most anterior portion. These processes extend gradually in caudal direction with progress of development. The neural crest material which remains dorsal to the spinal cord, begins to scatter at STAGE 27 as a first symptom of *dorsal fin formation*. This scattering process extends rapidly in a posterior direction at STAGE 28 (see further under "the development of the *skin*, etc.", on page 44).

Returning to the development of the *brain* a rapid further progress of internal organization can be observed during the period from STAGE 23 to STAGE 28, at which stage its general pattern has been mainly established. The segregation of *tel-* and *diencephalon* is indicated by a slight dorsal depression at STAGE 23. The ventricular walls get better modelled and the *cephalic flexure* increases markedly (STAGE 24). Its angle decreases from 135° at STAGE 22 to only about 90° at STAGE 25. The antero-lateral portion of the telencephalic wall thickens as a first indication of an *olfactory lobe formation* at STAGE 26 to 27. The posterior telencephalic and diencephalic roof becomes thinner and the *pineal body* begins to evaginate at STAGE 26. At STAGE 28 the *infundibular wall* begins to thin out and the *hypophyseal anlage*, derived from the dorsal edge of the stomodeal-hypophyseal anlage in the sensorial layer of the ectoderm and gradually segregating from it while migrating inwards, has already penetrated beyond the optic stalks. The *rhombencephalic roof* enlarges and thins out at STAGE 26. Now the ventricle is about twice as wide dorsally as ventrally. At STAGE 28 the rhombencephalic ventricle has become mushroom-shaped, while the roof plate now consists of a cubical to cylindrical, single cell layer.

With the progress of development the macroplacodes segregate from the sensorial layer of the ectoderm and invaginate.

The *ear placode* shows a first slight depression (underneath the flat epithelial layer of the ectoderm) at STAGE 23, a process which continues at the following stages. The ear vesicle so formed is closed at STAGE 27, at which stage it is however still connected with the sensorial layer of the epidermis. It becomes entirely separated from the epidermis at STAGE 28 (see further under "the further development of the *auditory organ*", on page 82).

The development of the *olfactory placode* proceeds much more slowly. The placode becomes distinct at STAGE 23 as a thickening of the sensorial layer of the ectoderm latero-dorsal to the prosencephalon, but remains flat till STAGE 29/30. At this stage the overlying epithelial layer of the epidermis becomes somewhat thinner and the contact between olfactory placode and telencephalon becomes less intimate (see further under "the further development of the *olfactory organ*", on page 74).

The development of the brain is accompanied by the formation of the *cephalic ganglia* and *cephalic nerves* from the *dorso-lateral* and *epibranchial placodes*. The *trigeminus ganglia* have already developed at STAGE 24. At this stage the *profundus ganglion* shows clear differentiation of ganglion cells. The *Gasserian ganglion* is located behind the eye. The large VII *placode* located ventro-cranial to the ear placode, and the IX *placode* located ventro-caudally show a "massive" cell proliferation; groups of cells proliferate from the epidermal matrix and sink inwards. This massive proliferation continues for some time (STAGE 25) but decreases gradually at STAGE 26 and STAGE 27. The placodes remain visible and active till STAGE 29/30. At STAGE 28 the *epibranchial placodes* segregate from the dorso-lateral placodes, the *epibranchial placode* 1 from the VII *placode* and the *epibranchial placode* 3 from the *common* IX and X *placodes*. The dorso-lateral placodes gradually disappear. At STAGE 31 *ganglion* VII is still connected with the dorso-lateral placode. The same holds for *ganglion* IX at STAGE 32. At STAGE 31 the *epibranchial placodes* 3 and 4 are still quite active and at STAGE 32 the *epibran-*

chial placodes 4 and 5 show massive cell proliferation. The *epibranchial placode* 5 is connected with *ganglion viscerale* x at the STAGES 33/34 and 35/36 and still with its lower end at STAGE 37/38, after which the last epibranchial placode disappears.

From STAGE 28 on, at which the *fibre tracts* begin to develop, the development of *brain* and *spinal cord* will be treated separately, just as the development of the *cephalic ganglia* and *nerves*, while the development of the *spinal ganglia* and *nerves* is combined with that of the *spinal cord* (see respectively under "the further development of the *brain*", on page 52, "the further development of the *cephalic ganglia* and *nerves*", on page 60, and "the further development of the *spinal cord, ganglia* and *nerves*", on page 67).

* *
*

THE EARLY DEVELOPMENT OF THE AXIAL SYSTEM up to STAGE 37/38

The initial development has been described under "the early development up to STAGE 15" on pages 23 to 27.

After the segregation of the *notochord* from the dorso-lateral mesoderm, a process which takes place at an early stage of development (STAGE 12½ to 13), the first indication of *somite* formation is found at STAGE 17. At this stage the *myocoelic cavity* is, however, still continuous. The process of somite segregation progresses in a cranio-caudal direction. The following numbers of somites have been individualized at corresponding stages: STAGE 18, three to four somites in segregation; STAGE 19, four to six somites; STAGE 20, six to seven somites; STAGE 21, eight to nine somites; STAGE 22, nine to ten somites distinct; STAGE 23, approximately twelve somites; STAGE 24, fifteen somites segregated; STAGE 25, sixteen somites distinct, but somite I (first head somite) being reduced; STAGE 26, seventeen somites, somite I disintegrated; STAGE 27, nineteen somites; STAGE 28, more than twenty somites; STAGE 29/30, twenty-four to twenty-five somites, the somites II and III (the second and third head somites) much reduced, the otic vesicle located at a level with head somite IV; STAGE 31, twenty-two to twenty-three postotic somites; STAGE 32, twenty-six postotic somites; STAGE 33/34, thirty-two postotic somites; STAGE 35/36, thirty-six postotic somites, and STAGE 37/38, about forty postotic somites. It must however be realized that in the course of development the otic vesicle acquires a more caudal position with respect to the axial system probably due to stretching processes of head and trunk. After its appearance the otic vesicle is located at the level of head somite I at STAGE 24. It is displaced to the level of head somite II at STAGE 26, to that of head somite III at STAGE 27 and to that of the anterior half of head somite IV at STAGE 28. The otic vesicle is situated opposite head somite IV at STAGE 37/38, and finally opposite the first trunk somite at approximately STAGE 45.

Simultaneously with the segregation of an increasing number of *somites* from the somitic mesoderm, their cellular differentiation proceeds in a cranio-caudal direction. The individual somites are separated by inter-myotomic fissures. At STAGE 20 the myocoelic cavities begin to disappear. However, the myotomes remain double-layered (an outer layer or *dermatome* and an inner layer, of which the dorsal portion represents the *myotome s.s.* and the ventral portion the *sclerotome*). In the most anterior somites the *myoblasts* are already spindle-shaped at STAGE 20, while at STAGE 21 *myofibrillae* are formed. These myofibrillae become arranged in fusiform bundles at STAGE 23. After the gradual liberation of the sclerotome and dermatome cells from the original somites (see below) the myotomes proper are individualized at STAGE 29/30. At the antero-lateral side of each trunk myotome a mass of cells ("*Urwirbelfortsatz*") projects ventrally, indicating the beginning of the formation of the *ventral somatic muscles* of the *trunk* (STAGE 31). Their cells spread more ventrad between the cutis and the somatopleure. This migration (STAGE 32 to 37/38) and subsequent detachment from the *dorsal somatic musculature* (beginning anteriorly at STAGE 37/38) also proceeds in a cranio-caudal direction.

Simultaneously the prechordal plate mesoderm and the notochord differentiate. The *prechordal plate* cell material is already dispersing into mesenchyme at STAGE 21, filling up the space between the floor of the brain and the roof of the future pharyngeal cavity. The *notochordal cells* become oriented perpendicular to the longitudinal axis at STAGE 20, while their vacuolization begins in the anterior half of the trunk at STAGE 23 to 24, spreading in cranial and caudal directions. The tip of the notochord which becomes curved ventralwards during the development of the cephalic flexure of the brain, begins to vacuolate separately at STAGE 24, at which stage also the *elastica externa* is for the first time indicated. This sheath does not develop as a distinct membrane before STAGE 28. The vacuolization of the notochord progresses rapidly, so that at STAGE 26 vacuolization has been achieved over the length of the first ten somites and is initiated in the area of the following seven somites. At STAGE 28 the notochord is not yet completely vacuolized, a process which still continues up to about STAGE 37/38. Simultaneously with the vacuolization, the nuclei of the notochordal cells migrate towards the periphery at STAGE 28, where they arrange themselves, thus forming the *notochordal epithelium* at STAGE 31. This process continues for some time up to STAGE 35/36.

From the ventro-medial (*sclerotomic*) portions of the somites the presumptive *axial mesenchyme* is liberated segmentally at STAGE 24. It maintains its segmental arrangement during several stages of development and is not definitively separated from the muscular portions of the somites before STAGE 29/30, at which stage also the future *cutis cells*, originating from the dorso-lateral wall of the somites (dermatome) become apposed to the inner surface of the epidermis (see under "the development of the *skin*, etc.", on page 44). From the region of the sclerotomic proliferation the mesenchyme cells gradually spread and surround the notochord and the spinal cord (the latter in a more scattered manner). Around the notochord their segmental arrangement is clearly noticeable at STAGE 28 when the *sclerotome cells* form ringlike concentrations. At STAGE 31 the mesenchyme begins to disperse intersegmentally, soon forming a continuous layer around the notochord. The mesenchymal sheath is strongly compressed by the bordering somites at STAGE 32. At this stage both notochordal sheaths, the *elastica externa* and *interna* can be distinguished, the latter being very thin. At STAGE 37/38 the mesenchyme forms a clearly discernible continuous *perichordal tube*.

Along the dorsal midline of the thin archenteric roof a slight depression indicates the future *hypochordal anlage* at STAGE 20. Its segregation from the dorsal entoderm begins at STAGE 24 and progresses in a cranio-caudal direction, a process which is more or less completed at STAGE 28. At STAGE 32 the hypochord is present as a dorso-ventrally flattened ribbon of yolk-laden cells, consisting of only a few cells in cross section. It begins to disappear at STAGE 41, where it is only a slender rod of cells, and it cannot be distinguished any longer at STAGE 43.

The caudal portion of the axial system which will take part in the formation of the *tail bud* develops relatively late. Although the tail bud itself becomes distinct for the first time at STAGE 24, its actual elongation begins somewhat later. After a slow initial phase this stretching process accelerates at STAGE 28. The tail bud contains, besides the spinal cord, notochord and somite anlagen, also the *postanal gut* and the *neurenteric canal* (see also under "the early development of the *alimentary system* and the presumptive *visceral skeleton* and *musculature*", on page 42). Due to the elongation of the tail the post-anal gut has been very much drawn out. The canalis neurentericus ceases to exist at STAGE 35/36 when the lumen disappears from the posterior part of the postanal gut. During the following stages the lumen of the postanal gut diminishes more and more. At STAGE 37/38 its posterior three fifths forms a solid strand of cells. At STAGE 39 its posterior part is disintegrating into a strand of loose cells, after which the postanal gut becomes reduced to a short appendage at STAGE 40 and disappears at STAGE 41.

The segregation process of the *myotomes* reaches the *tail region* at STAGE 29/30 and spreads further caudad in the following stages. The differentiation of the myotomes and notochord, described previously on pages 35 to 36, also spreads further caudad. At STAGE 32 the *myoblasts* are spindle-shaped at the base of the tail. The cells are still very rich in yolk particularly dorsally and ventrally. At the base of the tail the *chordal epithelium* and the *elastica externa* are formed, whereas behind the twenty fourth postotic somite the notochord is still at the stage of a "pile of coins". All its cells still contain yolk inclusions. At STAGE 33/34, with thirty two postotic somites, the notochord is vacuolated up to the level of the thirtieth postotic somite. This degree of differentiation has attained the level of the thirty fourth postotic somite at STAGE 35/36 and that of the thirty fifth postotic somite at STAGE 37/38, at which stages respectively thirty six and forty postotic somites have segregated. At STAGE 35/36 *myofibrils* become visible in the myoblasts of the anterior tail somites, while at STAGE 37/38 yolk begins to disappear from the anterior region of the tail notochord. The yolk is, however, still abundant behind the thirtieth somite.

From STAGE 37/38 on the development of the *axial system* will be successively described for the *head* region in the division "the further development of *skeleton* and *musculature* of the *head*", on page 93, for the *trunk* region in the division "the further development of *skeleton* and *musculature* of the *trunk*", on page 107, and for the *tail* region in the division "the further development of *skeleton* and *musculature* of the *tail*", on page 110, while the development of *shoulder girdle* and *forelimbs* and of *pelvic girdle* and *hindlimbs*, which arise later, will be described respectively in the divisions "the development of *skeleton* and *musculature* of *shoulder girdle* and *forelimbs*", on page 112 and "the development of *skeleton* and *musculature* of *pelvic girdle* and *hindlimbs*" on page 117.

* *
*

THE EARLY DEVELOPMENT OF THE ALIMEN-TARY SYSTEM AND THE PRESUMPTIVE VISCERAL SKELETON AND MUSCULATURE respectively up to STAGE 28 and STAGE 37/38

The initial development has been described under "The early development up to STAGE 15" on pages 22 to 28.

The development of the *archenteric canal* does not make much progress before STAGE 20. At about STAGE 20 to 21 the wide archenteric canal becomes dorso-ventrally compressed and narrowed in the future trunk region so that a subdivision into a *foregut*, a *midgut* and a *hindgut* with protruding entodermal mass (residue of the yolk plug) becomes manifest. The stretching of the embryo after STAGE 20 results in a stretching of the archenteric canal and in an elongation of the still undivided yolk mass. The yolk mass protrudes more and more into the median portion of the archenteric canal, reducing it to a narrow transverse slit at STAGE 22, while also the foregut and the caudal archenteron become dorso-ventrally compressed. The further stretching of the embryo at STAGE 25 leads to a further reduction of the medio-lateral and dorso-ventral dimensions of the foregut cavity.

The most characteristic feature in the development of the foregut is the formation of the *visceral pouches* and the *oral evagination*. The *first* and *second visceral pouches* are indicated as slight depressions in the lateral wall at STAGE 20. They deepen gradually at STAGE 21 and 22. The entodermal wall of the *first* visceral pouch evaginates and protrudes towards the simultaneously developing local thickening of the sensorial layer of the ectoderm. Both layers establish a first contact at STAGE 23 which contact becomes intimate at STAGE 24. In this way the *first visceral arch*, i.e. the *mandibular arch* is formed. The development of the *second* visceral pouch, separating the *hyoid arch* as *second visceral arch* from the following *branchial arches*, begins somewhat later. Here both layers approach each other at STAGE 24, make a first contact at STAGE 26 and establish an intimate contact at STAGE 27. The *third* visceral pouch is indicated as a slight depression of the latero-caudal wall of the foregut cavity at STAGE 23. The entodermal layer and the only slightly thickened sensorial layer of the epidermis approach each other at STAGE 26 and make first contact at STAGE 27, so that now the *first branchial arch*—i.e. *third visceral arch*—is separated from the rest of the branchial mesoderm and mesectoderm at STAGE 27. The *fourth* visceral pouch is indicated at STAGE 25, but is still a shallow groove at STAGE 27. At STAGE 29/30 the entodermal evagination protrudes as an actual fold into the branchial mesenchyme. The contact with the ectoderm is not established before STAGE 32, at which stage the anlage of the *fifth* visceral pouch is visible as a furrow in the ventro-lateral wall of the foregut cavity, penetrating only slowly into the underlying mesenchyme. The *second branchial arch*—i.e. *fourth visceral arch*—has thus been definitely formed at STAGE 32, while the segregation of the *third branchial arch*—i.e. *fifth visceral arch*—from the material of the *fourth branchial arch*—i.e. *sixth visceral arch*—does not occur before STAGE 35/36. Finally, the *sixth* visceral pouch is indicated at STAGE 40. The branchial cavity is more or less funnel-shaped, so that the anlagen of the fourth and fifth visceral arches are situated more medially than those of the third (see table).

VISCERAL SKELETON

mandibular arch	1st visceral arch				
			1st visceral pouch		
hyoid arch	2nd ,, ,,				
			2nd ,, ,,		
1st branchial arch	3rd ,, ,,				
			3rd ,, ,,	1st visceral cleft	
2nd ,, ,,	4th ,, ,,				
			4th ,, ,,	2nd ,, ,,	
3rd ,, ,,	5th ,, ,,				
			5th ,, ,,	3rd ,, ,,	
4th ,, ,,	6th ,, ,,				
			6th ,, ,,		

At STAGE 29/30 the cell material of the first two, and at STAGE 31 that of the first four visceral arches already shows a clear distinction between future *skeletogenic* and *myogenic elements*, of which the latter are now located more superficially. The skeletogenic elements are of neural crest origin. They were originally located superficially, but have already migrated into deeper layers. The myogenic elements, which are of mesodermal origin, are very rich in yolk material. At STAGE 33/34 the myogenic material of the hyoid arch has split up into a ventral and a lateral portion, out of which will develop respectively the *m. interhyoideus* and the *m. orbito-hyoideus* + *m. quadrato-hyoangularis*. At STAGE 35/36 the lateral myogenic cell group begins to split up again into the anlage of the *m. orbito-hyoideus* and that of the *m. quadrato-hyoangularis*, a process completed at STAGE 37/38.

At the same (transverse) level as the first visceral pouch a median depression has become visible in the oro-pharyngeal floor at a point opposite the cement gland at STAGE 33/34. The entodermal epithelium of this depression which represents the anlage of the *thyroid*, has thickened. From this thickened epithelium a finger-like outpocketing turns caudad between the two branches of the truncus arteriosus.

Synchronously with the formation of the first visceral pouches the *oral evagination* of the oro-pharyngeal cavity develops. It becomes visible for the first time at STAGE 20 and deepens gradually. Passing the successive stages of a distinct depression at STAGE 21 and a short funnel-shaped evagination at STAGE 22 the oral evagination penetrates into the entodermal protrusion located between forebrain and cement gland. The opposing *stomodeal-hypophyseal anlage*, indicated at STAGE 20 as a thickening of the sensorial layer of the epidermis just in front of the brain, at STAGE 21 acquires the form of a wedge-shaped protrusion, which makes contact with the oral evagination of the anterior pharyngeal wall. The stomodeal-hypophyseal anlage remains a solid cellular wedge from which the *hypophyseal anlage* gradually segregates at about STAGE 28, migrating between forebrain and dorsal pharyngeal wall. The *oral plate* itself does not become distinct before the epidermal protrusions begins to shorten around STAGE 28, at which time the *mouth anlage* becomes slightly concave.

The foregut is moreover characterized by the formation of the *liver diverticulum*, situated ventro-caudally. It is indicated very early, from STAGE 13½ onwards, and deepens gradually till STAGE 22. At about STAGE 20 it is dorso-ventrally compressed but now extends laterally and even dorso-laterally (from STAGE 22 onwards), so that it forms a narrow, transverse, crescent-shaped slit at the STAGES 22 to 25. In later stages the liver diverticulum becomes still narrower and longer, whereas the communication with the foregut remains funnel-shaped. At STAGE 28 the liver diverticulum opens by a very narrow slit into the sub-mesodermal space on the ventral surface of the entoderm-mass. At STAGE 33/34 the anterior wall of the liver rudiment bulges forward and the funnel-like primary hepatic cavity has become shorter. At this stage the liver rudiment as a whole has shifted in a cranial direction.

An elevation of the pharyngeal floor, situated above the heart rudiment, indicates the beginning of the segregation of the *pharyngeal cavity* from the *gastro-duodenal portion* of the *foregut* at STAGE 29/30. Before that stage both cavities are in broad communication with each other; the future gastro-duodenal cavity being represented only by the caudal part of the foregut which is dorso-ventrally extended but medio-laterally rather narrow. This gastro-duodenal portion of the foregut is in communication with the dorsally situated midgut represented by a narrow canal at STAGE 28. The transversal ridge marking the boundary between the pharyngeal and gastro-duodenal cavities protrudes more and more dorsalwards. Both cavities are sharply separated from each other at STAGE 35/36, at which stage the floor of the posterior part of the pharyngeal cavity has been raised still more and the ridge developing at the tip of this elevation nearly touches the roof of the gut cavity.

During the further elevation of the floor of the foregut just anterior to the liver diverticulum the wall of the gut forms a pair of lateral diverticula just over the liver rudiment, representing the anlagen of the *lungs*. At STAGE 31 they lie in front of the fifth branchial pouches but more ventrally. They are clearly distinguishable as lateral diverticula at STAGE 32. At STAGE 33/34 they are situated posterior to the transverse ridge marking the boundary between the oro-pharyngeal and the gastro-duodenal part of the alimentary canal.

The deep penetration of the lateral and ventral blastoporal grooves at the end of the gastrulation process have led to the formation of the *ectodermal proctodeum*. The thick entodermal mass representing the invaginated yolk plug, is retracted from the hindgut and its lumen becomes stretched at STAGE 21. The formation of the *tail bud* is, among other things, reflected in a caudal curving of the posterior portion of the archenteric canal at STAGE 22, foreshadowing the *postanal gut formation* and in the ventrad displacement of the anal opening at STAGE 22. With the further outgrowth of the tail bud the postanal gut anlage becomes stretched. Its lumen is in broad communication with the hind gut at STAGE 25. The postanal gut, clearly discernable at STAGE 26, extends to the tip of the tail bud at STAGE 28. At that stage it still communicates through the *canalis neurentericus* with the spinal canal. At STAGE 28 the narrow canal of the midgut broadens towards the posterior half of the body, and then bends ventralwards and cranialwards to meet the ectodermal rudiment of the cloaca, which opens by the anus on the postero-ventral surface of the embryo.

From STAGE 28 the development of the *oro-pharyngeal cavity* and that of the rest of the *alimentary canal* will be treated separately (see respectively the divisions, "the further development of the *oro-pharyngeal cavity*, etc.", on page 140 and "the further development of the *intestinal tract* and *glands*", on page 150), while the further development of the *hypophyseal anlage* is described in the division "the further development of the *brain*", on page 53.

From STAGE 37/38 the development of the *visceral skeleton* and *musculature* is described in the division "the further development of *skeleton* and *musculature* of the *head*", on page 91.

THE DEVELOPMENT OF THE SKIN, LATERAL LINE SYSTEM AND PIGMENTATION PATTERN

A. THE DEVELOPMENT OF THE SKIN

THE DEVELOPMENT UP TO METAMORPHOSIS

THE EARLY DEVELOPMENT OF THE PRE-LARVAL SKIN

At STAGE 16 the *cement gland* anlage begins to stand out from the rest of the epithelial layer of the ectoderm ventro-caudally of the anterior border of the neural plate. The first cellular differentiation is indicated by an elongation of its cells and a concentration of the embryonic pigment. During the following stages the cement gland anlage thickens more and more, and becomes a uniform oval thickening of the epithelial layer of the ectoderm at STAGE 19. At this stage a general maturing of the ectoderm is indicated by a displacement of the embryonic pigment. Whereas in the preceding stages the *embryonic pigment* was accumulated at the basal side of the cells of the epithelial layer, it now (STAGE 19) becomes dispersed over the entire cell body and subsequently (STAGE 20) accumulates at the apical side of the cells.—At the last-mentioned stage a median thickening of the sensorial layer of the ectoderm just in front of the brain indicates the *stomodeal-hypophyseal* anlage. At STAGE 21 this anlage has the form of a wedge-shaped protrusion of the sensorial layer which makes contact with the oral evagination of the foregut, which is visible for the first time at the preceding stage (see further under the early development of the *central nervous system, sense organs, ganglia* and *nerves,* on page 33, for the *hypophyseal anlage,* and under the early development of the *alimentary system* and the *presumptive visceral skeleton,* etc., on page 41, for the *mouth.* For the development of the *macroplacodes* for the *olfactory* and *auditory organs,* and of the *dorso-lateral* and *epibranchial placodes* for the *cephalic ganglia* and *lateral lines,* the reader is also referred to the division on the early development of the *central nervous system, sense organs, ganglia* and *nerves* on page 33).

From STAGE 23 on the *epidermis* becomes thinner, mainly by a stretching of the sensorial layer which now forms a thin epithelium of flattened cells. The primary differentiation of the epidermis has extended over the entire embryo except for the tail bud area at STAGE 26.—The *frontal gland area* is indicated as a thickening of the epithelial layer overlying the prosencephalon at STAGE 25. The frontal glands begin to differentiate at STAGE 26, when they become cylindrical.—At STAGE 29/30 the future *cutis cells* derived from the lateral portions of the somites (dermatomes) have become apposed to the inner surface of the epidermis.

THE FURTHER DEVELOPMENT OF THE PRE-LARVAL SKIN

The *cement gland* has fully differentiated and the secretion of its cells has started at STAGE 28. The embryonic pigment gradually disappears from the cement gland, a process which has been completed at STAGE 44. The cement gland, as a larval organ, begins to degenerate at STAGE 47 and has almost disappeared at STAGE 49. The last traces can still be found at STAGE 50.

The *frontal glands* also represent an early larval differentiation. They show signs of cellular degeneration at STAGE 44, when their embryonic pigment has completely disappeared. The frontal area is practically identical with the rest of the skin at STAGE 49.

The *pre-larval skin* does not show any characteristic changes during a long period of development (STAGE 29/30 up to STAGE 37/38). It consists of two layers of cells, an *outer epithelial layer* or *periderm* and an *inner sensorial layer* or *stratum germinativum*. The periderm contains much embryonic pigment which diminishes gradually during following stages until it has completely disappeared at STAGE 44. Both layers still contain many yolk platelets at STAGE 33/34. This yolky material is gradually consumed during the following stages and has almost disappeared at STAGE 40. From STAGE 44 on the *larval skin* is quite uniform over the entire surface of the larva.

Underneath the skin the first *chromatophores* differentiate in the head and trunk region at STAGE 33/34 (see further under the development of the *pigmentation pattern*, on page 50).

The first *connective tissue cells* differentiate underneath the epidermis and in the fins at STAGE 37/38. Hardly any changes occur up to STAGE 50. The skin is still composed of a double-layered epidermis of more or less flattened cells and a thin layer of connective tissue indicating the formation of the *corium* or *cutis*.

THE DEVELOPMENT OF THE LARVAL SKIN

At STAGE 50 vacuolisation starts in the cells of the sensorial layer of the *epidermis*, which begins to proliferate at the same time, so that a large number of *unicellular glands* develops, particularly in the fins and the adjacent skin of the trunk. These unicellular glands are also formed in the former frontal gland area on the upper jaw. Their number increases in all parts of the skin at STAGE 51 and following stages except for the skin of the tentacles where they do not develop. The unicellular glands of the larval skin have become very numerous at STAGE 56. The thin but compact layer of connective tissue underlying the epidermis at STAGE 51 has become somewhat thicker at STAGE 56.

THE DEVELOPMENT DURING METAMORPHOSIS

During the stages of metamorphosis pronounced changes take place in the skin. At STAGE 57 the *layer of connective tissue* thickens markedly and acquires a structure practically identical to that of the *stratum compactum* of the fully-differentiated skin. This holds for the entire skin, viz. for the areas where the adult skin develops from the stratum germinativum as well as for the areas where the larval skin will be broken down later on. The changes are less pronounced in the tail region and in the fins. In the *metamorphosing parts* of the *skin*, viz. in a broad girdle around the trunk, and areas on the upper and lower jaw and on the limbs, a strong proliferation occurs in the stratum germinativum. From the proliferating tissue a complete *new epidermis* is formed underneath the remainder of the larval epidermis, as well as the numerous glands below the epidermis. The layer between the epidermis and the stratum compactum develops into the *stratum spongiosum*, which contains the glands surrounded by some loose connective tissue in which blood and lymph vessels penetrate through openings in the stratum compactum. The draining tubes of the glands pass through the epidermis and open at its surface. Two types of *glands* can be distinguished, the so called "*granular*" and the "*mucous*" types, the former being the more common type at the earlier stages, the latter at the later stages. At STAGE 57 various developmental stages of the skin and skin glands can be found in the metamorphosing areas. The girdle around the trunk dorsally extends with a cranial protrusion up to the diencephalon, and caudally down to the base of the tail. Laterally and ventrally the girdle is much narrower and extends from a line directly behind the forelimbs and the heart down to a line somewhat anterior to the hindlimbs. The glands are most numerous and best developed on the back. Here melanophores, previously located underneath the stratum compactum, migrate into the stratum spongiosum where they settle immediately underneath the epidermis. Some even penetrate into the epidermis. In the area on the upper jaw only mucous glands are formed. The metamorphosing epidermis is thicker in this area than in other parts of the skin.

The following stages are characterized by a progressive development of the metamorphosing skin and by a gradual extension of the metamorphosing areas, while the interjacent larval skin gradually degenerates and shrivels up.

At STAGE 58 the *metamorphosing skin* of the forelimb, which has now broken through the larval skin, is caudally already continuous with the girdle of metamorphosing skin of the trunk. At STAGE 59 the area of adult skin on the upper jaw is almost completely developed. The mucous glands are very crowded so that they compress each other. Along the edge of the lower jaw a well developed epidermis is present at STAGE 59, but glands do not develop in this area. At this stage the metamorphosing areas are well circumscribed. They are separated from each other by strips of larval skin. The metamorphosing skin area of the trunk now extends partly over the telencephalon. Later the metamorphosing areas of the skin still increase somewhat in size. The trunk skin girdle covers the entire brain cavity and the caudal part of the heart at STAGE 61. During these stages the differentiation of the adult skin progresses. A large number of well developed glands can be found on the back at STAGE 58. The glands belong mainly to the granular type. At STAGE 60 numerous *horn papillae* begin to develop on the metamorphosing skin, the centre of each papilla formed by one cell of the stratum germinativum. At this stage a well defined *stratum corneum* has been formed only on the upper jaw in the neighbourhood of the olfactory organs. The metamorphosing skin is still covered with fragments of larval epidermis. The metamorphosis of the adult skin areas has been completed at STAGE 61, when a clear stratum corneum has been formed [1]).

In the stratum spongiosum underneath the metamorphosing areas of the skin, the *lymph sacs* begin to develop at STAGE 59. Most of them are already well developed at STAGE 60. In the strips of larval skin in between the metamorphosing areas the skin is connected with the underlying musculature by loose connective tissue, indicating in general the future *septa* between the lymph sacs.

The interjacent areas of *larval skin* show the first symptoms of degeneration at STAGE 59. The larval skin becomes thicker and the fibres and cells of the stratum compactum change their orientation as symptoms of shrinkage. This process starts in the area between the metamorphosing skin areas of the trunk and the hindlimbs. This strip of larval skin has already disappeared for the greater part at STAGE 60, while the other areas do not yet show clear symptoms of degeneration. The areas of larval skin on the head (between the metamorphosing areas of the upper and lower jaw and that of the trunk), which cover the entire filter apparatus, including the operculum, and surround the eyes, begin to shrivel at STAGE 62. This process is already far advanced at STAGE 63, when the area between the upper jaw and trunk fields has mostly disappeared, the operculum has become a shapeless mass of cell material, and only narrow strips of larval skin can be found around the eyes and between the lower jaw and trunk fields. At this stage the gill slits have disappeared. At STAGE 65 the skin has metamorphosed almost completely: the metamorphosed areas of the skin touch each other almost everywhere. A small area of very shrivelled larval skin still lies rostrally of the forelimbs. During the reduction of the *tail*, which starts at its tip at STAGE 58 and

---

[1]) Several small *lymphocyte accumulations* are found in the metamorphosed skin areas from STAGE 62 on. A constant large accumulation develops dorso-caudal to the forelimbs.

has made good progress at STAGE 63, the larval skin of the tail shrivels up and degenerates, so that only a small accumulation of pigment and degenerated cells represents the remnant of the tail at STAGE 66.

The skin of the *forelimb* develops out of the medial wall of the forelimb atrium. A basal membrane is formed at STAGE 51. Glands begin to develop in the skin of the basal portion of the forelimb at STAGE 53. Their number is still low at STAGE 54, but begins to increase at STAGE 55, when they appear in greater numbers in the outer surface of the limb, which is directed caudalwards. The hand and the inner surface are free of glands. At STAGE 57, when the forelimb has not yet broken through the body wall, the metamorphosis of the skin has hardly started in the forelimb. Dorsal, ventral and caudal to the base of the forelimb a small area of the wall of the atrium has the same structure as the skin of the limb and will also metamorphose. The rest of the larval skin of the atrium degenerates after the perforation of the forelimb at STAGE 58. (See also page 45.)

The skin of the *hindlimb* is somewhat more advanced than that of the forelimb, like the entire hindlimb anlage appears earlier than the forelimb anlage. Unicellular glands are already well developed at STAGE 51 in the proximal part of the outer surface of the limb. Also the metamorphosis of the skin of the hindlimb starts somewhat earlier than that of the forelimb skin and is already clearly indicated at STAGE 57. (See also page 45.)

B. THE DEVELOPMENT OF THE LATERAL LINE SYSTEM [1])

THE INTERNAL DEVELOPMENT

This development begins rather late and is nearly completed within a short lapse of time.

At STAGE 33/34 the anlagen of the *supra-* and the *infra-orbital lines* appear anterior, and the common anlage of the *trunk lines* posterior to the ear vesicle. They have already almost completely segregated from the other placodal anlagen at that stage. The *supra-orbital line* reaches a point above the eye at STAGE 35/36. At STAGE 40 this line has grown out to a point approximately midway between eye and olfactory organ, while its differentiation into individual sense organs has started. The *infra-orbital line* extends to a point below the eye at STAGE 35/36. The segregation into individual sense organs starts first in this line and is already indicated at STAGE 39. Probably as derivatives of the infra-orbital line, *hyomandibular sense organs* can be found in the form of cell proliferations to the left and right of the cement gland at STAGE 35/36. At STAGE 40 the hyomandibular line has extended posteriorly to the neighbourhood of the heart. At STAGE 45 the *parietal sense organs* are also present. The *dorsal* and *middle lines* of the *trunk* have extended beyond the pronephros at STAGE 35/36 without a

---

[1]) In this section the nomenclature given by ESCHER (1925) is followed.

sign of differentiation into individual sense organs. At STAGE 40 these lines have extended up to the base of the dorsal fin at the beginning of the tail. Their rostral portions have already segregated into individual sense organs, but their distal portions still form a continuous strip of cell material. At STAGE 37/38 the *occipital line* becomes visible opposite the caudo-ventral portion of the ear vesicle from where it extends first dorsad and then caudad at STAGE 40, at which stage it also segregates into individual sense organs. The occipital line has formed two to three sense organs dorso-caudally of the ear vesicle at STAGE 41. The *ventral line* of the *trunk* appears ventrally of the pronephros at STAGE 37/38. A strong proliferation of this anlage occurs at STAGE 39, at which stage it forms a more or less circular placodal cell mass. This ventral line has extended both ventrad and caudad, forming a continuous band of cells at STAGE 40. It runs from a point halfway between pronephros and ventral median line, first in a dorsal direction, then curves caudad underneath the pronephros, and terminates at some distance behind the pronephros. This line has extended further at STAGE 41, ventrally of the pronephros in a ventral direction and caudally of this organ in a ventro-caudal direction. At that stage it has partially segregated into individual sense organs. The ventral line of the trunk has reached the pre-anal fin at STAGE 43 and develops further as *"caudal line"* on the pre-anal fin at STAGE 44. At STAGE 45 it has formed one sense organ and the anlage of a second one. Although some lines continue to extend slightly after this stage no new formations occur.

At STAGE 41 all lateral lines have differentiated into individual, originally circular sense organs. The sense organs begin to stretch, forming elongated and later streak-shaped organs at STAGE 49, the oldest (proximal) ones being the most advanced. They form local thickenings of the epidermis with a large number of nuclei. At STAGE 51 they become more conical in form in cross section, their apical ends reaching to the surface of the skin.

THE EXTERNALLY VISIBLE DEVELOPMENT

The larval lateral line system becomes externally visible at the STAGES 43 to 45, and more clearly at STAGE 46, since the skin has become transparent.

The *supra-orbital line* runs rostrad from the ear, passes between eye and brain region and terminates at a point midway between the eye and the tentacle rudiment. The *infra-orbital line* running ventrally and caudally to the eye, and the *hyomandibular line* running backwards over the ventral surface of the head from a point midway between the eye and the ventral midline, first become visible at STAGE 43. The last-mentioned line can be followed at STAGE 46 posteriorly into the region of the heart and anteriorly towards the base of the tentacle rudiment, from where it extends mediad over the lower jaw, running parallel to its edge. The left and right lines will later on unite in the midline.

Groups of sense organs are found lateral to the telencephalic region, representing the *parietal lateral line organs*.

Dorsally the *dorsal* and *middle lateral lines* run along the trunk. The sense organs of the dorsal line are oriented roughly in the main axis of the line and those of the middle line roughly perpendicular, to the axis. Both lines run straight backwards from the region of the ear to a point dorsal to the pronephros, and from there in a dorso-caudal direction to the base of the tail from where they continue along the tail. From their anterior end the *occipital line* ascends dorsad over the ear at a short distance in front of the first muscle segment of the trunk, and turns backwards at a right angle near the median line.

More laterally the *ventral lateral line* runs over the abdomen, originating medially behind the heart, running first laterad and then caudad for a short while, then turning dorsad towards a point behind the forelimb, from where it runs ventro-caudad along the upper border of the abdomen until it reaches the pre-anal fin. On the pre-anal fin this line, now called the *caudal line*, runs to about the middle of the cloaca in a dorso-caudal direction. At STAGE 52 sense organs of the caudal line appear on the post-anal fin. They remain circular.

THE DEVELOPMENT OF THE LATERAL LINE SYSTEM OF THE ADULT

During metamorphosis only restricted changes occur in the lateral line system. It is taken up for the greater part into the adult skin. Some parts of the system lying in areas of larval skin wedged in between areas of adult skin (cf. the development of the *skin*, page 46) disappear gradually together with the larval skin at STAGE 61, so that parts of the *infra-orbital line* and the portion of the *hyomandibular line* between the corner of the mouth and the heart have disappeared at STAGE 64. The *sense organs* on the *tail* disappear during the reduction of the tail at STAGE 64. The shrinkage of the head leads to some displacements of the other lateral lines. The *supra-orbital sense organs* become crowded together along the dorsal edge of the eye by the degeneration of the larval skin around this organ. The *sense organs* on the *pre-anal fin* are possibly incorporated into the adult skin of the belly.

The individual *sense organs* in the adult skin consist of a series of three to five cone-shaped groups of sensorial elements, surrounded by epidermal cells.

C. THE DEVELOPMENT OF THE PIGMENTATION PATTERN

THE DEVELOPMENT UP TO METAMORPHOSIS

Melanophores appear for the first time dorsally on the head at STAGE 33/34[1]). They still contain yolky material and are only partially pigmented. Similar, but unpigmented elements can likewise be distinguished at this stage. Meanwhile melanophores appear on the pronephros and in a horizontal row along the trunk, beginning ventral to the pronephros and running along the ventral edge of the axial musculature. The melanophores spread from the dorsal surface of the head over the back, forming a narrow strip along the dorsal edge of the trunk musculature at STAGE 35/36 and become more heavily pigmented (black) at STAGE 37/38. At STAGE 39 they are arranged at two different levels, viz. in a *superficial layer* immediately underneath the dorsal skin and the base of the dorsal fin, and in a *deeper layer* upon the dorsal surface of the central nervous system. Anteriorly melanophores also appear around the nasal pits and posteriorly they appear superficially along the ventral edge of the musculature of the tail at STAGE 39. During the following stages (STAGES 40 to 43) the superficial melanophores spread from the dorsal row ventrad over the area of the axial musculature of the tail (except in its rostral part). They become arranged in irregular, more or less parallel longitudinal rows. The deeper layer of melanophores gradually covers the entire dorsal surface of the central nervous system, showing a sharp lateral boundary at STAGE 43. Melanophores also appear underneath the olfactory placodes, in the filter apparatus, around the eyes and in the peritoneum during these stages. During the following stages (STAGES 44 to 46) only minor changes occur.

At STAGE 46 *xanthophores* appear in the peritoneum and on the outer layer of the eye cup. At STAGE 48 they form a continuous layer in the peritoneum and on the ventral part of the eye. At STAGE 50 they also appear in the pericardium.

At STAGE 47 many *melanophores* seem to migrate from their superficial position towards the deeper layer. From that stage till STAGE 54 the superficial pigmentation is much less dense than before in the ventral part of the axial musculature area. Melanophores appear around the n. octavus at STAGE 48 and around the other cephalic nerves at STAGE 49. At the same stage they appear on the thymus gland and in the dorsal and ventral fins near the tip of the tail. The pigmentation of the fins proceeds gradually in a rostral direction during the following stages. At STAGE 54 there is a marked increase in the number of melanophores in the tail, where they begin to arrange themselves in longitudinal rows and along

---

[1]) Stages at which processes of pigmentation occur are only approximate, since these processes are to some extent dependent on external conditions and growth rate.

the muscle segment borders. At that stage the number of melanophores begins to increase also in the head region, while their size decreases, and they acquire a more homogeneous distribution over the head. The overall pigmentation in head and tail regions increases further during the next stages.—The pigmentation of the blood vessels starts at STAGE 49 in the head region. In the tail melanophores are also found at STAGE 53 around the dorsal branch of the caudal vein and at STAGE 54 around the arteria caudalis. They have appeared around all blood vessels of the tail at STAGE 56.—Melanophores appear on the hindlimb buds at STAGE 51 and on the forelimbs at STAGE 54. On the hindlimbs, later in development, pigmentation is lacking on the joints of the toes at STAGE 56. With the development of an isolated area of pigmentation on the ventral fin near the cloaca at STAGE 56 the larval pigmentation pattern has now been fully differentiated.

THE DEVELOPMENT DURING METAMORPHOSIS

During metamorphosis several changes take place in the pigmentation pattern. The beginning of the transformation of larval into adult skin is indicated by a change of colour into a greenish shade and by the appearance of small, round, whitish dots, representing the openings of the adult skin glands. This begins in patches on the dorsal surface of the trunk, on the upper jaw and on the hindlimbs at STAGE 57. These patches of adult skin expand during the following stages (see under the development of the *skin*, page 45). Besides some local differentiations such as a pigmentfree spot which appears above the "Stirnorgan" at STAGE 57, and irregular dark spots formed in the adult skin on the back by a local expansion of the *melanophores* at STAGE 59, *guanophores* appear on the belly and the thighs, giving these areas a silvery colour. They appear for the first time at STAGE 58 and increase rapidly in number at STAGE 59, at which stage they also spread to the adult skin areas of the forelimbs. At STAGE 60 they appear on the ventral surface of the lower jaw.

With the fusion of the adult skin areas the definitive pigment pattern of the young frog has been established at STAGE 66.

* *
*

THE FURTHER DEVELOPMENT OF THE BRAIN

The earlier development up to STAGE 28 has been described in the division "The early development of the *central nervous system, sense organs, ganglia* and *nerves*", on pages 29 to 33.

THE DEVELOPMENT OF THE BRAIN FROM STAGE 28 TO ABOUT STAGE 40

In the following period of development, ranging from approximately STAGE 28 to approximately STAGE 40, the yolky material is consumed, the fibre tracts and later the commissures develop, and the segregation of the various brain structures continues.

The *yolky material*, still abundant at STAGE 28, is gradually consumed, so that practically no yolk is left at STAGE 40, and the last platelets have disappeared at STAGE 41.

At STAGE 28 a narrow zone of *nerve fibres* develops along the ventro-lateral portion of mes- and rhombencephalon and simultaneously along the anterior portion of the spinal cord (see page 67). This *marginal layer*, which is still very thin at STAGE 29/30 and STAGE 31, becomes broader and extends somewhat more lateralwards in the mesencephalic area at STAGE 32. At STAGE 33/34 a well developed marginal layer can be found in the diencephalon, especially in its hypothalamic part, and in the mesencephalon, but less pronounced in the ventro-lateral part of the myelencephalon. The marginal layer thickens and extends laterad. At STAGE 37/38 it is also present in the lateral telencephalon. The lamina terminalis, however, does not contain distinct *white matter* before STAGE 41.

A *chiasmatic ridge* develops as a thickening of the floor of the forebrain at STAGE 29/30. It is rostrally separated from the thick *lamina terminalis* by a shallow groove. During the following stages the ridge becomes higher and well separated from the lamina terminalis by the deepening *preoptic recess* (STAGE 31 and 32). At STAGE 33/34 the first *optic fibres* appear basally in the middle of the semicircular chiasmatic ridge. Simultaneously the preoptic recess broadens and forms small lateral recesses, while a thick commissural plate develops in front of it. During the following stages the number of optic fibres increases rapidly in the chiasmatic ridge till the definitive number is acquired. Near the borderline of pros- and mesencephalon some commissural fibres develop in the region of the *caudal commissure* (*commissura posterior*) around STAGE 37/38. Their number increases rapidly, so that many fibres are present at STAGE 40. At STAGE 47 this commissure has greatly increased in size. In the commissural plate in front of the preoptic recess some fibres of the *anterior commissure* develop at STAGE 39; they become more abundant at STAGE 40. At this stage some fibres are also crossing in the *habenular commissure*, connecting the now distinct habenular thickenings. The commissure becomes more distinct at STAGE 41. The *cerebellar plate* broadens somewhat at STAGE 37/38, but does not become thicker than the mesencephalic roof lateral to the mid-dorsal line before STAGE 40. Fibres of the *cerebellar commissure* first appear at STAGE 42.

During this period of development the brain acquires more and more its general form and structure. At STAGE 29/30 the anterior portion of the *prosencephalon* is still strongly bent. During the following stages it straightens gradually, but is not completely straightened before STAGE 43. At STAGE 37/38 the *prosencephalic ventricle* has increased in size and the thin prosencephalic roof rostral to the epiphysis has elongated. Simultaneously the *dorsal hypothalamus* protrudes further into the third ventricle; a similar development of the *ventral hypothalamus* can be seen at STAGE 42. The thin-walled *lateral recesses* of the *preoptic recess*, visible for the first time at STAGE 33/34, have been well developed by STAGE 37/38. At this stage tall columnar ependymal cells develop in the region of the caudal commissure, representing the anlage of the *subcommissural organ*.

The *pineal body*, which has begun to evaginate at STAGE 26, is gradually pinched off. The anlage is relatively large, semicircular in form and contains a large lumen in broad communication with the ventricle of the brain at STAGE 29/30. It becomes mushroom-shaped with a still rather wide but already somewhat constricted opening of its lumen into the ventricle at STAGE 31. The pineal body has become more massive with only a small lumen dorsally and with an extremely small orifice communicating with the ventricle at STAGE 32. At STAGE 33/34 its lumen has become still smaller, while its base shows a small pit which is the remnant of the orifice. At STAGE 37/38 it has a slit-like lumen running through the massive organ. The pineal body becomes further flattened and its lumen almost disappears. At STAGE 39 the anlage is rather elongated, massive and with a somewhat pointed rostral end, lying on the now thin prosencephalic roof, while its caudal end lies upon the somewhat thicker *pars intercalaris* of the diencephalon. Basally it sometimes still shows a very small lumen and orifice. These seem to be lacking at STAGE 40, at which the pineal body forms a large, flat, rather massive cell-cluster.

During this period the *infundibular wall* thins out gradually and the *hypophyseal cell plate*, segregating from the *stomodeal-hypophyseal anlage*,

becomes gradually displaced towards its definitive position in front of th
tip of the notochord. At STAGE 29/30 the infundibular wall is still rathe
thick, but thinner than the rest of the prosencephalic wall. This re main
the case at STAGE 31, although the roof of the prosencephalon becomes
thinner. At this stage the infundibular wall has markedly flattened in the
midplane. At STAGE 29/30 the hypophyseal cell plate is not yet very distinct.
At STAGE 31 it reaches to the caudal border of the chiasmatic ridge, while
its rostral end is still in distinct connection with the deeper, sensorial layer
of the ectoderm. At STAGE 32 this connection is interrupted. The hypophyseal cell plate now gradually shifts caudad. At STAGE 37/38 the main portion
of the hypophyseal plate lies ventral to the caudal two-thirds of the
chiasmatic ridge, reaching caudalwards just beyond this ridge. At this
stage the front part of the hypophyseal cell plate has loosened. At STAGE 39
the rostral end of the hypophyseal plate lies caudal to the rostral end of
the chiasmatic ridge. At STAGE 40 the rostral end of the *pars distalis* of the
hypophysis is situated at the middle of the chiasmatic ridge, while its
caudal end tapers out. At STAGE 41 it reaches from just caudal to the
middle of the chiasmatic ridge to the ventro-rostral side of the notochordal
tip, while its caudal end has rounded off. At STAGE 42 the anlage of the
pars distalis of the *adenohypophysis* reaches finally from the caudal end of
the optic chiasma to the ventral side of the notochordal tip. Until this
stage no differentiation occurs in the anlage.

There is already a very distinct *pros-mesencephalic groove*, but no distinct
mes-rhombencephalic one at STAGE 29/30. At STAGE 31 the latter still only
forms a very shallow depression. At STAGE 32 it becomes more distinct
in the dorsal midline between the mesencephalic roof and the future
cerebellar plate, which now in the midline is just as thin as the mesencephalic roof. This groove deepens gradually at STAGE 33/34 and extends
laterad and ventrad at STAGE 37/38. On the ventricular side the mes-
rhombencephalic border bulges inward and finally cuts deep into the
ventricle at STAGE 40. Simultaneously the border between pros- and
mesencephalon protrudes also middorsally as a distinct ridge into the
ventricle.

During this period of development the *roof* of the ivth *ventricle* gets
rapidly thinner, a process already clearly indicated at STAGE 31. It has
become very thin at STAGE 32, while the caudal border of the thickening
cerebellar plate becomes very distinctly demarcated against it at STAGE
37/38.

At STAGE 41 the anlage of the *pia mater* becomes visible at the level of
the mesencephalon. At STAGE 42 the *dura-endocranial membrane* begins to
differentiate also at the level of the mesencephalon. At STAGE 43 the
dura-endocranial membrane, which develops further, becomes heavily
pigmented dorsal to the brain. It does not surround the entire brain

before STAGE 46. Again at the level of the mesencephalon, at the lateral sides, the membrane separating the *ecto-* and *endomeninx* is splitting off from the dura-endocranium at STAGE 48. This membrane becomes much more distinct alongside the mes- and rhombencephalon at STAGE 49, while delamination likewise occurs at the level of the prosencephalon, a process completed at STAGE 50.

At STAGE 43 some few *pial vessels* start entering the brain. Their number does not increase very much during the following stages. The blood supply becomes, however, more voluminous and the vessels penetrate into deeper layers of the brain at STAGE 46 (see also under "the development of the *heart* and *vascular system*, etc." on page 123).

THE FURTHER DEVELOPMENT OF THE BRAIN

The *prosencephalon* straightens during the first stages of this period of development, a process completed at STAGE 43. At this stage the *cerebral hemispheres* begin to develop. Their formation proceeds slowly, although at STAGE 46 the hemispheres are already much more developed. This process continues at later stages. At STAGE 50 the *olfactory bulbs* begin to fuse, a process not completed before STAGE 58. The rostral ends of the *lateral ventricles* now reach into this fused part of the hemispheres.

At the dorsal borderline of tel- and diencephalon the *choroid plexus* differentiates. The first signs of this differentiation occur at STAGE 43, when the cells in the midline of the prosencephalic roof just become cylindrical. At STAGE 44 invagination starts but is still very slight. This invagination continues during the following stages. At STAGE 47 the plexus, which is large and has invaginated more deeply, looks like a bunch of grapes. At STAGE 49 *bilateral telencephalic choroidal invaginations* begin to develop at both sides of the *median plexus*. They are more clearly visible at STAGE 50, when the median plexus has strongly developed and has become more complicated in appearance.

The *paraphysis* develops rather late and can be distinguished as a small tube-like anlage only at STAGE 49. This begins to ramify, forming some small tubules at STAGE 50. At STAGE 51 many well developed paraphyseal tubules have been formed, protruding in a dorsal direction. No significant changes occur at later stages.

In the diencephalon the *habenular region* differentiates further by the formation of spherical masses of white matter (STAGE 42) connected by the *habenular commissure*. The commissure is already well developed at STAGE 44 while the *habenulae* become large and well-differentiated at STAGE 46. The habenular region greatly increases in size during the following stages. At STAGE 51 a well developed habenular commissure extends over some length in the midplane rostro-caudally. At this stage nuclear differentiation has made good progress in the diencephalon.

The *pineal body* does not change much in the STAGES before and after 40. It forms a flattened, elongated body, sometimes with some very large lacunae, mainly in its basal portion, and an orifice which is tiny or absent (STAGE 41 to 42). At STAGE 42 its rostral end, containing a small lumen, begins to segregate from the rest of the pineal body, indicating the anlage of the "Stirnorgan". Further development of the latter proceeds very slowly, so that little progress has been made at STAGE 45. At STAGE 46, however, the "Stirnorgan" has developed much further and has a triangular shape. It is situated in between the epidermis and the dura-endocranium, still caudal to the prosencephalic choroidal invagination, and is as yet broadly connected with the rostral end of the *epiphysis*. At STAGE 47, its rostral end reaches in between the hemispheres. It remains broadly connected with the epiphysis during the STAGES 47 and 48. This connection, however, becomes much more constricted at STAGE 49 and decreases to a thin nervous strand at STAGE 50. At that stage the "Stirnorgan" is oval and rather elongated, containing a lumen in its dorsal half. The entire organ has shifted into a more rostral position, so that its centre now lies dorsal to the paraphyseal anlage. From STAGE 51 on the "Stirnorgan" begins to decrease in size, becoming more spherical. For some time it looks rather solid (STAGE 51) but a lumen appears again at STAGE 52. The dorsal cell layer becomes thin and the basal one thicker. Later on it diminishes still more in size and becomes solid again at STAGE 55 and more spherical at STAGE 56. At STAGE 60 it is a tiny spherical organ. From STAGE 50 on the "Stirnorgan" shifts more and more rostrad. Its caudal border lies at the level of the paraphysis at STAGE 52 and at that of the lamina terminalis in front of the paraphysis at STAGE 53. At STAGE 55 the organ lies well rostral to the lamina terminalis and much nearer to the level of the olfactory bulbs. At STAGE 56 it lies just caudal, and at STAGE 58 dorsal to the fused olfactory bulbs. At STAGE 60 its position is already rostral to the latter and at STAGE 62 it finally lies in front of the rostral end of the brain capsule, dorsal to the nasal septum. A nervous connection with the rostral end of the epiphysis remains intact, probably extending to the habenular commissure.

The *epiphysis* itself in which some larger lacunae and sometimes a small orifice were found at STAGE 49, acquires a very elongated and flattened shape with a large lumen and a caudally situated distinct orifice just caudal to the habenular commissure at STAGE 50. Its form is rather variable, containing either many lacunae or one large lumen at STAGE 51. At later stages no further changes occur, except that the orifice becomes somewhat wider at STAGE 58.

The *subcommissural organ* represented by a columnar ependyma ventral to the caudal commissure (comm. posterior) develops slowly. The cells of the ependyma gradually become taller, the organ protruding deeper into the ventricle at STAGE 50.

In the ventral portion of the diencephalon changes occur in the lateral recesses of the preoptic recess and particularly in the infundibulum and hypophysis. The blind ends of the *lateral recesses* of the *preoptic recess* run in a somewhat caudal direction at STAGE 44.

At STAGE 41 large *lateral infundibular recesses* develop. The anlage of the *adenohypophysis* becomes more compact, its largest mass of cells lying just in front of the notochord at STAGE 45. The differentiation of the hypophysis starts in the *pars intermedia* at STAGE 47 when the distal part of the adenohypophysis has grown still larger. At this stage the infundibular epithelium is still flat. The *infundibular process* of the *neurohypophysis* begins to develop at STAGE 49 by a differentiation of the infundibular cells. A small number of *fibres* appear in the base of the infundibulum. Simultaneously the first anlagen of the *tuberal parts* of the *adenohypophysis* become manifest. From STAGE 50 on the latter and the infundibular process of the neurohypophysis differentiate rapidly. The tuberal parts become very distinct and well developed at STAGE 51, extending in a rostro-caudal direction. The differentiation of the *pars intermedia* also proceeds at STAGE 52. The infundibular process increases in size, especially laterally, and a greater number of nerve fibres can be found in the infundibular floor at STAGE 51 to 52, so that all parts of the hypophysis have been well developed at STAGE 53; at later stages they only increase in size and show a further cytological differentiation. The infundibular process of the neurohypophysis particularly increases in length at STAGE 57.

In the mesencephalon the *optic ventricles* begin to develop at STAGE 41. First they extend lateralwards, forming large lateral optic-ventricular recesses in the caudal part of the tectum at the STAGES 42 and 43, and later on at STAGE 44 in a somewhat more caudo-lateral direction, so that the tectal optic ventricles begin to protrude into both caudal tectal poles at STAGE 48. This is still more pronounced at STAGE 50, at which the tectal lobes themselves protrude caudalwards. The optic ventricles reach into the posterior tectal poles at STAGE 51. The peripheral nuclear layer of the *optic tectum* begins to develop at about STAGE 47, and is well differentiated especially in its rostral part at STAGE 48. At STAGE 49 the lamination of the optic tectum has made good progress; at least five nuclear layers are present now, including the ependyma. In the caudal part of the tectum the white matter has hardly appeared at STAGE 48. During the following stages the tectal laminar differentiation progresses rapidly proceeding in a caudal direction, so that the white matter is also present in the midline in the caudal part of the optic tectum at STAGE 52. At STAGE 55 tectal differentiation seems to have reached its final stage. At STAGE 50 a constriction begins to appear between the caudal part of the optic tectum and the ventrally situated *tegmental portion* of the mesencephalon.

After STAGE 42 at which, laterally, the *corpora cerebelli* have thickened and the first fibres have appeared in the *cerebellar commissure* near the deep mes-rhombencephalic groove, the *cerebellar plate* in the mid-dorsal line develops rather slowly. It is not very much thicker than the mesencephalic roof at STAGE 44, although the number of commissural fibres increases. At STAGE 46 the white matter in the cerebellar plate is very distinct, while at STAGE 48 the amount of white matter in the midplane is about equal in volume to the ependymal layer. It continues to increase in thickness during the following stages, so that in the midplane it far exceeds the gray matter (i.e. the ependymal layer) at STAGE 50. The white matter is strongly developed at STAGE 51, at which the cerebellar plate protrudes rostralwards in the midplane. In the laterally located *corpora cerebelli*, which have begun to thicken at STAGE 42, *cells of Purkinje* start differentiation at STAGE 46. The layer of Purkinje cells is much more prominent at STAGE 49. The Purkinje layers of both corpora cerebelli extend to and fuse in the midline at STAGE 53, a situation already nearly achieved at STAGE 52. At STAGE 49 a few cells of the future *molecular layer* can be found in the white matter. Their number slowly increases during the following stages (STAGE 54 and beyond). In the cerebellum the *granular layer* develops from the ependyma at STAGE 53, increasing much in thickness during the following stages (STAGE 56 and later on). The structural development of the cerebellum is practically finished at STAGE 56, after which it simply increases in size both relatively and absolutely. Finally at STAGE 59, the cerebellum has become very massive in midplane sections, its three layers being well developed. The ependymal layer still forms granular cells.

A *nucleus isthmi* may just be distinguished at STAGE 53 and is clearly visible at STAGE 55.

At STAGE 43 the cells in the midline of the rostral part of the myelencephalic roof have just become cylindrical. They form the first anlage of the *myelencephalic choroid plexus*. At STAGE 44 invagination of this anlage starts along the mid-dorsal line, becoming more distinct at STAGE 46. Formation of the myelencephalic choroid plexus first proceeds along the mid-dorsal line in a rostro-caudal direction. This is clearly visible at STAGE 47 in comparison with former stages. Only the most caudal part of the roof of the IVth ventricle stays membraneous. Then, plexus differentiation starts likewise from the mid-dorsal line in a lateral direction. At STAGE 48 this lateralward extension of the choroid plexus has not yet quite reached the taeniae of the roof of the ventricle. The lateral border of the myelencephalon is actually reached by the choroid plexus at STAGE 49. Owing to the irregularity of the process of invagination the myelencephalic plexus appears to show evaginations. At STAGE 49, in this way, choroidal sacs, seemingly strongly protruding in a dorsalward direction, are present over the entire length of the roof of the ventricle, the very thin membranous posterior part excepted.

In general, all parts of the *brain* are already well developed at STAGE 53, so that later development consists mainly of growth and some further cytological differentiation. In the period of metamorphosis only topographical changes take place, particularly in the position of the *"Stirnorgan"* which is simultaneously reduced to a mere rudiment.

* * *

THE FURTHER DEVELOPMENT OF THE CEPHALIC GANGLIA AND NERVES

The early development of the *cephalic ganglia* and *nerves* has been described up to STAGE 37/38 in the division "The early development of the *central nervous system, sense organs, ganglia* and *nerves*", on page 33.

THE DEVELOPMENT OF THE CEPHALIC GANGLIA AND NERVES

At STAGE 32 the first fibres of the *olfactory nerve* pass from the nasal placode to the closely associated prosencephalon. There is a rapid increase in the number of these fibres during the following stages, so that by STAGE 35/36 they have already formed a thick bundle. The nasal organs begin to withdraw from the hemispheres at STAGE 45, lengthening the olfactory nerves. The nerve, which is still very short at STAGE 45, is first directed laterad. At STAGE 46 the direction is latero-craniad but at STAGE 47, at which the nerve has lengthened considerably, it directs craniad.

The second cephalic nerve, the *optic nerve*, begins to develop at STAGE 33/34, at which the first optic fibres appear in the optic stalks, which still contain a wide lumen. The interretinal part of the optic nerve becomes distinct at STAGE 35/36. The nerve, when leaving the eye, bends ventrocaudad and runs for some distance along the outer surface of the tapetum at STAGE 39.

The *oculomotorius nerve* appears at STAGE 33/34. It branches in the loose mesenchyme around the eye at STAGE 35/36 and becomes connected with the undifferentiated eye muscle anlagen at STAGE 39, which differentiate at STAGE 41, at which also the *ganglion ciliare* begins to appear. The n. opticus and n. oculomotorius first run through a common foramen in the crista trabeculae at STAGE 47 but become separated by cartilage at STAGE 51.

The *nervus trochlearis* has reached the m. obliquus superior at STAGE 35/36, so that the eye muscle innervation develops at a rather early stage. The n. trochlearis runs through a separate narrow canal in the crista trabeculae at STAGE 53.

At STAGE 24 the first nerve fibres can be seen in the root of the vth *cephalic nerve*, and some fibres also emerge from the *profundus ganglion* as well as ventrad from the *Gasserian ganglion*. The first part of the *n. maxillomandibularis* becomes distinct at STAGE 25 and spreads in a distal direction, so that the nerve can be followed up to the neighbourhood of the cement gland at STAGE 26, and reaches its inner side at STAGE 28. At STAGE 27 the first part of the *n. ophthalmicus profundus* becomes distinct. The nerve can be followed up to the neighbourhood of the nasal placode at STAGE 28. It finally reaches the nasal placode at STAGE 33/34. The Gasserian ganglion begins to be displaced dorsalwards at STAGE 39. The n. ophthalmicus profundus and the n. maxillomandibularis become separated by the processus ascendens palatoquadrati at STAGE 42.

The *nervus abducens* can be distinguished only very late and can be followed from its origin in the rhombencephalon to the ganglion v-vii at STAGE 49 and finally to the m. rectus posterior at STAGE 50.

From *ganglion* vii nerves begin to grow out at STAGE 31. At that stage the *n. hyomandibularis* vii is already present in the form of a cellular strand. It becomes connected with the mesoderm of the hyoid arch at STAGE 33/34 and reaches the skin and the lateral line organs of the lower jaw at STAGE 35/36. Connection has been established between the *main branch* and the m. interhyoideus at STAGE 39. A *lateral branch* of the n. hyomandibularis vii becomes visible at STAGE 47, running also to the m. interhyoideus. The *ganglion dorsolaterale* vii is still connected with the supraorbital lateral line anlage at STAGE 32. The *n. buccalis* vii which appears at STAGE 32 makes connection with the infraorbital lateral line at STAGE 33/34 just as the *n. ophthalmicus superficialis* vii with the supraorbital lateral line. The *n. palatinus* vii has reached the roof of the mouth cavity at STAGE 33/34.

Ganglion viii can be distinguished medial to the anterior part of the ear vesicle at STAGE 31 and extends along the ventral wall of the ear vesicle at STAGE 33/34. It sends a branch ventro-craniad along the wall of the ear vesicle at STAGE 42. At STAGE 47 ganglion viii sends a branch caudad along the ductus utriculo-saccularis.

The *nerves of ganglion* ix develop later. A distinct root of ix can be found at STAGE 32. The *n. glossopharyngeus* is connected with the mesoderm of the first branchial arch at STAGE 33/34 and has reached the ventral end of this arch at STAGE 39, at which the *ramus pharyngeus* has reached the roof of the pharynx. At this stage the *g. superius* and the *g. petrosum* become distinctly separated. The n. ix branches below the floor of the pharynx at STAGE 40 and innervates the m. levator arcuum branchialium at STAGE 42.

The *ganglion dorso-laterale* x is still connected with the lateral line anlage of the trunk at STAGE 33/34. The *n. lateralis* x emerges posteriorly from the g. dorso-laterale x at STAGE 35/36. The roots of x become distinct at STAGE 39, at which the *n. recurrens-intestinalis* x and *n. lateralis inferior* x grow out. The latter has reached the level of the future forelimb at STAGE 40 and that of the ventral lateral line at STAGE 41. The *n. branchialis* x, developed as a cellular strand, has become connected with the mesoderm of the second and third branchial arches at STAGE 35/36. The n. branchialis x^2 has extended to near the pericard at STAGE 40 and sends branches to the m. lev. arc. branchialium and to the roof of the pharynx at STAGE 43. The *n. auricularis* x develops at STAGE 47 while branches of the *n. lateralis* x become distinct at STAGE 49.

The *n.n. occipitales* are growing out to the occipital lateral line organs at STAGE 39.

Hardly any changes occur after STAGE 49, at which the cephalic ganglia and nerves have fully developed.

THE DESCRIPTION OF THE CEPHALIC GANGLIA AND NERVES AT STAGE 49

The *n. olfactorius* emerges from the rostral end of the telencephalon hemisphere with a medial and a lateral root. The lateral root originates rather dorsally from the lateral side of the hemisphere, the medial root emerges more ventrally. The nerve runs craniad over the ethmoid plate; a division into a medial and lateral fibre tract may be indicated over a great part of its length. It reaches the ventromedial side of the nasal organ and branches into the various parts of this organ.

The *n. opticus*, surrounded by a meningeal sheath, runs from the chiasma in the bottom of the diencephalon laterad over the basicranial plate, passes through the optic foramen, runs laterad through the orbita, makes a ventral bend and reaches the eye.

The *n. oculomotorius* emerges a short distance behind the infundibulum from the bottom of the mesencephalon, brushes past the lateral edge of the infundibulum and runs craniolaterad over the basicranial plate. It emerges through the foramen nervi oculomotorii, which at this stage is not yet separated from the optic foramen (at STAGE 50 to 51 both foramina are entirely separated by cartilage). On the other side of this foramen it crosses the n. ophthalmicus profundus v, runs dorsal to the origin of the m. rectus superior and rectus posterior, ventral to the arteria optica, and sends a branch to the m. rectus superior. Then it passes between the optic nerve and the m. rectus inferior, innervates this muscle and sends a branch to the very small *ganglion ciliare*. Then it divides into 2 branches: one of these runs on the dorsal side of a blood vessel (branch of the optic vein) to the m. rectus anterior, the other one on the ventral side of this vein to the m. obliquus inferior.

The *n. trochlearis* emerges from the lateral side of the mesencephalon rather posteriorly. It runs in a craniolateral direction over the dorsal margin of the crista trabeculae somewhat behind the foramen nervi oculo-motorii (at STAGE 53 it passes through a very narrow canal in the dorsal part of this wall). Then it runs forward through the orbita dorsal to the optic nerve and all the eye muscles, and reaches the m. obliquus superior.

The common root of *n.n. trigeminus* and *facialis* emerges rather caudally and dorsally from the lateral side of the medulla oblongata, runs ventro-craniad along the inner side of the auditory capsule, spreads fanwise into a flat band and passes into the common trigeminus-facialis ganglionic complex, which is situated in the foramen proöticum. The ganglion trige-mini, which occupies the middle part of this complex consists of large clear ganglion cells. From this ganglion 2 nerves arise: 1) The *n. ophthalmicus profundus*, from the lower part of ganglion v, runs rostrad ventral to the processus ascendens palatoquadrati, then enters the orbita through the fenestra subocularis, runs ventral to the origin of the m. rectus superior, dorsal to the origin of the m. rectus posterior, crosses the n. oculomotorius and n. opticus on the dorsal side (at this place the ganglion ciliare is situ-ated), runs craniad through the orbita, dorsal to the m. obliquus superior, and sends a medial branch to the nasal organ and lateral branches to the skin laterodorsally to this organ; 2) The *n. maxillomandibularis*[1]), from the upper part of ganglion v, runs craniad dorsal to the processus ascendens palatoquadrati, passes ventral to all eye muscles and the optic nerve, runs over the m. adductor mandibulae posterior forward to the lateral edge of the mouth, where it sends branches to the skin. One or two terminal branches (*n. mandibularis*) run medioventrad on the anterior side of Meckels' cartilage and innervate the cement-gland and the m. intermandibularis.

The *n. abducens* emerges rather caudally from the bottom of the medulla oblongata, runs craniad over the basicranial plate and reaches the lower side of the trigeminus-facialis ganglionic complex. Here it joins the n. ophthalmicus profundus, runs with this under the proc. ascendens pala-toquadrati into the orbita, then passes on towards the m. rectus posterior.

The *trigeminus-facialis ganglionic complex* consists of:

a) The upper part of the trigeminus-facialis ganglionic complex, consist-ing of rather small ganglion cells, represents the *ganglion dorso-laterale* vii. From its lateral side a strong nerve, the *n. lateralis anterior*, is given off, which immediately splits into two branches: 1) The *n. ophthalmicus super-ficialis*, runs craniad lateral to the crista trabeculae, and innervates with small branches the supraorbital lateral line; 2) The *n. buccalis*[2]) runs lateroventrad, first on the dorsal surface of the thymus, then on the outer side of the m. lev. arc. branchialium, immediately behind the eye, where it innervates the postorbital lateral line, then beneath the eye along the infra-orbital lateral line, and is continued to near the lateral edge of the mouth.

b) The lower part of the trigeminus-facialis ganglionic complex repre-sents the ganglion vii. Its dorsal half, the *ganglion ventrolaterale* vii, consists

---

[1]) This nerve has been mistaken for the N. palatinus vii by KOTTHAUS (1933), whereas WEISZ (1945) calls it the dorsal palatine branch of the facial.

[2]) By KOTTHAUS (1933, fig. 10) this nerve is indicated as r. maxillaris v. WEISZ (1945) erroneously has taken the n. buccalis for the n. hyomandibularis vii.

of big clear ganglion cells, whereas the ventral part, the *ganglion geniculi*, is composed of smaller and darker cells. From the former, the n. hyomandibularis arises, from the latter the n. palatinus. The *n. hyomandibularis* VII [1]) runs laterocraniad on the ventral side of the proc. ascendens palatoquadrati, then beneath the bottom of the orbita. It gives off a lateral branch beneath the eye to the m. interhyoideus. Then it runs laterocraniad along the posterior margin of the pars palatina palatoquadrati towards the articulation between the pars palatina and the hyomandibular. It passes ventrad immediately along the lateral surface of this articulation, on the inner side of the thick levator hyoidei muscle which surrounds it, then breaks through this musculature and reaches the skin and lateral line sense organs of this region with many branches. A terminal branch runs mediad along the lower side of the hyomandibular towards the m. interhyoideus and the skin and lateral line sense organs of the lower jaw. The *n. palatinus* [2]), arising from the lower part of ganglion VII, runs craniad, with a slight bend to the medial side, on the roof of the pharynx and mouth cavity, and innervates the taste organs in the anterior part of this roof. It ends on the lower side of the ethmoid plate near the choana.

The root of the *n. acusticus* emerges immediately behind that of the n. facialis. The *ganglion acusticum* is situated on the medial side of the sacculus. Its anterior part is somewhat separated from the rest; the nerve arising from this part innervates the macula utriculi and the ampullae of the can. semicircularis anterior and lateralis. The greater posterior part of the ganglion sends branches towards the papilla amphibiorum, macula lagenae, papilla basilaris and the ampulla of the can. semicircularis posterior. Between the two parts, a separate branch arises which runs to the macula sacculi.

Four roots of the *n. glossopharyngeus*, emerging separately from the lateral side of the medulla oblongata at some distance behind the n. acusticus and a rather great distance apart, join and run in a posteroventral direction medial to the auditory capsule towards the jugular foramen, where they form the common *glossopharyngeus-vagus ganglionic complex*. The upper and middle portion of this complex mainly consist of large clear ganglion cells; the lower part is composed of smaller and darker cells.

The *n. glossopharyngeus* arises from the most anterior root, and forms a small *ganglion superius* IX against the posterior wall of the auditory capsule. Then a fibre tract runs first laterad, then craniad in a groove on the lateroventral side of the otic capsule, and swells once more into the big elongated *ganglion petrosum*. From the anterior end of this ganglion, the nerve continues forward and splits immediately into 2 branches: a) The lateral branch, the *ramus posttrematicus*, continues forward along the lateral

---

[1])   WEISZ (1945, fig. 2) calls this nerve the "mandibular branch of the profundus".
[2])   This nerve has been called by WEISZ (1945) the "ventral branch of the facial".

side of the thymus and proc. ascendens palatoquadrati, and reaches the anterior part of the m. levator arcuum branchialium; b) The medial branch, the *ramus pharyngeus*, passes ventromedial to the thymus and proc. ascendens, sends branches to the m. lev. arc. branchialium, and innervates the taste organs in the posterior part of the roof of the pharynx. Apparently it anastomoses with the n. palatinus vii.

The following nerves arise from the *vagus ganglion complex*:

(1) The *n. auricularis*, from a little group of small ganglion cells in the upper portion of the vagus complex, the general cutaneous *ganglion* x, runs along the posterior side of the auditory capsule, first caudad, then dorsad and splits into 2 branches: The first runs craniad ventrolateral to the auditory capsule, on the outer side of the m. lev. arc. branchialium, bends ventrad behind the thymus and innervates the skin behind the eye; The second runs mediocraniad dorsal to the auditory capsule and innervates the skin in this region.

(2) The *n. lateralis* x, from the upper portion of the vagus complex, the *ganglion dorso-laterale* x, runs caudad on the lateral side of the myotomes, and splits into the following branches: a) *rami occipitales*, to the occipital lateral line; b) *n. lateralis superior* to the upper lateral line; c) *n. lateralis medius*, runs caudad dorsal to the pronephros and innervates the middle lateral line of the trunk; and d) *ramus ventralis*, to the skin in the neighbourhood of the pronephros.

(3) The *n. lateralis inferior* originates from the middle portion of the vagus complex, the *ganglion ventro-laterale* x. Immediately after its origin from the ganglion, it pierces the origin of the m. lev. arc. branchialium from the proc. muscularis capsulae auditivae, then runs lateroventrad along the peritoneum dorsal to the lung, behind the branchial skeleton (this part probably includes the *r. branchialis* x[4]) and in front of the pronephros and forelimb. It crosses the cell strand connecting the forelimb rudiment with the skin, gives off branches to the peritoneum, then continues on the outer side of the abdominal muscles, gives off a strong branch to the skin ventrolateral to the limb, then bends caudad and innervates the lower lateral line of the trunk and the skin both dorsal and ventral to this line.

(4) The *n. recurrens-intestinalis*, arising from the lower, small-celled portion of the vagus complex, the *ganglion viscerale* x, runs mediocraniad upon the glottis musculature and gives off branches to the oesophagus and glottis, to the larynx musculature and the wall of the lung.

(5) The *n. branchialis* x, arising from the most lateroventral part of the *ganglion viscerale* x, gives off immediately a small branch which runs craniad ventral to the aorta toward the posterior part of the roof of the pharynx. The main branch runs first laterad, then craniad behind the n. glossopharyngeus, then ventrad under the skin in the roof of the lateral part of the branchial cavity behind the aortic arch. It sends branches to the m.

lev. arc. branchialium, a branch passing ventrad through the third branchial arch, the *r. branchialis* x³, then the main branch, the *r. branchialis* x², bends ventrad around the branchial cavity, immediately behind the aortic arch and 2nd branchial cartilage, and ends in the floor of the pharyngeal cavity and the musculature of this region.

THE DEVELOPMENT OF THE CEPHALIC GANGLIA OF XENOPUS IN COMPARISON WITH RANA

The development of the cranial ganglia in *Xenopus* mainly corresponds to that in *Rana* as described by KNOUFF (1928). Some minor differences are present, however, which may be indicated here:

(1) In *Rana* there is a distinct *trigeminal placode* dorsocaudal to the eye. In *Xenopus* this is much less clear; at most a diffuse proliferation of cells from the ectoderm in this region may be observed.

(2) In *Rana* there are separate *maxillary* and *mandibular branches* of the *n. trigeminus*. In *Xenopus* both are united into a common *n. maxillomandibularis*. In conjunction with this, no separate *maxillary* and *mandibular ganglia*, as in *Rana*, can be observed in *Xenopus*. The derivation of the components of the *trigeminal complex* from the neural crest and placodes, respectively, is much less clear in *Xenopus* than it is in *Rana* according to KNOUFF's description.

(3) In *Rana*, the *acustic ganglion* consists of two histologically distinct ganglionic elements, which KNOUFF calls *acusticum anterior* and *posterior*, respectively. The former is derived from the lateralis placode, the acusticum posterior appears later in development and arises by a proliferation of cells of the ventromedial wall of the ear vesicle. In *Xenopus*, no similar distinction between two parts can be made at early stages, and no evidence has been obtained of a proliferation of ganglion cells from the wall of the ear vesicle.

(4) In *Rana*, separate *dorsolateral* and *epibranchial* IX *placodes* are present. In *Xenopus* there is a common IX *placode* which extends from the ventro-caudal side of the ear placode towards the upper margin of the first branchial pouch.

(5) In *Rana*, the upper part of the IX ganglion is connected with a subotic lateral-line primordium; consequently it is called *ganglion lateralis* IX by KNOUFF. In *Xenopus* no subotic lateral line primordium has been observed, and no lateralis nerves arise from *ganglion superius* IX.

(6) In *Rana* the *dorsolateral* x *placode* is separated from the IX *placode*; in *Xenopus* both placodes are continuous. The dorsolateral placode thickening in the vagus region exists for a short time only (STAGE 28); at later stages, no dorsal placodal contribution to the vagus ganglion can be observed.

* * *

DIVISION VIII

THE FURTHER DEVELOPMENT OF THE SPINAL
CORD, GANGLIA AND NERVES

The early development has been described in the division "the early development of the *central nervous system, sense organs, ganglia* and *nerves*", on page 32 up to STAGE 31, at which the *neural crest* has been entirely withdrawn from the *neural tube*.

THE DEVELOPMENT FROM STAGE 31 TO APPROXIMATELY STAGE 39

The separate description of the further development of *spinal cord* and *ganglia* begins with the first appearance of a thin lateral crust of *nerve fibres* in the extreme anterior end of the cord at STAGE 28. This crust of nerve fibres extends rapidly along the spinal cord, so that at STAGE 29/30 nerve fibres can be followed along 50 to 60% of the length of the cord; at STAGE 31 a thin crust extends over 70 % of the cord; at STAGE 32 the nerve fibres terminate just anterior to the level of the cloaca; at STAGE 33/34 some nerve fibres extend already into the tail, while at STAGE 35/36 a few nerve fibres extend almost to the caudal end of the cord.

Simultaneously with the development of the fibre tracts, the first cellular differentiations occur in the formation of *Rohon-Beard cells*, which first appear in trunk levels at STAGE 33/34. They appear in anterior levels of the tail at STAGE 37/38, and well back into the tail at STAGE 39. When they appear they are still small in size, but grow larger with progress of development. Concurrently the *roof plate* thickens and the dorso-ventral extent of the *neurocoel* is reduced. During this period of development the *neural crest cells* migrate laterally and ventrally at more and more caudal levels. At STAGE 32 some crest cells have migrated ventrad of the cord in levels nearly as far caudad as the cloaca, while the lateral and ventral migration can be noted in the tail region except in the last few sections. The subsequent aggregation of ganglion cells is indicated at STAGE 37/38.

In the anterior segments of the trunk, *spinal nerves* originating from the ventral side of the spinal cord can be followed up to the ventral side of the somites at STAGE 33/34. The development of the spinal nerves progresses caudalwards, so that in more and more caudal segments spinal nerves become distinguishable at following stages, while they simultaneously extend further towards the lateral plate.

THE DEVELOPMENT FROM STAGE 39 TO APPROXIMATELY STAGE 50

In the following period of development the lateral layer of nerve fibres— *lateral white matter*—increases in thickness and a thin crust of white matter appears on the ventral surface of the cord at STAGE 39. At this stage the lateral white matter is at least one third as thick as the lateral *gray matter* at the anterior end of the cord, which proportion has increased to one half at STAGE 40. At STAGE 42 the white matter reaches dorsally almost to the midline.

At STAGE 39 another process of cellular differentiation starts, viz. a few cells from the central gray matter bulge into the white matter. At this stage these cells have not quite detached themselves from the gray matter, but detachment at the extreme anterior end of the cord has been achieved by STAGE 41. At this stage the first *motor neurons* begin to enlarge at the edge of the gray matter. The segregation of motor neurons from the central gray matter extends rapidly along the cord. At STAGE 44 the motor cells stand out but are not yet completely separated from the central gray matter in the tail region. At STAGE 45 they stand out in posterior trunk levels as well as in the tail, but not in anterior trunk levels. Cells of the *mesial motor column* may be completely separated from the central gray matter, and two such cells may be seen together at STAGE 48. At this stage the *median ventral fissure* appears. It becomes deeper at later stages.

During this period of development the *spinal ganglia* begin to differentiate. They are still inconspicuous and composed of aggregates of six to ten cells each at STAGE 39, but they grow larger during the following stages. At STAGE 44 they already contain twenty to thirty cells. At STAGE 46 some ganglion cells begin to enlarge and already equal Rohon-Beard cells in size in the tail region. The same degree of enlargement is achieved at mid-trunk levels at STAGE 47, at which stage all trunk ganglia are sixty μ long or longer. They are generally located dorsally to the level of the notochord. The differentiation of the ganglion cells progresses gradually, so that some ganglion cells already exceed the largest of the Rohon-Beard cells in size at anterior trunk levels at STAGE 49.

At STAGE 40 the first ten spinal nerves and ventral roots are well developed; they run along the ventral surface of each somite to the lateral plate. The more caudal spinal nerves are less well developed. From STAGE 44 on the *dorsal roots* become distinguishable; these are well developed by STAGE 47.

The *spinal nerves* originally run approximately laterad after reaching the ventral surface of the somites. At STAGE 40 the spinal nerves 2 and 3 begin to deviate caudad, whereas spinal nerve 4 still runs approximately laterad. Consequently the nerves begin to converge. The convergence and caudal displacement starts for the spinal nerves 8, 9, and 10 at approximately STAGE 44. At that stage spinal nerve 8 runs slightly caudad, spinal nerve 9 runs approximately laterad and spinal nerve 10 runs slightly craniad. The gradual caudal extension of the pronephros and of the coelomic cavity at the STAGES 45 and 46 leads to a more and more pro-

nounced caudad displacement and convergence of the spinal nerves innervating the limb buds, since these run respectively along the caudal side
of the pronephros and around the caudal poles of the coelomic cavity on
both sides of the rectum.

The forelimb rudiment appears at STAGE 46 at a level with trunk somite
3, the hindlimb rudiment at STAGE 44 at a level with trunk somite 9. At
STAGE 48 the spinal nerves 2 and 3 and problably also 4 reach the forelimb
region, running partially very close together. A relative caudalwards
displacement of the limb rudiments with respect to the axial system leads
to a further caudad deviation of the spinal nerves innervating the limbs.
At STAGE 48 the forelimb rudiment is situated at the level of trunk somite 4,
and the hindlimb rudiment at the level of trunk somite 10. At STAGE 49
the spinal nerves innervating the forelimb run first latero-caudad around
the caudal pole of the pronephros, and then turn craniad, where the
plexus brachialis is formed. The spinal nerves innervating the hindlimb run
over some distance nearly caudad and then around the caudal poles of
the coelomic cavity, where they begin to form the *plexus lumbo-sacralis*.
The spinal nerves, particularly those innervating the limb buds, increase
in size and are already well developed at STAGE 49.

Meninges begin to appear around the spinal cord at STAGE 44, at which
the *dura-endocranial membrane* is being formed around its most anterior
region. The membrane is forming along the spinal cord of the entire trunk
at STAGE 45, and has become quite distinct in the trunk and anterior tail
region in between STAGE 46 and 47.

THE FURTHER DEVELOPMENT FROM STAGE 50 ON

In the following period of development the larval *Rohon-Beard cells*
disappear, and their function is taken over by the *spinal ganglia*. At STAGE 50
they begin to decline. Their number and size are reduced, a process which
continues up to STAGE 55, when only a few, with pycnotic nuclei, are
present. At STAGE 50 the *brachial ganglia* have a few large cells, but the
lumbar ganglia have none. The enlargement of the ganglion cells at lumbar
levels begins at STAGE 52. This process of differentiation proceeds slowly
at first, but later accelerates. At STAGE 54 the cells of the lumbar ganglia
are still beginning to enlarge, at STAGE 55 their cells are mostly moderate
in size with a few large cells. At STAGE 56 the lumbar ganglia have cells
of all sizes, including many large ones.

In this period the cells of the *lateral motor columns* differentiate. At STAGE
50 they are barely distinguishable in lumbar levels, showing slightly
elongate or ovoid nuclei. Not until STAGE 51 do they become distinguishable
at brachial levels. Nucleoli begin to appear in the *lumbar motor column* cells
at STAGE 52 and in the cells of the *brachial motor column* at STAGE 53. Simultaneously the nuclei of the lumbar motor column cells enlarge. Cytoplasm

can be distinguished at the narrower ends of the egg-shaped nuclei of the lumbar motor column cells at STAGE 53, while a few nuclei of the brachial motor column have cytoplasmic rims at STAGE 54. At STAGE 54 many nuclei in the lumbar motor column have cytoplasmic rims, a situation not achieved for the brachial motor column until STAGE 55. The cells of the motor columns spread laterally at STAGE 54 and their nuclei enlarge gradually. At STAGE 55 the lumbar motor column nuclei frequently have twice the cross section area of adjacent non-motor cells, an enlargement which continues at later stages. The brachial motor column has about thirty cells per side (per section of ten μ), whereas the lumbar motor column has fewer than twenty cells.

The *dorsal column* or *dorsal horn* is beginning to appear in postbrachial levels at STAGE 52 and is for the first time present in prebrachial levels at STAGE 53. At STAGE 54 the dorsal columns are present throughout trunk levels, and persist even a short distance into postsacral levels. During this period of development the *mesial motor column* cells have increased, so that as many as four can be found on each side (per section of ten μ) at STAGE 52.

The enlargement of the nuclei of the *motor columns* continues, while simultaneously their number per section decreases. At STAGE 57 the largest *lumbar motor column* cells have nuclei three to four times as large as the average of the adjacent non-motor nuclei (in cross section area). This value has increased to five times for the lumbar column at STAGE 59 and to six times at STAGE 60, while at STAGE 59 the value is up to four times for the largest nuclei of the *brachial motor column*. The lumbar motor column usually contains fewer than twelve cells per side (per section of ten μ) at STAGE 58, while the brachial motor column generally shows fewer than twenty cells per side at STAGE 59.

The *forelimb nerve* is entering the forelimb anlage at STAGE 50 and has already invaded the anlage at STAGE 51. At STAGE 52 the forelimb nerve is bifurcated, a lateral branch running to the forelimb anlage s.s. and a medial branch to the future shoulder girdle. At STAGE 53 the latter is again divided into a dorsal and a ventral branch. The *hindlimb nerve* is entering the hindlimb bud at STAGE 50 and has invaded its proximal portion at STAGE 51. At STAGE 52 the hindlimb nerve is bifurcated, a lateral branch running to the hindlimb anlage s.s. and a medial branch to the future pelvic girdle. During the further development of the limbs the limb nerves extend and split up into branches to the various muscle anlagen.

In association with the development of the hind limbs and pelvic girdle and the processes of metamorphosis, the *lumbar ganglia* descend with respect to their roots. At STAGE 53 the lumbar ganglia lie already caudad to them. At STAGE 56 the root of ganglion 8 overlaps part of ganglion 7 and the root of ganglion 9 all of ganglion 8. Ganglion 6 also descends and overlaps the lumbar level of the cord at STAGE 57. At STAGE 61 ganglion 8 has descended to a level well posterior to the dorsal root of ganglion 9, and except for a few sections all of ganglion 9 is posterior to the lumbar level of the spinal cord. At STAGE 63 the root of ganglion 9 enters the spinal cord just posterior to the level of ganglion 7. The root of ganglion 7 appears anterior to ganglion 6 for the first time at STAGE 64. At STAGE 65 ganglion 8 has moved so far caudad that the lumbar level of the spinal cord ends at the level of this ganglion. Finally, the dorsal root of ganglion 9 enters the spinal cord at the level of ganglion 7 at STAGE 66. Concurrently with their descent, the *lumbar ganglia* increase in size, and simultaneously the *spinal cord* reduces at postsacral levels. The greatest cross section area of ganglion 9, being almost as large as the adjacent spinal cord at STAGES 59 to 62, becomes larger than that at STAGE 63. At later stages this becomes even more pronounced. At STAGE 64 ganglion 9 has a greater cross section diameter than the adjacent spinal cord for nearly one half of the length of the ganglion. At STAGE 65 this holds for about two-thirds of its length, while finally at STAGE 66, in some cross sections, the total area of ganglion 9 and its roots is up to two times as great as the area of the cord, and in nearly all sections the ganglion is larger than the adjacent cord.

Delamination starts in the *dura-endocranial membrane* in the most anterior region of the trunk at STAGE 49, a process subsequently spreading in a posterior direction. At STAGE 50 the *ecto-* and *endomeninx* have segregated in the anterior trunk region down to the level of the pronephros. This process of delamination has reached the tail region at STAGE 51 and is completed at STAGE 52 or 53, after which both membranes only increase in thickness.

The *sympathetic chain ganglia* begin to appear as tiny groups of small ganglion cells along the medio-ventral side of the myotomes, anteriorly at STAGE 49, more posteriorly at STAGE 50. They increase considerably in size during the following stages. At STAGE 51 the *connecting nerves* become manifest and a well developed chain of sympathetic ganglia can be found along the medio-ventral side of the myotomes. They are connected cranially with the *ganglion jugulare*. The anlage of this ganglion, originally stretched out along the lateral side of the notochord at the level of the first myotome, is displaced ventralwards and occupies more or less its definitive position at the medio-ventral side of the first myotome at STAGE 50. The chain of sympathetic ganglia extends caudally nearly down to the end of the mesonephric region at STAGE 51 to 52. At mid-trunk levels the *nervus splanchnicus* originates from the sympathetic chains and runs along the arteria coeliaco-mesenterica. At later stages the system increases in size, but there is otherwise little change.

As a general feature the *floor plate* of the spinal cord thickens, as the *central canal* decreases in dorso-ventral extent at STAGE 54. At STAGE 56 the roof plate increases further in thickness. This is more pronounced at STAGE 58, when the neurocoele becomes greatly shortened dorso-ventrally, so that it is more or less circular at STAGE 60, its size being no larger than that of the largest nuclei of the motor column cells. At STAGE 62 it is reduced to a size considerably less than that of the largest of the motor column nuclei.

THE DESCRIPTION OF THE SPINAL CORD AND GANGLIA

The successive developmental processes described in the previous pages have finally led to the following general structure of the *spinal cord*, *spinal ganglia* and *nerve roots* at STAGE 66.

At the level at which the *fourth entricle* terminates, the *central canal* is large, extending over at least one fifth of the dorso-ventral extent of the cord. The *ventral median fissure* is shallow, whereas the *dorsal median sulcus* is deep. A *dorsal column* is evident, but a *ventral column* is lacking. Passing caudad, the dorsal sulcus quickly fills in, until it is virtually obliterated, while the ventral fissure increases in depth. In the distance of about one mm. the central canal constricts very markedly, mainly from the dorsal side. It becomes a short, narrow slit, elongated dorso-ventrad. The *ependyma*, some three to five nuclei thick at the extreme anterior end, becomes reduced to a thickness of but one or two nuclei.

Mesial and *lateral motor columns* can be seen in the *brachial region*. The lateral column appears at the level of ganglion 2, and extends to about half a mm. caudad of ganglion 3. This is about one fourth of a mm. caudad of the most posterior motor rootlet of nerve 3. *Ganglion* 2 lies completely anterior to the entrance of its dorsal root into the cord. *Ganglia* 3 and 4 have dorsal roots which run directly mediad to the cord. All of the more posterior *ganglia* (5-9) lie posterior to the level of entrance of their dorsal roots into the cord. The more posterior the ganglion, the more elongate is its dorsal root. Caudad of ganglion 3, the *spinal cord* shows a reduction in its total cross section area.

The *lumbar region*, like the brachial, is marked by an enlargement in the cross section area of the cord. The most anterior motor cells of the lumbar region (*lateral motor column*) appear about three fourths of a mm. anterior to the entrance of dorsal root 7, and about one tenth of a mm. anterior to the exit of the first of the motor rootlets of spinal nerve 7. The cells of the lateral motor column terminate opposite ganglion 8. Posterior to ganglion 8 the *spinal cord* diminishes sharply in size, becoming nearly circular in cross section, and with a circular, slightly enlarged central canal. The last of the lumbar ganglia—ganglion 9—is found at this level, and with the fibres of its root it may be twice as large in cross section area as the cord at the same level. At the level of ganglion 9, and at more caudal levels, the spinal cord flattens considerably. The ventral median fissure is nearly obliterated. The central canal also becomes flattened.

The paired fibres of *Mauthner's cells* are quite prominent throughout all of the cord anterior to ganglion 9. At the level of this ganglion, or caudal to it, they become indistinguishable from other large fibres in the ventral portion of the cord.

The motor cells of the *lateral motor columns* of both the *brachial* and *lumbar* enlargements are largest in the middle of their distributions. The cells at both anterior and posterior ends of the two distributions are distinctly smaller in size.

The *spinal ganglia* occupy the intervertebral foramina, and the adjacent spaces lateral to the neural arches. The large lumbar ganglia, 8 and 9, impinge slightly upon the space of the spinal canal.

* * *

THE FURTHER DEVELOPMENT OF THE OLFACTORY ORGAN

The early development of the *olfactory placode* has been described up till STAGE 29/30 under "the early development of the *central nervous system, sense organs, ganglia* and *nerves*", on page 33.

THE PRIMARY SEGREGATION OF THE OLFACTORY ORGAN

The *olfactory placode* develops from the sensorial layer of the ectoderm, forming a thickening, gradually increasing in dimensions. At STAGE 29/30 the overlying epithelial layer of the ectoderm becomes somewhat thinner at the point of the future *nasal pit*, so that a slight depression of the outer surface becomes visible at STAGE 31. It deepens gradually during the following stages. The long, bottle-shaped olfactory cells end radially in the bottom of the pit at STAGE 35/36. At STAGE 37/38 the olfactory organ begins to segregate into an antero-medially situated, smaller lobe, the *organon vomero-nasale* or *Jacobson's organ* and a latero-caudal larger lobe, the *olfactory organ s.s.* The median organon vomero-nasale has a separate, narrow lumen at STAGE 40, while the lumen of the lateral lobe is an open, flat hollow. The median lobe, containing a slit-like lumen, has become narrower at STAGE 41 and has shifted medio-ventrad underneath the forebrain, approaching the pharyngeal roof. Its lumen widens at STAGE 42, while that of the lateral lobe hardly alters. It becomes compressed against the cartilaginous plate of the pharyngeal roof at STAGE 43. At this stage narrow cell strands lead from the lateral nasal pits to the epithelium of the pharyngeal cavity, where small pits appear in the pharyngeal roof, representing the future *choanae*. At STAGE 44 these pits deepen, and the cell strands become more massive at STAGE 44 and 45. At the latter stage also the lumen of the nasal pit forms a protrusion which reaches into the beginning of the cell strand.

THE FURTHER DEVELOPMENT OF THE OLFACTORY ORGAN

The *Jacobson's organ* becomes a long tube with a narrow lumen at STAGE 46. It runs along the medial side of the nasal pit with which it is broadly connected. At STAGE 47 it also extends craniad, so that it becomes a massive tube, closed at both ends but connected through a side-opening with the open nasal pit. Its caudal end begins to branch, forming the *Jacobson's glands*. At STAGE 47 up to three glands can be distinguished. The entire organ increases in size and the cranial portion forms a swelling at STAGE 50.

The number of Jacobson's glands has increased only slightly; only three to four branches can be distinguished. This number has increased to four to five at STAGE 51 and to about ten at STAGE 52, while fifteen to twenty glands can be distinguished at STAGE 55, reaching further caudally than the organon vomero-nasale itself. At later stages the number of Jacobson's glands increases still more. They are situated dorsal and caudal to the organon vomero-nasale. The *organon vomero-nasale* does not change very much.

The *olfactory organ s.s.* forms a medio-caudal swelling in the form of a closed pouch at STAGE 46. At STAGE 50 this pouch reaches more caudally than the organon vomero-nasale. The cavity of the nasal pit becomes funnel-shaped at STAGE 49, its tip approaching the pharyngeal epithelium in the area of the future choana. The *choanae* have perforated at STAGE 50, at which the main lumen of the nasal organ has become more extensive. Each choana is closed by a valve, which is covered by the pharyngeal mucous membrane. The valve opens cranially. This situation is maintained during several stages and represents the larval olfactory organ (see Abb. 4, FÖSKE, 1934).

CHANGES IN THE OLFACTORY ORGAN DURING METAMORPHOSIS

At the beginning of metamorphosis further changes occur. At STAGE 55 the cranial extremities of the two nasal pits approach each other. The main lumen of the *nasal pit* has widened in a medio-lateral direction and has formed a pouch dorso-medial to the organon vomero-nasale. The left and right pouches approach each other. The organa vomero-nasalia, situated near the oral mucous membrane, are connected with the antero-medial side of the olfactory organ s.s. (see Abb. 5, FÖSKE, 1934). At STAGE 56, at which the *nostrils* have become smaller by an outgrowth of their caudal and lateral edges, a semicircular ridge has been formed in the lumen of the olfactory organ, dividing it into a larger cranio-medial ("*Haupthöhle*", FÖSKE, 1934) and a smaller caudo-lateral diverticulum ("*Nebenhöhle*", FÖSKE, 1934). From the larger diverticulum, at the medial side of the semicircular ridge, a funnel-like protrusion leads into a duct which turns in a ventro-caudal direction and branches into three to five glandular ducts, forming the anlage of the *glandula oralis interna*. The gland lies dorsal to the Jacobson's gland. At the antero-ventral side of the smaller caudal diverticulum ("*dorsaler Abschnitt der Nebenhöhle*", Abb. 9, FÖSKE 1934) another banana-shaped diverticulum ("*ventraler Abschnitt der Nebenhöhle*, Abb. 9, FÖSKE 1934) has formed at STAGE 56. Finger-like branches have been formed from the cranio-lateral wall of the lateral diverticulum ("Nebenhöhle") of the olfactory organ at STAGE 61. Their lumina are in connection with the lumen of the diverticulum. Their number has still increased at STAGE 63.

At STAGE 56 the *naso-lachrymal* duct becomes visible, running from the lateral wall of the central lumen of the lateral diverticulum of the olfactory organ to the skin in the neighbourhood of the nasal pit. Only its beginning in the olfactory organ contains a lumen; its distal portion running to the skin, is still a solid cord. During the following stages the naso-lachrymal duct extends in the direction of the eye. It has a bilobed ending at two thirds of the distance from nasal pit to eye at STAGE 57, and very close to the eye at STAGE 62, at which stage a *tear furrow* leads to the eye. Meanwhile its lumen extends distalwards into the anlage and opens to the exterior by two apertures from STAGE 60 on. At STAGE 63 the naso-lachrymal duct opens at the base of the lower eyelid at the tip of a papilla with a slit-shaped opening.

During metamorphosis the *choanae* have moved caudad. They are situated at the level of the cranial margin of the eyes at STAGE 64. The *naso-lachrymal ducts* have therefore shortened. The entire *olfactory organs* have turned; their longitudinal axes now direct cranio-caudally. The *Jacobson's glands* have also moved caudad. The most caudal glands reach down to the level of the cranial margin of the eye. The glands end in the immediate vicinity of the n. olfactorius and protrude around the nerve into the ventral portion of the cartilaginous nasal capsule.

THE DESCRIPTION OF THE OLFACTORY ORGAN

At STAGE 64 the development of the olfactory organ has nearly reached its definitive state and shows the following structure. Along the caudal margin of the *nostril* a circular fold has developed, extending to its lateral side. The lumen of the *olfactory organ* consists of a larger median cavity and a smaller lateral one. The cavities are connected with the nostril through a narrow duct, which can be closed by the *valve-like skin fold*. The lateral cavity is in communication with a number of accessory cavities, which have branched several times. The median and lateral cavities are separated by a fold which reaches nearly up to the nasal opening, thus dividing the nostril into two ducts leading into separate cavities. The *organon vomero-nasale* has thick walls and a narrow lumen which is rather small in comparison with the other diverticula of the olfactory organ. The duct of the *glandula oralis interna* begins near the cranial end of the floor of the median cavity. The gland consists of only a few glandular ducts. The *naso-lachrymal duct* now leads from the base of one of the diverticula of the lateral cavity to a ventral thickening (papilla) of the lower eyelid. The *choanae* have become wide and are stretched in a longitudinal direction continuing caudad into two broad furrows in the pharyngeal roof. There are no real valves; the cranial margins are merely protruding.

* * *

THE FURTHER DEVELOPMENT OF THE EYE

The initial development up to STAGE 22 has been described in the division "the early development of the *central nervous system, sense organs, ganglia* and *nerves*" on page 30.

THE EARLY DEVELOPMENT OF THE EYE TO APPROXIMATELY STAGE 39

At STAGE 23 the ventral and lateral walls of the *primary eye vesicle* become thicker in comparison with its dorso-medial wall. This relative increase in thickness becomes more pronounced during the following stages. The primary eye vesicles have been fully developed at STAGE 25, but do not show any sign of invagination before STAGE 26, at which local protrusions develop on the inner surface of the lateral wall. The actual process of invagination starts at the antero-dorsal margin at STAGE 27 and continues during the following stages. At STAGE 28 the *retinal layer* is already very thick and the *tapetum* is a single cell layer, which is very thin in its dorso-median portion. This thick retinal layer invaginates first particularly along its outer margin, the central portion being still convex at STAGE 29/30. At STAGE 31 the retinal layer has become concave. The *eye cup* deepens during the following stages and the margins of the optic cup forming the *choroid fissure* have made contact with each other at STAGE 32. The cavity of the primary optic vesicle rapidly diminishes in size with the invagination of the retinal layer. At STAGE 32 it has disappeared except in the region where the *optic stalk* (which has begun to segregate from the optic cup at STAGE 31) passes into the optic cup. Simultaneously the walls of the optic stalk have become thinner. The optic stalk still contains a wide cavity in open communication with the ventricle of the forebrain at STAGE 32. This cavity does not become obliterated before STAGE 35/36. The cells of the optic stalk have lost their regular arrangement and their number has decreased at STAGE 37/38. The optic cup does not change very much. At STAGE 33/34 its outer and inner layer make contact with each other except in the vicinity of the optic stalk, where a remnant of the cavity of the primary optic vesicle is still present. Although decreasing in size, this remnant can still be found up to STAGE 39. The edge of the optic cup bounding the aperture of the future *pupil*, has become thinner at STAGE 35/36. The proximal part of the choroid fissure has closed at STAGE 37/38. At STAGE 39 only its distal part is still open, and this does not close before STAGE 46.

Simultaneously with the beginning of invagination of the retinal layer of the eye cup, the overlying sensorial layer of the ectoderm, which is in intimate contact with the retinal layer, begins to thicken as a first sign of *lens* formation. The lens rudiment is still continuous with the sensorial layer of the ectoderm, and is enclosed by the edge of the optic cup at STAGE 31. It has become detached from the ectoderm at STAGE 33/34, at which the lens anlage has become approximately spherical. Although at STAGE 33/34 the lens has sunk deeper into the optic cup, it no longer entirely fills up the latter, leaving a slit between lens and retina into which mesenchyme cells have penetrated. This slit widens in later stages (STAGE 39). A *lens cavity* begins to appear between the thin outer and the thickened inner wall of the lens, and the cells of the outer wall begin to form a *lens epithelium* at STAGE 33/34. The lens cavity embraces the thickened inner wall of the lens at STAGE 35/36 and becomes reduced to a narrow slit at STAGE 37/38. It has disappeared at STAGE 41.

During the STAGES 32 to 39, in which the primary morphogenesis is completed, the first cellular differentiations appear. The first *pigment granules* become visible in the thin outer layer of the optic cup at STAGE 32. This pigment increases in the following stages, so that the cells of the pigmented layer of the eye are filled up with pigment granules at STAGE 39. *Nerve fibres* appear in the optic stalks at STAGE 33/34, increasing in number in following stages. The retinal layer begins to differentiate at STAGE 35/36. Nuclear and fibre layers first become arranged in the center of the *pars optica retinae* at STAGE 35/36, so that three *nuclear layers* have formed, in between which single inner and outer *plexiform layers* are foreshadowed at STAGE 37/38. In this central part of the retina the *visual cells* have begun to differentiate; from some of the nuclei of the external nuclear layer a protoplasmic bud containing an achromatic vacuole bulges radially towards the pigment layer at STAGE 37/38. This process continues during the following stages. The various retinal layers can be well distinguished at STAGE 40, when nerve fibres converge towards the *optic papilla*, whence they can be followed through the pars optica retinae, forming the *fasciculus opticus*. The primary differentiation of the visual cells has extended peripherally and nearly reaches as far as the *pars caeca retinae*, while centrally the first outer segments of the visual cells begin to appear at STAGE 40. At STAGE 41 the thinner pars caeca retinae has formed a flat ring. At STAGE 42 ordinary *rods* and *cones* can be distinguished. The thin outer and the thicker inner plexiform layers have extended gradually. At STAGE 46 they have reached the periphery of the pars optica retinae nearly as far as the pars caeca retinae except for the region of the just closed choroid fissure.

THE DEVELOPMENT OF THE EYE FROM APPROXIMATELY STAGE 39 TO STAGE 45

At STAGE 39 the cells of the proximal wall of the *lens* have begun to lengthen and at STAGE 40 the degeneration of the nuclei of the central *lens fibres* has started, while the cells of the most distal part of the lens epithelium have flattened.

The *arteria hyaloidea* is visible outside the optic cup in the neighbourhood of the choroid fissure at STAGE 37/38. It reaches the interior of the eye by way of this fissure at STAGE 39, and can be followed through its remnant into the slit-shaped cavity between the pars optica retinae and the proximal side of the lens at STAGE 40; it forms small branches against the inner surface of the retina at STAGE 41.

The surrounding mesenchyme becomes arranged in a delicate layer around the pigmented layer of the optic cup at STAGE 39 and continues as a thin *cornea primitiva mesodermalis* or *inner cornea* at STAGE 40, lying freely between the front side of the lens and the corneal part of the skin, or *outer cornea*. At STAGE 42 the inner cornea becomes attached to the outer one in the center of the corneal area. The mesenchymal coat of the eye shows the first indications of a segregation into an inner thin, pigmented *choroid coat* and an outer, still delicate *scleral coat* at STAGE 44. The latter is continuous with the cornea primitiva mesodermalis or inner cornea. The inner thin choroid coat contains darkly pigmented cells at STAGE 45. The scleral coat becomes more clearly visible at STAGE 49, although it is still only one cell layer thick. At STAGE 42 some mesenchyme cells are present in the corner between the inner cornea and the outer surface of the pars caeca retinae. They have begun to arrange themselves into a *pectinate ligament* at STAGE 43.

The *eye muscles* become outlined as separate cell masses at STAGE 39, but are much better defined at STAGE 40. They become attached to the delicate outer scleral coat at STAGE 44.

THE FURTHER DEVELOPMENT OF THE EYE

During further development differentiation proceeds. At STAGE 45 the majority of the *visual cells* have their final structure. At STAGE 47 to 48 the appearance of the rods and cones and their relative numbers are the same as in the adult, although they are not yet full-grown. Their final growth starts at STAGE 49 and lasts until beyond STAGE 66. At STAGE 52 delicate strands of mesenchymal cells bridge the space between *inner* and *outer cornea*. The area of contact between the two layers spreads from the center of the cornea outwards at STAGE 55. At STAGE 59 this fusion has extended nearly as far as the point of attachment between the ligamentum pectinatum and the inner cornea. Cartilage cells appear in the *sclera* in the neighbourhood of the optic nerve at STAGE 55. This very thin cartilaginous shell extends gradually in later stages. At STAGE 58 it already covers more than half of the eye surface and thickens at the nasal side of the eye at STAGE 63. The sclera has not yet become entirely cartilaginous at STAGE 66.

For the development of the *ductus nasolacrimalis* the reader is referred to the division "the development of the *olfactory organ*" on page 76.

Two further differentiations appear. Anterior to the eye a small rudiment of the *gland of Harder* becomes visible at STAGE 60. At STAGE 62 the gland has grown further into the interior and lies ventral to the musculus rectus internis. It finally discharges into the *conjunctival sac* formed by the *lower eyelid* at the nasal side of the eye. This lower eyelid, representing the second late differentiation of the eye, has begun to develop as a small fold of the skin at STAGE 62 and has grown over the lower portion of the cornea at STAGE 66.

THE DESCRIPTION OF THE EYE

At the end of this development (STAGE 66) the eye has the following structure:

Near the lateral edge of the dorsal side of the head the *eyes* bulge out of their orbital capsules. The shape of the eye is almost spherical. The vaulting of the *sclera* continues nearly unchanged into that of the *cornea*. The stratified epithelium of the cornea increases in thickness towards the corneal centre. The fibrous layer of the cornea is composed of a very delicate outer stratum (*tunica propria cutanea*) and a thicker inner one (*tunica propria sclerotica* or the original *cornea primitiva mesodermalis*). Both layers have fused. The tunica propria sclerotica continues into the fibrous part of the sclera. For the greater part the sclera consists of cartilage. On its inner and its outer surface the cartilaginous shell is covered with a delicate fibrous layer. It extends to the outer side of the eye as far as the insertion of the *extrinsic eye muscles*. In this stage the cartilaginous shell is still one or two cells in thickness except for its nasal part, where the cartilage is somewhat thicker. Inside the sclera can be distinguished a thick, dark-pigmented membrane, which is homologous to the *tunica vasculosa* of the *chorioidea*. Locally big vessels are seen in the tunica vasculosa. Where they are quite surrounded with pigmented cells, as is mostly the case, the pigmented membrane splits into two layers, enwrapping the vessels. On the inner surface of the tunica vasculosa a *lamina choriocapillaris* is present as a delicate membrane. Whether it can be distinguished or not depends upon the filling of the capillaries with blood. The large cells of the pigment layer of the retina are crammed with dark brown pigment granules. The cells preserve their character along the *pars ciliaris retinae* and along the *iris* to the edge of the *pupil*. The various strata of cells and fibres of the inner layer of the *pars optica retinae* are well developed. The *rods* and *cones* have their adult appearance and their size does not differ much from that of the full-grown elements. The fibres of the ganglion-cell layer can easily be followed into the *fasciculus opticus* which forms a long strip at this stage. The cell layers of the pars optica retinae pass into a one-layered columnar epithelium on the internal surface of the *ciliary body*. These cells contain no pigment. The internal surface of the *iris* shows a simple pigmented epithelium. The tissue covering the outside of the pars caeca retinae shows the same dark pigmentation as the chorioidea with which it is continuous. *Muscle fibres* could be distinguished neither in the ciliary body nor in the iris. *Ciliary processes* are not visible. Just in the angle between the ciliary body and the tunica propria sclerotica of the cornea a *pectinate ligament* is present consisting of a dense trabecular tissue. It extends halfway up the iris. The *lens* is nearly spherical and shows no peculiarities in its structure. It is held in position by the *ciliary zonule* attaching to the root of the ciliary body. An *opthalmic artery* with two *ciliary branches* can be followed piercing the sclera not far from the fasciculus opticus. The finer pattern of its branches cannot be determined without special preparations of the vessels. A *hyaloid artery*, however, can be easily distinguished. It reaches the interior of the eye by way of a remnant of the *choroid fissure* in the ciliary body. The artery ramifies on the outer surface of the *vitreous body*. Some *veins* leave the eyeball. One of them can be followed discharging into the *internal jugular vein*.

The eye has a *lower eyelid*. A *gland of Harder* is to be found at the nasal side of the eye; it lies between the eye muscles. It discharges into the *conjunctival sac* formed by the lower eyelid likewise at the nasal side of the eye. A *nasolachrymal duct* ends ventral to the eye at the outer surface of the lower eyelid.

Besides the four *musculi recti* and the two *musculi obliqui*, attached to the cartilaginous part of the sclera, a *musculus retractor bulbi* is inserted onto the proximal part of the sclera.

* *
 *

THE FURTHER DEVELOPMENT OF THE AUDITORY ORGAN

The initial development up to STAGE 28 has been described in the division "The early development of the *central nervous system, sense organs, ganglia* and *nerves*" on page 33.

THE EARLY DEVELOPMENT OF THE EAR VESICLE UP TO STAGE 46

At STAGE 29/30 the isolated *ear vesicle* has greatly increased in size; this has led to the thinning out of its walls, particularly laterally. The following development consists of the formation of the anlage of the *ductus endolymphaticus* at the dorso-caudo-medial side of the *ear vesicle (s.s.)* at STAGE 32. The anlage has become more distinct at STAGE 33/34 and has grown out further at STAGE 35/36. It has become a tube-like evagination at STAGE 41.

At STAGE 33/34 the ventro-medial epithelium of the expanding ear vesicle begins to thicken as a first indication of the formation of the *sensorial epithelium*. This sensorial epithelium, which has more clearly developed at STAGE 35/36, splits into a large medio-caudal and a smaller latero-cranial portion at STAGE 39. At STAGE 40, at which the expanding ear vesicle has become slightly compressed and flattened medio-laterally and dorso-ventrally, the latero-cranial sensorial anlage has split up again into a more medial and a more lateral portion, the latter representing the *crista externa*. The *crista posterior* splits off from the medio-caudal sensorial anlage at STAGE 41, while at STAGE 43 the *crista anterior* forms from a part of the cranial sensorial anlage, the rest forming the *macula utriculi*. The medio-caudal sensorial anlage splits finally into two separate anlagen at STAGE 44.

Meanwhile the modelling of the auditory vesicle makes further progress. At STAGE 41 some lateral protrusions appear, while at STAGE 43 *septa* are formed for the *canalis lateralis*. At STAGE 44 the ear vesicle is divided into a *pars superior* and a *pars inferior*, representing respectively the anlagen of the *utriculus* and the *sacculus*. The *macula utriculi* is situated at the border of both anlagen. The canalis lateralis has been partially separated from the utriculus by septum formation at STAGE' 46. At this stage the distal end of the ductus endolymphaticus has enlarged, forming the *saccus endolymphaticus*.

THE DEVELOPMENT OF THE EAR VESICLE FROM STAGE 46 TO 49

At STAGE 47 [1]) the utriculus has formed the three semicircular canals. The *canalis lateralis* has been separated over a more extensive area from the main lumen than the *canalis posterior*, whereas the *canalis anterior* has only been formed over a small section. The three semicircular canals have been completed at STAGE 49. In the canals the *cristae* have become better demarcated at STAGE 47; actual *ampullae* have not been formed before STAGE 49.

At STAGE 47 the utriculus and sacculus are still broadly communicating with each other. This connection has narrowed at STAGE 50, during which the *utriculus* has been divided into a *pars anterior* and a *pars posterior*. In the *sacculus* the *maculae lagenae* and *sacculi* and the *papillae amphibiorum* and *basilaris* are formed at STAGE 47. At the same stage separate *recessi* for the papilla amphibiorum and the papilla basilaris are being formed. They are already well developed at STAGE 49. The *papilla amphibiorum* gets subdivided into two portions at STAGE 53. The macula lagenae is still situated with the macula sacculi in one and the same diverticulum at STAGE 47. The macula lagenae lies in a separate diverticulum at STAGE 49.

The *n. octavus* enters the ear capsule at the medial side at STAGE 47; its *ramus anterior* sends branches to the cristae anterior and externa and to the macula utriculi. The macula sacculi is innervated by the *intermediate portion* of the n. octavus, while its *ramus posterior* terminates at the macula lagenae, the papilla amphibiorum, the papilla basilaris and the crista posterior.

THE FURTHER DEVELOPMENT OF THE AUDITORY ORGAN

The further development of the auditory organ consists mainly of the extension of the saccus endolymphaticus within the brain capsule, the formation of the perilymphatic system and the development of the cartilaginous ear capsule.

At STAGE 47 the *ductus endolymphaticus* opens into the auditory vesicle at the border of utriculus and sacculus. The ductus forms a long and narrow canal which extends outside the auditory capsule as the *saccus endolymphaticus*. The latter, very much lobed, is apposed dorso-laterally against the plexus chorioideus of the fourth ventricle of the brain. At STAGE 49 the saccus endolymphaticus increases in size, becomes still more lobed, and extends more mediad. The widened ductus endolymphaticus opens now into the sacculus near its communication with the utriculus. During later stages the saccus endolymphaticus extends craniad and caudad and splits craniad into a *ventro-lateral* and a *dorso-lateral lobe*. At STAGE 51 the rostral end of

---

[1]) At STAGE 47, which is apparently much older than STAGE 46, the endolymphatic system has developed much further.

the ventro-lateral lobe extends up to the mesencephalon, and the caudal extremity of the saccus reaches beyond the plexus chorioideus of the myelencephalon. The saccus also extends more mediad. At STAGE 54 the ventro-lateral lobe reaches up to the diencephalon. The left and right sacci come to lie against each other above the plexus chorioideus and extend caudad down to the first spinal nerve. The ventro-lateral lobe reaches up to the telencephalon at STAGE 55, at which the saccus extends caudad up to the second spinal nerve. The saccus reaches caudad to the third spinal nerve at STAGE 56. At this stage the dorso-lateral lobe extends up to the diencephalon and reaches to the front end of the telencephalon at STAGE 57. The caudal extension of the saccus finally reaches the fourth spinal nerve at STAGE 58.

The *perilymphatic system* is for the first time indicated at STAGE 47, when the *recessus papillae basilaris* of the *perilymphatic system* has been formed. The system is still mainly developed as loose connective tissue. The *saccus perilymphaticus* begins to be formed at STAGE 48. It extends into the foramen jugulare at STAGE 49. At STAGE 50 the *ductus perilymphaticus* has been formed together with the *spatium sacculi* and the *recessus papillae amphibiorum* of the *perilymphatic system*. The saccus perilymphaticus extends finally into the fossa condyloidea at STAGE 51.

For the development of the *ear capsule* and the *middle ear* see below.

THE DESCRIPTION OF THE AUDITORY ORGAN

By the end of metamorphosis (STAGE 66) the *auditory system* has become quite complex. It is here given a brief systematic description.

The *utriculus* consists of a *pars anterior* and a *pars posterior*. In the ventral wall of the pars anterior, rostral to the *foramen utriculo-sacculare*, lies the *macula utriculi*. The *ampulla* with *crista* of the *canalis lateralis* (*crista lateralis* or *externa*) is situated at a short distance rostrally and laterally of the *macula utriculi*. Still somewhat more rostrad the *ampulla* with *crista* of the *canalis anterior*, and on the same level but most caudad the *ampulla* with *crista* of the *canalis posterior* are located. The canalis anterior and posterior are medially connected with the utriculus through the *crus commune*. At the border between utriculus and sacculus the *ductus endolymphaticus* opens out at the medial side into the sacculus.

The *sacculus* consists of the *sacculus s.s.* with the *macula sacculi* located rostro-ventro-medially, the *lagena* with the medio-caudal *macula lagenae*, the *pars basilaris* with the *papilla basilaris*, located caudal to the lagena, and finally the *pars papillae amphibiorum* with the *papilla amphibiorum* [1]), situated dorso-caudal to the pars basilaris and close to the opening of the ductus

---

[1]) In the literature the *papilla amphibiorum* has often been confusingly described as *papilla neglecta*. It is, however, not homologous with the papilla neglecta of the mammals, which is located in the utriculus.

endolymphaticus. The papilla amphibiorum consists of two portions separated by a thinner area.

The *ductus endolymphaticus* forms a rather wide canal. It passes into the *saccus endolymphaticus* which is located inside the brain capsule where it is apposed to the brain. The left and right sacci do not communicate. They are very much lobed, and extend from the fourth or fifth spinal nerve up to a point beyond the telencephalon.

The *perilymphatic system* consists of the *cysterna perilymphatica*, which is mainly located at the lateral side of the sacculus and against the foramen ovale, and is connected with the *saccus perilymphaticus* through the *ductus perilymphaticus*, running from the dorso-lateral to the medial side. The saccus perilymphaticus has two recessi, the *recessus partis papillae basilaris* and *partis papillae amphibiorum*, with which it is respectively apposed to the recessus papillae basilaris and recessus papillae amphibiorum of the endolymphatic system, separated from these by thin membranes, the *membranae tympani*, which are located opposite the corresponding papillae. A part of the saccus perilymphaticus protrudes into the brain capsule on the medio-caudal side, where it is apposed to the brain.

The *ganglion acusticum* lies outside the ear capsule. The *n. acusticus* emerging from the ganglion splits into three parts. The *ramus anterior* innervates the cristae of the canalis anterior and horizontalis and the macula utriculi. The *intermediate portion* which runs together with the ramus posterior ends at the macula sacculi. Finally the *ramus posterior* terminates at the macula lagenae, the papillae basilaris and amphibiorum and the crista of the canalis posterior.

For the development of the *cartilaginous* and *bony ear capsule* the reader is referred to the division "the further development of *skeleton* and *musculature* of the *head*" on pages 88, 94 and 98 and for more details concerning the *middle ear* to the division "the further development of the *oro-pharyngeal cavity*, etc." on page 148.

* * *

THE FURTHER DEVELOPMENT OF SKELETON AND MUSCULATURE OF THE HEAD

The early development of the *visceral arches* and *musculature* has been described up to STAGE 37/38 in the division "the early development of the *alimentary system* and the *presumptive visceral skeleton* and *musculature*" on page 39.

A. THE DEVELOPMENT UP TO METAMORPHOSIS

THE PRIMORDIAL DEVELOPMENT OF THE CRANIUM AND VISCERAL SKELETON

After STAGE 35/36 the curvature of the brain decreases, leading to pronounced *topographical changes* in the future *cranium* and *visceral skeleton*. At STAGE 35/36 the tip of the notochord lies at the same transversal level as the recessus opticus, while the oral plate of the mouth anlage is situated caudal to this level. The presumptive anlagen of the cartilagines Meckeli are situated still further caudally, viz. at a level with the infundibulum. The mesenchyme underlying the lamina terminalis of the telencephalon and representing the future ethmoid-trabecular plate, is lying entirely anterior to the oral plate in a more or less horizontal plane ventral to the level of the notochord. During the period from STAGE 35/36 to about STAGE 44 the line connecting the tip of the notochord with the recessus opticus gradually turns upwards, and finally comes to lie in prolongation of the longitudinal axis of the trunk. The now cartilaginous ethmoid plate is situated at the same horizontal level as the trabeculae and the notochord. Meanwhile the visceral skeleton stretches strongly and extends more and more craniad, so that the cartilagines Meckeli finally form the most cranially projecting portion of the skeleton, displacing the mouth opening to the dorso-rostral side of the head. The strands of mesenchyme on both sides of the notochord, originally directed caudad (up to STAGE 37/38), begin to turn around the tip of the notochord. At STAGE 40 they make an angle of 45° with the notochord and at STAGE 42 they are directed perpendicularly to the notochord, towards the eyes. This rotation still continues during the chondrification of the basal plate, so that finally at STAGE 47 the two wings of the basal plate make an angle of 135° with the notochord. They are directing latero-craniad towards the eyes, which have also moved forwards and lateralwards.

The *mesenchyme* around the notochord and the brain, which is mainly of mesodermal origin, shows a first orientation at STAGE 31. From the tip of the notochord the mesenchyme is directed latero-caudad to caudad on both sides of the notochord and craniad on both sides of the prosencephalon, the two cranial wings approaching each other in front of the forebrain. The mesenchyme of the future cranium becomes more and more condensed, so that at STAGE 40 the various parts can be well distinguished as mesenchymatous condensations, viz. the presumptive *basal plate*, the two *trabeculae* and the *trabecular plate*, of which the latter still consists of two strands of mesenchyme approaching each other underneath the front end of the forebrain. Both halves fuse at STAGE 41, at which the basal plate, the trabeculae and the trabecular plate have become procartilaginous, enclosing the *fenestra basicranialis*. Anterior to the trabecular plate the *ethmoid plate* begins to appear at STAGE 40.

The formation of the visceral pouches has divided the mesectoderm and mesoderm into separate cell masses at STAGE 33/34, except for the most dorsal and the most ventral cell material which remains continuous. Dorsally this is particularly evident between the first and the second arches, where the palatoquadratum will be formed. Medio-ventrally the mesenchyme of the arches communicates, particularly between the second and following arches, where the basihyobranchiale and the hypobranchial plate will be formed. In the following period the material of the arches is further condensed. At STAGE 39 the *copula* between the future left and right cartilagines Meckeli and the *basihyale* are individualized as mesenchymatous formations. At the next stage also the *cartilagines Meckeli* and the *palatoquadratum*, with *processus muscularis*, *commissura quadrato-cranialis anterior* and *arcus subocularis*, as well as the *ceratohyale* and the *branchial arches* are individualized as mesenchymatous formations, so that the primordial anlage of the visceral skeleton has been laid down at STAGE 40. The palatoquadratum, with processus muscularis and arcus subocularis, has become procartilaginous at STAGE 41, just as the cartilagines Meckeli, the ceratohyale and the greater part of the *basihyobranchiale*. The caudal part of the latter and the connection of the arcus subocularis with the basal plate are still mesenchymatous. The commissura quadrato-cranialis anterior has become procartilaginous at STAGE 42, at which the palatoquadratum with the processus muscularis, the lateral parts of the ceratohyalia, the basihyobranchiale and the *plana branchialia* begin to chondrify.

THE FURTHER DEVELOPMENT OF THE CRANIUM

The REGIO OCCIPITALIS: The two wings of the *basal plate* are connected underneath the tip of the notochord by chondrifying mesenchyme at STAGE 43. From the caudal side of the basal plate the *parachordalia* begin to grow out at STAGE 45. They develop in between the notochord and the parachordal musculature (head somites), which is pushed laterad. Posteriorly the parachordalia are connected with each other underneath the notochord as the *hypochordal commissure*, which is formed at STAGE 45 to 46 as a ventral chondrification of the perichordal tube. Thus the *planum basale* is formed.

At STAGE 43 some mesenchyme cells have accumulated along the floor and the front and hind wall of the ear vesicle. This mesenchyme increases gradually. It is partially in a procartilaginous state at STAGE 45 and begins to chondifry at STAGE 46, forming the anlage of the *ear capsule*. At that stage the ear capsule consists of a bowl-shaped floor-plate and two dorsally directed arches, a rostral one, the *cupula anterior* and a caudal one, the *cupula posterior*. Their basal portions are flat and broad, while their apical portions are thick and narrow. The *crista otica* is formed along the outer surface of the cupula anterior and is connected with the trabecula by a mesenchymatous ligament at STAGE 45, which becomes procartilaginous at STAGE 46 and chondrifies at STAGE 47. The *trabecula* subsequently forms a *processus oticus trabeculae* which fuses with the extending crista otica in between STAGE 46 and 47, whereby the *foramen prooticum* is formed. The cupula posterior forms the *processus muscularis capsulae auditivae* in between STAGE 46 and 47, which is already well developed at STAGE 47, at which the anterior crista otica has developed along the anterior surface of the ear capsule. In the lateral wall of the blastematous ear capsule another center of chondrification appears at STAGE 46, so that at STAGE 47 only the *foramen ovale* remains in the lateral wall of the ear capsule. At STAGE 47 the ear capsule becomes connected with the parachordalia, which connection is still procartilaginous, but has become cartilaginous at STAGE 48.

The large medial window of the ear capsule is gradually reduced in size by the outgrowth of its edges, which makes particular progress at STAGE 50. The remaining window is divided into two separate windows by a strip of cartilage, descending from the dorsal edge and running to the caudal edge. The dorso-caudal window, the *foramen endolymphaticum*, forms the passage for the ductus endolymphaticus, while the ventrocranial window forms the *foramen acusticum* at STAGE 50. This foramen is split into two separate openings by a cartilaginous bridge which develops between the anterior and the posterior branch of the n. acusticus at STAGE 52.

The originally bean-shaped ear capsules change markedly in form by the formation of the *semicircular canals*. Inside the ear capsule cartilaginous ridges develop on both sides of the semicircular canals of the ear vesicle, a process which has started between STAGE 47 and 48. The ridges gradually approach each other, then fuse, after which they extend in two directions. The *canalis semicircularis anterior* and *lateralis* of the ear capsule are formed at STAGE 49 and the *canalis semicircularis posterior* at STAGE 50. A horizontal ridge, the *crista parotica* (KOTTHAUS, 1933) develops ventro-laterally along the outer wall of the canalis semicircularis lateralis between the STAGES 49 and 51, overarching gradually the foramen ovale. The *crista parotica* becomes confluent anteriorly with the branchio-cranial commissure, while laterally a triangular lobe and caudally a bilobed plate is formed, separated from the processus muscularis of the posterior wall of the ear capsule by a deep incision.

The *occipital arches* are indicated as mesenchymatous formations in between STAGE 46 and 47. They have become procartilaginous at STAGE 47 and begin to chondrify at STAGE 50 to 51. Chondrification is completed at STAGE 53. Ventrally the occipital arches become confluent with the *atlas arches* by the formation of the *occipito-vertebral tissue*. This is changing into procartilage at STAGE 49 and is chondrified at STAGE 52. It differentiates partially into two halves at STAGE 54, the cranial half fuses with the occipital arch and the caudal half with the atlas arch. In the middle region, where the *synovial cavity* will be formed, the intervertebral tissue remains unchondrified, whereas ventrally the atlas arch remains confluent with the occipital arch. The dorsal parts of the occipital arches form the *cristae occipitales laterales*, which extend obliquely forwards and upwards, where they meet a median dorsal roof, the *tectum posterius*, at STAGE 55. The latter is formed by outgrowths of the dorso-medial edges of the cristae oticae, which have fused at STAGE 51. The anterior extension of the occipital arch is limited laterally by the *foramen jugulare*, through which the glossopharyngeus and vagus nerves pass. This foramen has been formed at STAGE 47 by the formation of a procartilaginous connection between the occipital arch and the ear capsule, which connection chondrifies at the STAGES 48 to 49.

The REGIO ORBITO-TEMPORALIS: At the level of the anterior trabecular plate the *palatoquadratum* is connected with the neurocranium by the *commissura quadrato-cranialis anterior*, representing a latero-ventral protrusion from the trabecular plate. This commissure is still procartilaginous at STAGE 43. At this stage the palatoquadratum extends caudalwards as a thin mesenchymatous ligament, the future *processus ascendens* of the palatoquadratum, which is connected with the latero-ventral protrusion of the trabecula, the *pila antotica*, by the *arcus subocularis palatoquadrati*. The latter is already for the greater part cartilaginous at STAGE 44, but is not completed before STAGE 46, when it leads to the formation of the *fenestra subocularis*.

In the further development, in which the regio orbito-temporalis stretches markedly, the *trabeculae* become more and more massive, while the connections with the basal plate, the ethmoid plate and the palatoquadratum broaden considerably. Cartilaginous *cristae* develop along the medial and dorsal sides of the trabeculae, a process which begins in between STAGE 46 and 47. At the base of the trabeculae the *foramen caroticum*, the passage for the arteria carotis interna, has been formed at STAGE 46.

The medial outgrowth of the trabeculae and the simultaneous posterior extension of the anterior trabecular plate lead to reduction of the *fenestra basicranialis*. At STAGE 49 this fenestra still extends from the recessus opticus up to the caudal border of the infundibulum. At STAGE 50 it only encloses the floor of the infundibulum, and at STAGE 51 it has disappeared completely. The dorsal outgrowths of the trabeculae form the lateral walls of the cranium. These *cristae trabeculares* develop more rapidly caudally than cranially. They enclose the nervus opticus and nervus oculomotorius in between STAGE 46 and 47, leaving originally a single foramen for both nerves. At STAGE 51 the *pila metoptica* is formed, dividing this foramen into a *foramen opticum* and a *foramen oculomotorium*. Finally at STAGE 53 the n. trochlearis passes through a narrow canal in the crista trabecularis. The *foramen prooticum*, through which several branches of the trigeminus and facialis nerves pass, is formed in between STAGE 46 and 47 by the connection of the pila antotica or processus oticus trabeculae with the antero-dorsal wall of the ear capsule.

Underneath the floor of the cranium a membrane bone, the *parasphenoid*, is formed at STAGE 51. It forms a very narrow bone lamella extending from the caudal half of the ear capsules up to the middle of the eyes. Dorsal to the cavum cranii a pair of membrane bones, the *frontoparietalia*, are formed. They appear as mesenchymatous condensations at STAGE 50 and begin to ossify caudal to the eyes at STAGE 51. Ossification proceeds up to STAGE 55, at which long but narrow bone lamellae are formed on both sides of the dorsal midline, covering the *fenestra fronto-parietalis*.

The REGIO ETHMOIDALIS: At STAGE 43 the *ethmoid plate* forms a large rectangular cartilaginous plate ventral to the olfactory organs. Caudally it is connected with the trabecular plate and laterally with the palatoquadratum by the still procartilaginous commissura quadrato-cranialis anterior. The ethmoid plate extends forwards, particularly its antero-lateral edges, which grow out laterad around the olfactory organs as *ethmoid flanges*, which together with the ethmoid plate and the *cartilago tentaculi* form the *cartilagines labiales superiores*. This outgrowth begins at STAGE 44. The cartilagines labiales superiores have reached the base of the tentacles at STAGE 47.

The olfactory nerves originally pass freely over the ethmoid plate. At STAGE 50 a median cartilaginous ridge is formed, lateral outgrowths of which later overarch the two nervi olfactorii in the form of the *tectum anterius* and fuse with ridges developed lateral to the olfactory nerves at STAGE 53, thus forming the two *canales olfactorii* through which the olfactory nerves run. The orbital cartilages, developed out of the trabeculae, are connected antero-dorsally with the thin tectum anterius. In the latter the *orbitonasal foramina* are formed by the outgrowth of the tectum and the subsequent enclosure of the nasal branch of the n. ophthalmicus profundus.

THE FURTHER DEVELOPMENT OF THE VISCERAL SKELETON

The PALATOQUADRATUM: This element, running approximately parallel to the trabecula, forms a pyramidal protrusion at its anterior end, the *processus muscularis palatoquadrati*. At the level of this processus the palatoquadratum is connected with the cranium through the *commissura quadrato-cranialis anterior*. Caudally the palatoquadratum extends as *processus ascendens*, which forms the *arcus subocularis* after its connection with the latero-ventral protrusion of the trabecula. All these structures are present as mesenchymatous condensations at STAGE 40. The palatoquadratum with the processus muscularis and processus ascendens has become procartilaginous at STAGE 41 and begins to chondrify at STAGE 42. This chondrification is more or less completed at STAGE 43, at which the commissura quadrato-cranialis anterior has become procartilaginous. This commissura has finally chondrified at STAGE 46. The caudal part of the arcus subocularis palatoquadrati has become procartilaginous at STAGE 44 and has been completely chondrified at STAGE 46. The *processus ventro-lateralis palatoquadrati* appears as a mesenchymatous condensation on the ventro-lateral side of the processus ascendens palatoquadrati at STAGE 47. It has become procartilaginous at STAGE 49 and has formed a well defined cartilaginous processus at STAGE 50. The commissura quadrato-cranialis anterior is connected with the ethmoid plate by a *ligamentum cornu-quadratum mediale* (VAN EEDEN, 1951). The ligament is indicated at STAGE 46 and develops further in between the STAGES 48 and 51. Dorso-laterally to the articulation of the cartilago Meckeli with the palatoquadratum, at the base of the processus muscularis, the *processus cornu-quadratus lateralis* is formed. At STAGE 45 it is still mainly mesenchymatous, only the base forming a procartilaginous protrusion of the palatoquadratum. The proximal portion has become cartilaginous at STAGE 46. Chondrification has extended markedly in between STAGE 46 and 47 but has not yet extended beyond the base of the tentacle at STAGE 49.

The MANDIBULAR ARCH: Anteriorly the palatoquadratum articulates with the *cartilago Meckeli*. The cartilagines Meckeli of both sides are connected through a *copula* or *cartilago labialis inferior*. Both are laid down as mesenchymatous condensations at STAGE 40. The anlagen of the cartilagines Meckeli have become procartilaginous at STAGE 41 and cartilaginous at STAGE 43, at which the copula is still procartilaginous. Chondrification of the copula is not completed before STAGE 46. Along the medial surface of the cartilago Meckeli a membrane bone, the *goniale*, is formed at STAGE 53 to 54.

The HYOID ARCH: The *ceratohyale* articulates with the ventral surface of the palatoquadratum. This second visceral arch is the most massive one. The ceratohyale is present as a mesenchymatous condensation at STAGE 40 It is procartilaginous at STAGE 41 and begins to chondrify at STAGE 42, while chondrification is completed at STAGE 43. At STAGE 44 the *ceratohyalia* are situated in prolongation of each other in a horizontal plane, approximately perpendicular to the longitudinal axis. Medially their upwards curving ends nearly touch each other. The median end of the ceratohyale forms a caudal protrusion, the *processus hyalis posterior*. The two processus have a mesenchymatous connection with the *hypobranchial plate* at STAGE 44, which connection changes into a ligament at STAGE 46. Laterally the ceratohyalia curve caudad towards the palatoquadrata. The rostral end of the basihyobranchiale becomes enclosed by the upwards curving median extremities of the ceratohyalia. The *basihyobranchiale* has been laid down as a mesenchymatous formation at STAGE 40, has become partially procartilaginous at STAGE 41 and has chondrified completely at STAGE 43. In between STAGE 46 and 47 it has fused with the hypobranchial plate, which seems to have a paired origin.

The BRANCHIAL SKELETON: These elements have already been laid down at STAGE 40 in the form of mesenchymatous condensations. The anlagen of the *plana branchialia* are procartilaginous at STAGE 42. They are chondrified and are fused with the hypobranchial plate at STAGE 44. The chondrification of the *arcus branchiales* 1, 2 and 3 has started in the region of the branchial clefts at STAGE 43. At STAGE 45 the arcus of the three anterior branchial arches are already chondrified to a great extent and are connected laterally by the cartilaginous *commissurae terminales*. At this stage the fourth arcus branchialis begins to chondrify, a process completed at STAGE 46. At STAGE 46 also the entire basihyobranchiale, the plana branchialia and the last commissura terminalis are chondrified, so that the entire branchial skeleton with the paired branchial chambers has been formed. The two *branchial chambers* show a wide dorsally directed concavity perforated by three *branchial clefts* which are located in between the four *ceratobranchialia*. During further growth the arcus branchiales broaden markedly, particularly those of the first and the fourth arch, the branchial clefts diminishing in a relative sense. Cartilaginous *branchial rays* develop along the inner side of the arcus branchiales in between STAGE 46 and 47—particularly along the second and third arcus—supporting the filaments of the filter apparatus (cf. the further development of the *oro-pharyngeal cavity*, etc., on page 141). The lateral extremities of the second and fourth arcus branchiales are synchondrotically attached to the crista parotica and the processus muscularis capsulae auditivae (SEDRA and MICHAEL, 1955) or processus branchialis of the crista parotica (KOTTHAUS, 1933) at STAGE 49 to 50. The branchial chambers have now become slightly constricted.

At STAGE 45 to 46 mesenchyme begins to accumulate between the laryngeal epithelium and the mm. constrictores laryngis, foreshadowing the formation of the ARYTENOID ANLAGEN. These anlagen are well defined at STAGE 48. During the following stages the mesenchyme extends further and further round the larynx in its more posterior region, forming the anlage of the *annulus cricoideus*. At STAGE 53 chondrification starts in the arytenoids, as well as dorsally in the annulus cricoideus.

THE FURTHER DEVELOPMENT OF THE MUSCULATURE

Although the segmental structure of the head is still clearly visible at STAGE 40, all muscles are already recognizable and begin to differentiate. The myogenic elements, which have already begun to elongate at STAGE 39, have the form of long cylinders at STAGE 43. Myofibrils appear at STAGE 45 and cross striation can be seen in the m. levator mandibulae anterior at STAGE 46 and in the other muscles at STAGE 47.

At STAGE 40 the *m. levator mandibulae anterior*, the *m. levator mandibulae posterior* and the *m. intermandibularis* can already be distinguished in the region of the mandibular arch. At STAGE 41 the *m. tentaculi* segregates from the m. levator mandibulae anterior. In the region of the hyoid arch the *m. quadrato-hyoangularis*, individualized at STAGE 35/36, has extended into the mandibular arch at STAGE 40. At STAGE 41 also the *m. orbito-hyoideus* and the *m. interhyoideus* are clearly discernable. The *m. genio-hyoideus* has been formed at STAGE 40. In the branchial region the *mm. constrictores branchiales*, the *mm. subarcuales*, the *mm. transversi ventrales* and the *m. levator arcuum branchialium* can be distinguished. The *m. cucullaris* is present at STAGE 55 and later develops further.

At STAGE 40 the anlage of the *m. dilatator laryngis* is very faintly delimited from the mesenchyme surrounding the larynx. It can be followed over some distance caudo-laterad at STAGE 41. At that stage the anlagen of the *mm. constrictores laryngis* are recognizable. The anlagen of both sides meet each other dorsal and ventral to the larynx at STAGE 42.

B. THE DESCRIPTION OF SKELETON AND MUSCULATURE OF THE HEAD at STAGE 55

THE CHONDROCRANIUM

The OCCIPITO-AUDITORY REGION: The *occipital arch* is confluent with the *atlas* vertebra, except dorsally. The free parts of this arch are represented by a pair of *cristae occipitales laterales* which extend obliquely forwards and upwards and meet a median dorsal roof, the *tectum posterius*. The anterior extent of the occipital arch can be limited laterally by a pair of *foramina jugularia*, which are concealed by the auditory capsules and through which the glossopharyngeal and vagus nerves pass. The dorso-medial edges of the auditory capsules anterior to the region of the tectum posterius, represent the *taeniae tecti marginales*. The floor of this region is composed of the *planum basale* which surrounds the notochord posteriorly, while anteriorly there is no cartilaginous floor for the notochord.

The *auditory capsules* are well developed, each possesses a conspicuous *crista parotica* which is prolonged, all over its length, into a broad plate-like *processus muscularis capsulae auditivae*. This is perforated by a small foramen for a branch from the glossopharyngeal nerve and by a larger one for a branchial artery. On the side wall of each capsule lies a wide *fenestra ovalis* which is spanned with fibrous connective tissue and is masked externally by the shield-like processus muscularis capsulae auditivae. The medial capsular wall is pierced with the *foramina acustica anterius* and *posterius* and also with the *foramen endolymphaticum* which lies at a higher level than and nearly between the two acoustic foramina. The posterior region of each capsule is confluent, through the *foramen perilymphaticum inferius* with the subcapsular *condyloid fossa*. The floor of this fossa is fused with the occipital arch posteriorly and is broadly confluent with the capsular floor anteriorly. The fossa itself is spanned posteriorly with dense fibrous tissue, leaving a small laterally directed aperture for the outlet of the glossopharyngeal and vagus nerves. The antero-lateral wall of each capsule is confluent with the *larval otic process* which is broadly continuous with the processus muscularis capsulae auditivae posteriorly, and with the so called processus ascendens quadrati anteriorly. Just in front of each capsule lies the *foramen prooticum* through which the n. ophthalmicus superficialis VII, n. maxillo-mandibularis v, n. ophthalmicus profundus v, n. palatinus VII, n. hyomandibularis VII and the lateral head vein pass.

The ORBITO-TEMPORAL REGION: The floor of this region is formed by the *basis cranii* which represents the fused *trabeculae*. Below this lies the cuneiform *parasphenoid*, the posterior end of which underlies the notochord. Dorsally, there is a wide *fronto-parietal fenestra* which is spanned with a darkly pigmented connective tissue membrane. The fenestra is covered with a pair of developing *fronto-parietal bones* which are thin medially and are bound with connective tissue fibres. Laterally, the side walls of the cranium are represented by the *orbital cartilages* which are thick and long.

In the temporal region, the *palatoquadrate* cartilage is suspended to the neurocranium through a broad *processus ascendens quadrati* which is fused with the *pila antotica* as well as with the orbital cartilage. Laterally, this process extends into the ventrally directed *processus ventro-lateralis quadrati*. The latter sends a thin medial commissure which is fused with the processus muscularis capsulae auditivae. This commissure acts as a support for the *thymus foramen* laterally. The processus ascendens quadrati is pierced with two nerve tunnels which leave it dorsally above the foramen prooticum. The anterior tunnel is for the outlet of the n. maxillo-mandibularis v and the posterior one serves for the passage of the n. ophthalmicus superficialis VII. The two tunnels proceed anteriorly into a common groove. The palatoquadrate cartilage proceeds forwards and slightly downwards as a slender bar, which bounds a narrow *subocular fenestra*. This fenestra is spanned by a layer of fibrous connective tissue.

Ventrally, the *foramen prooticum* proceeds into a groove which lodges the n. palatinus VII, n. hyomandibularis VII, n. ophthalmicus profundus V and the lateral head vein. This groove is supported ventrally by a connective tissue layer. Medial to this groove the basis cranii is pierced with the *foramen caroticum primarium* for the passage of the arteria carotis interna. More anteriorly, the orbital cartilage possesses two nerve tunnels, a narrow dorsal *trochlear tunnel* and a much wider ventral *oculomotor tunnel*. The former serves for the exit of the trochlear nerve and the latter for the passage of the oculomotor nerve and the arteria ophthalmica magna. The oculomotor tunnel opens externally at the anterior level of the processus ascendens quadrati. A short distance in front of the oculomotor tunnel lies the *optic foramen* which separates a relatively short *pila metoptica* from a much longer *pila praeoptica*. Anterior to the foramen caroticum primarium the basis cranii is pierced by a narrow *foramen cranio-palatinum* through which passes the palatine branch of the arteria carotis interna.

The ETHMOID REGION: The floor of this region is formed by the *basis cranii* which extends anteriorly into a broad *ethmoid plate*. This plate is pierced with a pair of foramina for the passage of the medial nasal branches of the profundus nerve. The ethmoid plate extends laterally into the *ethmoid flanges*.

Posteriorly, the ethmoid region possesses a broad *tectum anterius* which forms a roof for this area and connects the orbital cartilages. Both the tectum anterius and the basis cranii are confluent together through a broad vertical *septum* which separates the two *olfactory tunnels* for the passage of the olfactory nerves. At the anterior level of the tectum anterius lies the *orbitonasal foramen* which is limited anteriorly by a thin, weakly developed, *lamina orbitonasalis*. The tectum anterius flows anteriorly into a median crest which separates the two nasal cavities from each other.

The PALATOQUADRATUM: In the ethmoid region the *palatoquadrate bar* has two outgrowths, a medial *commissura quadrato-cranialis anterior* and a lateral *processus muscularis quadrati*. The former is broadly confluent with the side wall of the cranium and is prolonged forwards into a triangular *processus cornu-quadratus medialis*. This is connected with the side of the ethmoid plate, anterior to the choana, through a ligamentous band, the *ligamentum cornu-quadratum mediale*. The processus muscularis quadrati is short and thick. It is directed upwards in its posterior half, while anteriorly it is more or less horizontal. This process is connected with the commissura quadrato-cranialis anterior through a dense ligamentous band. Anteriorly the processus muscularis quadrati proceeds into a slender cartilaginous bar, the *processus cornuquadratus lateralis*, which gains fusion with the lateral edge of the ethmoid flange. Just at this area, it gives off a *tentacular cartilage* to support the corresponding tentacle. This tentacular cartilage is a long and slender process. Just medial to the base of the processus cornu-quadratus lateralis, the *palatoquadrate* gives off a small process which is the larval *pars articularis quadrati* for the reception of the articular region of the lower jaw. Slightly posterior, the palatoquadrate cartilage possesses an articular facet for the ceratohyale.

The MANDIBULAR ARCH: The lower jaw is simple in being composed of a pair of *Meckels' cartilages* fused together through a median *inferior labial cartilage* which is curved with its concavity facing downwards. An osseous element on the medial surface of Meckels' cartilage represents the primordium of the *goniale*.

The HYOID ARCH: The *ceratohyalia* are massive and well developed. They extend obliquely parallel with the lower jaw. Ventrally, each ceratohyale possesses a ridge for the insertion of the hyoid muscles. Dorsally, there is a slight protuberance for articulation with the overlying palatoquadrate. The postero-dorsal border of each ceratohyale possesses a concavity for lodging the anterior region of the corresponding branchial chamber. Medially, each ceratohyale is extended into a postero-medial process. The *basihyale* does not exist separately, but is continuous with the *basi-branchiale* into a median *basihyobranchiale* (see below).

The BRANCHIAL APPARATUS: The *branchial arches* on either side are completely confluent with each other, forming a branchial chamber which is cup-shaped with a wide concavity directed dorsally. Each chamber is perforated ventrally by three gill slits, separating the four *cerato-branchialia* from each other for some distance only. The second and third cerato-branchialia are comparatively narrower than the first and fourth ones. The dorsal rim of the whole chamber is curved internally into a ledge. The area of the branchial chamber lateral to the gill slits represents the *commissura terminalis*. This is synchondrotically attached to the processus muscularis capsulae auditivae in two regions; one opposite to the second ceratobranchiale (the *second cranio-branchial commissure*) and the other opposite the dorso-medial edge of the fourth ceratobranchiale (the *fourth cranio-branchial commissure*). Opposite to the third branchial arch, there is the third cranio-branchial process which is syndesmotically connected with the processus muscularis capsulae auditivae. The ledge projecting from the medial region of the branchial chamber is broad and extends as a horizontal shelf. This shelf is more developed opposite the second and third ceratobranchialia forming the second and third branchial processes. The outer surface of the whole branchial chamber is smooth. However, its inner surface is ragged with a very large number of *branchial rays*. The rays present on the first and fourth ceratobranchialia are much shorter than those on the second and third ones. Each of the latter arches possesses one long ramified arboreal ray which is continuous with the second and third branchial processes respectively.

The *hypobranchial plate* is a small piece which is confluent with the bases of the first pair of ceratobranchialia. It is pierced with a pair of foramina for the thyreopharyngeal arteries. The *basihyobranchiale* is a well developed structure with a broad, deep ventral keel. It is confluent posteriorly with the hypobranchial plate as well as with the anterior region of the bases of the first pair of ceratobranchialia. It extends anteriorly to the base of the larval *hyoglossal sinus*. Loose cartilage cells join the medial sides of the ceratohyalia with the basihyobranchiale. The postero-medial processes of the ceratohyalia form a peg-socket articulation with the dorsal surface of the basihyobranchiale. This articulation is strengthened by a ligament binding both structures.

The LARYNX: The only structures developed are the primordia of the *arytenoids* which are oval and formed of juvenile cartilage cells. The *annulus cricoideus* is only feebly developed dorsally, and is composed of a median cartilagenous piece fringed with dense mesenchyme cells.

THE MUSCULATURE

The mandibular muscles are represented by the three *mm. levatores mandibulae* and the *m. intermandibularis*, of which the *m. levator mandibulae anterior*, which is the largest, originates more posteriorly on the palatoquadratum and splits into three bundles, of which two, the *pars medialis* and *pars intermedius*, insert onto the Meckels' cartilage, and one, the *pars lateralis*, onto the tentacle. The *m. levator mandibulae posterior* originates more anteriorly on the palatoquadratum and inserts onto the Meckels' cartilage, while the poorly developed *m. levator mandibulae externus* originates from the processus muscularis and inserts onto the Meckels' cartilage. The left and right *m. intermandibularis posterior* originate from the Meckels' cartilages and meet in a medio ventral raphe.

The hyoid muscles consist of three *levator muscles* and a *m. interhyoideus*. The *m. orbito-hyoideus* originates from the processus muscularis and inserts onto the ceratohyale; the *m. suspensorio-hyoideus* originates from the palatoquadratum and inserts onto the ceratohyale, and the *m. quadrato-hyoangularis* originates from the articular surface between the palatoquadratum and the ceratohyale and inserts onto the Meckels' cartilage. The left and right *m. interhyoideus* originates from the ceratohyalia and meet in a medioventral raphe.

The branchial musculature consists of a larger number of muscles. The *mm. constrictores branchiales* form a continuous sheet which originates from the processus ascendens, the processus ventro-lateralis palatoquadrati, and the crista parotica processus muscularis capsulae auditivae and inserts onto the branchial chamber. The *m. subarcualis rectus* 1 originates on the first arcus branchialis and inserts onto the ceratohyale. The *m. subarcualis rectus* 2 originates and inserts onto the second arcus branchialis. The *mm. subarcuales recti* 3 and 4 originate and insert onto the third arcus branchialis.

The *mm. transversi ventrales* originate partially from the left and right second ceratobranchiale and partially from the left and right processus muscularis capsulae auditivae and meet in a medio-ventral raphe. The *m. levator arcuum branchialium* originates also from the processus muscularis capsulae auditivae and inserts onto the fourth ceratobranchiale. Finally the *m. cucullaris* originates from the scapula and inserts onto the fourth arcus branchialis.

The hypobranchial spinal muscles include only the *m. genio-hyoideus*, which originates from the first ceratobranchiale and inserts onto the inferior labial cartilage.

The *m. dilatator laryngis* originates from the processus muscularis capsulae auditivae and inserts onto the arytenoids, while the *mm. constrictores larynges* originate from the dorsal and ventral raphe and insert onto the arytenoids.

C. THE DEVELOPMENT DURING METAMORPHOSIS

THE DEVELOPMENT OF THE SKELETON

During the stages of metamorphosis the following changes occur:

In the REGIO OCCIPITALIS the *occipito-atlantal joint* is formed at STAGE 58, at which the *occipito-prootic ossification* has started. At STAGE 60 the *processus muscularis capsulae auditivae* becomes syndesmotically confluent with the fourth *commissura cranio-branchialis*. The latter begins however to disappear and is already absent at STAGE 61. All branchial commissures have disappeared at STAGE 62, while the processus muscularis capsulae auditivae has disappeared at STAGE 63. The adult *processus oticus* appears at STAGE 64 and fuses with the auditory capsule at STAGE 65.

In the area of the foramen ovale of the ear capsule the *operculum fenestrae ovalis* and the *pars interna plectri* appear at STAGE 59, while at STAGE 61 the *pars media* and *externa plectri* and the *annulus tympanicus* begin to appear as mesenchymatous condensations. At STAGE 63 the pars interna and media plectri have become cartilaginous. The pars media and externa plectri have become confluent at STAGE 64, at which the cartilaginous annulus tympanicus has become sickle-shaped. At STAGE 65 the pars externa plectri has begun to chondrify, while ossification has started in the pars media plectri. At STAGE 66 the plectral apparatus has been completed and the operculum fenestrae ovalis is confluent with the wall of the ear capsule.

In the REGIO ORBITO-TEMPORALIS the *frontoparietalia* are fused at STAGE 56, leaving only a *parietal foramen*. The floor and the side walls of the cranium begin to show erosion phenomena at STAGE 61. The *dorsal outlets* of the *foramen prooticum* have become confluent at STAGE 62. At STAGE 63 the *foramen metopticum* or *oculomotorium* and the *foramen caroticum primarium* have become confluent, while at STAGE 65 both have become confluent with the *foramen prooticum* by the disappearance of the processus ascendens palatoquadrati and the progressive decay of the basis cranii. The pair of *foramina cranio-palatina*, described at STAGE 55, are occluded at STAGE 63. At STAGE 66 the side wall of the cranium shows the following foramina: *for. opticum, for. trochlearis,* the *common for. prooticum, metopticum* and *caroticum primarium,* and the *dorsal outlet* of the *for. prooticum.*

At STAGE 63 the *post-palatine commissure* appears. Meanwhile the *parasphenoid* extends underneath the *septum nasi.* At STAGE 66 the postpalatine commissure is well developed and the *orbitosphenoid* has become partially fused with the parasphenoid, while the frontoparietalia have become better developed.

In the ETHMOIDAL REGION important changes occur. The *nasal capsules* are rapidly formed. At STAGE 58 the *septum* and *tectum nasi* appear, just as the *crista intermedia* and the *cartilago alaris.* The *nasale* appears at the same stage. The crista intermedia sends a *lamina inferior* to the ethmoid plate; both fuse at STAGE 59, while at STAGE 60 the *cartilago obliqua* and the *processus praenasalis superior* appear, the latter connecting the cartilago alaris with the developing *solum nasi.* The cartilago obliqua becomes connected with the crista intermedia through the lamina inferior at STAGE 60. The rostral part of the *ethmoid plate* and the *ethmoid flanges* begin to diminish at STAGE 60. The latter have already disappeared at STAGE 61. The *lamina orbito-nasalis,* which has become confluent with the processus cornu-quadratus medialis of the palatoquadratum at STAGE 60, forms a *processus maxillaris anterior* at STAGE 61. On the upper jaw the *maxillae* appear at STAGE 58, while the *praemaxillae* and *septomaxillae* are formed at STAGE 60. *Teeth* develop on the praemaxillae and maxillae at STAGE 60 (cf. the further development of the *oropharyngeal cavity,* etc., on page 146).

Very pronounced changes occur in the PALATOQUADRATUM. After the fusion of the *processus cornu-quadratus medialis* with the lamina orbitonasalis at STAGE 60, erosion starts at STAGE 61 in the *processus muscularis,* the *processus ascendens,* the *processus cornu-quadratus lateralis* and in the *commissura quadrato-cranialis anterior,* after which the palatoquadratum starts turning and shifting backwards. At STAGE 61 the palatoquadratum makes an anterior angle of only 50° with the floor of the cranium. This angle increases to 65° at STAGE 62, 85° at STAGE 63, 105° at STAGE 64 and finally 115° at STAGE 66, at which the palatoquadratum has become syndesmotically connected with the post-palatine commissure. During this turning the *processus muscularis* and the *ligamentum cornu-quadratum mediale* have disappeared at STAGE 62, as well as the *cartilago tentaculi,* while the *processus cornu-quadratus lateralis,* the *processus ascendens,* the larval *processus oticus* and the *medial portion* of the *commissura quadrato-cranialis anterior* have disappeared at STAGE 63. The *processus pterygoideus* and the *processus maxillaris posterior* become differentiated at STAGE 63. As membrane bones there appear on the palatoquadratum the *squamosum* at STAGE 62, and the *pterygoid*, around the processus pterygoideus, at STAGE 63.

The LOWER JAW or *cartilago Meckeli* is markedly displaced caudad. The *dentale* and the *outer lobe* of the *goniale* appear at STAGE 60. The inner and outer lobes of the goniale fuse at STAGE 62. Dentale and goniale finally cover the greater part of the cartilago Meckeli at STAGE 66.

The HYPOBRANCHIAL APPARATUS undergoes a strong reduction. Erosion starts in the keel of the *basihyobranchiale* at STAGE 61, while the *craniobranchial processes* have disappeared at STAGE 62. The *branchial chambers* become reduced at STAGE 63 and have disappeared at STAGE 64. As new formations the *thyroid processes* appear at STAGE 62 and the *alae* as the postero-lateral wings of the hypobranchial apparatus at STAGE 63, while the *hyoglossal sinus* is transformed into the *hyoglossal foramen* at STAGE 65. This foramen has become wider at STAGE 66, at which the ceratohyalia, alae and thyroid processes extend from the *corpus hyoideum*.

The LARYNGEAL CARTILAGES, which consist of the primordia of the *arytenoids* and the roof of the *annulus cricoideus* at STAGE 55, have developed gradually. The arytenoids have chondrified completely at STAGE 60 and the floor of the annulus cricoideus has formed at STAGE 61. The *bronchial processes* appear at STAGE 64. The larynx is finally continuous with the thyroid processes of the corpus hyoideum through a *hyocricoid connection* at STAGE 66. At the end of metamorphosis the hypobranchial apparatus and larynx of male and female are still identical.

THE DEVELOPMENT OF THE MUSCULATURE

During the stages of metamorphosis the following changes occur:

At STAGE 58 the *m. intermandibularis anterior* appears without a median raphe. It is attached from its early appearance to the cartilagines Meckeli. At STAGE 66 the *m. intermandibularis posterior* becomes attached to the goniale. The *pars lateralis* of the *m. levator mandibulae anterior*, previously inserting onto the tentacle, becomes reduced at STAGE 61 and loses its individuality at STAGE 63. The *pars intermedia* of the *m. levator mandibulae anterior* is fused with the *m. levator mandibulae posterior* at STAGE 64 and has become attached to the palatoquadratum only. Together with the *pars lateralis* of the *m. levator mandibulae anterior* it forms the adult *m. levator mandibulae posterior*, which originates from the frontoparietale and the auditory capsule and is broadly inserted onto the goniale at STAGE 66. At this stage the larval *pars medialis* of the *m. levator mandibulae anterior* forms the *m. levator mandibulae anterior* of the adult, originating from the frontoparietale and inserting by a long tendon onto the goniale. The *m. levator mandibulae externus* finally originates from the squamosum, the annulus tympanicus and the quadratum and is broadly inserted onto the outer surface of the goniale at STAGE 66.

The *m. orbitohyoideus*, previously attached to the processus muscularis palatoquadrati, takes its origin from loose connective tissue fibres posterior to the eroding processus at STAGE 61. The muscle fuses with the *m. suspensorio-hyoideus* and acquires a new origin on the auditory capsule at STAGE 63. The compound muscle becomes inserted onto the posterior end of the lower jaw at STAGE 65. The previous muscles and the *m. quadrato-hyoangularis* form the adult *m. depressor mandibulae* at STAGE 66. This muscle originates from the auditory capsule and is inserted onto the posterior end of the lower jaw. The *m. interhyoideus* becomes attached to both the ceratohyale and the palatoquadratum at STAGE 61 and is finally attached to the palatoquadratum only at STAGE 66. A new muscle, the *m. petrohyoideus*, appears at STAGE 62. It originates from the auditory capsule and meets the muscle of the other side in a median raphe ventral to the annulus cricoideus at STAGE 66.

The *mm. constrictores branchiales*, the *mm. subarcuales recti* and the *m. transversus ventralis* of the second ceratobranchiale are represented by loose muscle fibres only at STAGE 63 and have disappeared at STAGE 64.

The *m. cucullaris* changes its origin onto the auditory capsule at STAGE 66.

The *m. geniohyoideus* becomes separated into a *pars medialis* and a *pars lateralis* at STAGE 60. The pars medialis finally originates from the annulus cricoideus at STAGE 66, while the pars lateralis arises from the ala. Both insert onto the Meckel's cartilage.

The *m. dilatator laryngis* gains a new origin at the cricoid at STAGE 65, so that it finally originates from the thyroid process and the annulus cricoideus, and inserts onto the arytenoids. The *mm. constrictores laryngis* of the adult are identical with those of the larva.

* * *

THE FURTHER DEVELOPMENT OF SKELETON AND MUSCULATURE OF THE TRUNK

A. THE DEVELOPMENT OF THE AXIAL SKELETON

The early development up to STAGE 37/38 has been described in the division "the early development of the *axial system*", on page 35.

At STAGE 37/38 the *notochord*, which is now completely vacuolated, extends cranially to a point between the eyes (see also under the development of the *cranium*, on page 87). The *elastica interna* is still very thin. At STAGE 42 the tip of the notochord reaches up to the infundibulum. At STAGE 46 the tip begins to disintegrate, so that at STAGE 47 the notochord extends only to the front margin of the auditory capsule. The anterior end is further reduced at STAGE 49 to 50. After the development of the neural arches and the chondrification of the vertebral portions of the perichordal tube (see below), the arches exert pressure upon the notochord, so that the notochordal sheaths are much thinner underneath the arches than intervertebrally. The cranial portion of the notochord is much reduced at STAGE 55 (see also under the development of the *cranium*, on page 93).

From STAGE 56 on the *notochord* gradually degenerates. The intervertebral cartilage formations of the perichordal tube are much thicker than the vertebral portions at STAGE 56, so that the notochord becomes compressed intervertebrally. At STAGE 64 the atrophying notochord runs ventral to the vertebral column. The remaining notochordal tissue is shrunken and becomes separated from the notochordal sheaths which are in the process of resorption. At STAGE 65 only vestiges of the notochord are present along the vertebral column. Remnants of the notochord still persist ventral to the vertebral column at STAGE 66.

The development of the *axial skeleton* proceeds in a cranio-caudal direction, except for the *atlas*, which shows an aberrant development, and the first *two vertebrae*, which are retarded in development with respect to the following vertebrae. The vertebrae develop out of the *neural arches*, the *perichordal tube* and the *hypochord* (in the urostyle region).

The future *neural arches* appear as condensations of mesenchyme opposite the myocommata. Between the anlagen of the consecutive neural arches the spinal cord is surrounded by scattered mesenchymal cells. The *perichordal tube*, which forms a clearly discernible but very thin continuous mesenchymal layer around the notochord in between the STAGES 37/38 to 50, begins to differentiate at STAGE 51.

The *atlas arches* appear as condensations of mesenchymal cells posterior to the occipital arches and in front of the second trunk myotomes at STAGE 47. Their basal portions are procartilaginous and their dorsal portions in a blastematous state at STAGE 48. The line of demarcation between the basal portions of the occipital and the atlas arches is not distinct at this stage. At STAGE 49 to 50 the bases of the atlas arches start to chondrify, while the remaining portions are still blastematous. Between the atlas and the occipital arches a dense *occipito-vertebral tissue* which is changing into procartilage is formed, representing an *interdorsal arch*. Chondrification of the basal portions of the atlas arches is completed at STAGE 51. Chondrification has extended dorsalward up to the area above the spinal cord at STAGE 52. Ventrally the bases of the atlas arches are in cartilaginous continuity with those of the occipital arches as a result of the chondrification of the occipito-vertebral tissue. The rest of the occipito-vertebral tissue is chondrified at STAGE 53, but differentiated into a cranial half, which now fuses with the occipital arches, and a caudal half which unites with the atlas arches. In between, where the future *synovial cavity* will be formed, the tissue remains unchondrified. No indications of a joint are present yet between the occipital and the atlas arches at STAGE 56.

In the further development the increasing occipito-vertebral cartilage pushes the basal portions of the atlas arches further backwards. The first indications of an *occipito-vertebral joint* appear at STAGE 60. It is completely established at STAGE 62, arising partly as a result of degeneration and resorption of cartilage. The remaining cartilage forms the *occipital condyles* and the *cups* on the anterior surface of the atlas arches. A part of this intervertebral tissue is added to the centrum of the atlas, formed out of the chondrifications of the perichordal tube (see below). This situation represents more or less the final state.

The development of the *vertebrae* proceeds in a characteristic way. The condensations of mesenchyme opposite the myocommata become procartilaginous at STAGE 47. The bases of the third and fourth *neural arches* are procartilaginous at STAGE 48. They are chondrified except for the parts dorsal to the spinal cord at STAGE 51. Chondrification is completed at STAGE 52 in the fourth arches and nearly completed in the third arches. The dorsal ends of the first and the second arches are still connected by connective tissue at STAGE 54. Here chondrification is only completed at STAGE 55. The neural arches are completely chondrified from the fourth to the ninth vertebra at STAGE 52. The 10th pair of arches is chondrified at STAGE 54 and the 11th pair at STAGE 55. The 12th pair of arches is chondrified at STAGE 56. Except for the 13th and the 14th pair of arches which chondrify partially, the rest of the spinal cord remains enveloped only by connective tissue arches.

Perichondral ossification appears on the outer and inner surfaces of the *neural arches* at STAGE 54. Meanwhile destruction of cartilage as well as calcification of the cartilage occurs, particularly in the basal portions. The outer perichondral bone has become much thicker at STAGE 55, while at the same time a large amount of cartilage as well as inner perichondral bone has been resorbed. In the neural arches of the first three vertebrae almost all the cartilage dorsal to the spinal cord has been resorbed at STAGE 60, so that the arches now consist of only the outer perichondral bone and a thin inner strip of calcified cartilage. At that stage the basal portions of the arches, which consist of calcified cartilage, contain *primary marrow cavities*. The inner bony lamella as well as the cartilage are then further resorbed, so that the dorsal and dorso-lateral portions of the first to the eighth vertebra consist of only an outer perichondral bony lamella at STAGE 62. Nearly the whole neural arch is formed by the outer perichondral bony layer; at STAGE 66 only the bases of the arches contain enchondral bone.

The 13th pair of arches is in a procartilaginous state at STAGE 55. The 10th pair of arches fuses with the 11th pair at STAGE 57. The 11th and 12th pairs of arches fuse at STAGE 58. At STAGE 60 the cranial portion of the 13th pair of arches, which is now completely chondrified, fuses also with the 12th pair. Now even the basal portions of the 14th pair of arches have become procartilaginous. At STAGE 61 the 11th pair of arches is partly ossified perichondrally. At STAGE 63 perichondral bone is not yet present on the cartilage of the 12th pair of arches. At this stage the cartilage of the 13th pair of arches starts degenerating, after which resorption sets in. In the *urostyle region* the neural arches together with the dorsal perichordal cartilage form a hollow cartilaginous cylinder perforated by the spinal nerves. At STAGE 65 the cartilage of the 13th pair and the procartilage of the 14th pair of arches have been converted into connective tissue and are in the process of resorption. At STAGE 66 the vertebrae of the urostyle region consist of calcified cartilage, while some further resorption still takes place, except for the 10th vertebra in which the whole space between the inner and outer perichondral bony layers of the arches is occupied by marrow cavities.

The still very thin *perichordal tube* has become thicker at STAGE 52, its ventral portion being now in a procartilaginous state. In its cranial portion chondrification takes place in the lateral walls at STAGE 52. At STAGE 53 the perichordal tube begins to chondrify *vertebrally*, i.e. underneath the arches, so that the arches now appear to rest almost directly upon the elastica externa. *Intervertebrally* the perichordal tube has become thicker on its dorsal side at STAGE 53. All chondrifications extend in a cranio-caudal direction. *Ventrally* the chondrification reaches to the region of the fourth vertebra at STAGE 53 and to that of the sixth vertebra at STAGE 54. It extends finally over the entire length of the trunk at STAGE 55. Anteriorly the ventral chondrification of the perichordal tube is continuous with the hypochordal commissure of the basal plate. The *lateral* portions of the perichordal tube are partly chondrified in the atlas region at STAGE 51. The lateral chondrification extends down to the intervertebral region between the second and the third vertebra at STAGE 52 and then stops.

The vertebral portions of the *perichordal tube* dorsal to the notochord have changed into procartilage down to the region of the fourth vertebra at STAGE 54. They are chondrified down to the sixth pair of arches at STAGE 55, whereas the intervertebral portions have formed cartilage down to the third vertebra at that stage, thus forming anteriorly a complete cartilaginous perichordal tube down to the third vertebra. The vertebral chondrification extends to the region of the 9th vertebra at STAGE 56, while the intervertebral cartilage formation has then reached the region of the 5th vertebra. At STAGE 56 perichondral bone formation appears dorsally on the vertebral chondrifications of the perichordal tube, while calcification of the cartilage matrix has set in. The vertebral chondrification of the perichordal tube has reached the 10th vertebra and the intervertebral cartilage formation the 8th vertebra at STAGE 57. The latter forms thickenings compressing the notochord intervertebrally. Both vertebral and intervertebral chondrifications extend posteriorly to the 10th vertebra at STAGE 58. Chondrification and perichondral ossification of the dorsal aspects of the vertebral portions of the perichordal tube have extended down to the 12th pair of neural arches at STAGE 60, while intervertebrally the chondrification has reached the 11th pair. At this stage the ventral cartilage formation continues over some distance into the tail. The vertebral centra, except those in the urostyle region, consist of calcified cartilage with primary marrow cavities. At STAGE 62 the intervertebral cartilages begin to calcify in the pre-urostyle region.

Degeneration of the ventral cartilage formation in the *perichordal tube* starts anteriorly at STAGE 63. At the next stage the entire ventral cartilage formation has been converted into loose connective tissue, while simultaneously the lateral cartilage formations (only present anteriorly) are in the process of resorption. The dorsal, epichordal chondrifications and ossifications remain, forming the centra of the atlas and the other vertebrae and the intervertebral cartilage. At STAGE 65 the intervertebral cartilages become divided into two portions, forming an *anterior socket* which is added to the hind end of the preceding centrum, and a *posterior ball* added to the front end of the following *opisthocoelous centrum*. Between socket and ball *intervertebral joints* are formed.

Ventral to the perichordal tube a rod of connective tissue appears in the region of the future *urostyle* at STAGE 49, thickens at STAGE 50 and becomes rectangular in cross section at STAGE 51. It increases gradually in thickness during the following stages. It starts to chondrify at STAGE 56 to 57, forming a ventral ridge, and is now called the *hypochord*. At STAGE 58 it extends from the intervertebral region between the 9th and 10th vertebrae down to a point beyond the 12th vertebra, from where it continues as ordinary connective tissue. It becomes perichondrally ossified on its ventral surface at STAGE 60 to 61. The chondrification extends still somewhat more caudad and reaches the 13th pair of neural arches at STAGE 63. Meanwhile some of the cartilage on its dorsal surface is being destroyed in the anterior region of the hypochord. At STAGE 64 the urostyle is completely chondrified anteriorly and ventrally, but not posteriorly and dorsally, so that at STAGE 66 the urostyle is ossified anteriorly and chondrified laterally and ventrally, but remains fibrous in the posterior dorsal region. Concomitantly with the reduction of the notochord the front end of the hypochord approaches the dorsal, intervertebral cartilage between the 9th and the 10th vertebrae at STAGE 65 and almost touches it at STAGE 66.

On the dorso-lateral side of the spinal cord and dorsal to the spinal ganglia the consecutive neural arches are connected by strips of mesenchyme at STAGE 51, representing the anlagen of the *zygapophyses*. The cartilage of the arches spreads into these mesenchymal strips at STAGE 52, whereby the zygapophyses are actually established. At the location of the future joints the tissue remains unchondrified. At STAGE 54 the zygapophyses are well developed and *synovial cavities* arise in the connective tissue between the *pre-* and *post-zygapophyses* in the region of the 4th to the 9th vertebra. In the region of the first three vertebrae they are less developed. Between the first and the second vertebra a cartilaginous continuity develops and the development of a joint is not yet indicated at STAGE 55. At this stage, at which the tissue surrounding the zygapophyseal synovial cavities between the 4th to the 9th vertebra chondrifies, well developed zygapophyses are also present between the 9th and the 10th pair of arches. The first indications of synovial cavities between these arches do not appear before STAGE 57.

Small cartilaginous *diapophyses* develop on the neural arches of the second to the fourth vertebra at STAGE 55. They appear on the 5th pair of neural arches at STAGE 57. Short cartilaginous diapophyses are present on the 5th to the 9th vertebra at STAGE 60.

"*Transverse processes*" arise partly as direct chondrifications of tissue in the myocommata and are not merely outgrowths of the neural arches. They should be called *pleurapophyses* since they are vertebro-costal in origin. The "transverse processes" of the 5th to the 9th vertebra are present as condensations of mesenchyme in the proximal portions of the myocommata at STAGE 58. Those of the 9th vertebra are much longer and more advanced than the other ones. At STAGE 60 those of the 9th vertebra are chondrified medially and are fused with the diapophyses. At STAGE 62 they expand laterally. However, they have not yet reached the ilia at STAGE 66. At STAGE 64 they are medially ossified perichondrally, while at that stage those of the 5th to the 8th vertebra show the first signs of direct ossification. The "transverse processes" of the 5th to the 8th vertebra have not yet attained their full length at STAGE 66.

The *ribs* appear for the first time at STAGE 52 as condensations of connective tissue in the myocommata opposite the second, third and fourth pairs of arches. The chondrification only proceeds slowly. The distal portions of the first pair are only in a procartilaginous state, those of the second pair are chondrified and those of the third pair are still blastematous at STAGE 56. At STAGE 57 the first and second pairs of ribs are chondrified and attached to the diapophyses of the second and third vertebrae respectively by means of dense connective tissue. The second rib is much longer than the first and reaches laterally to the suprascapula. The distal portions of the third pair of ribs are chondrified at STAGE 57 and these ribs are completely chondrified at STAGE 58. They are longer than the second pair and are curved backwards. The attachments of the first and second pairs of ribs—these ribs are now ossified perichondrally—are now changing into procartilage, whereas those of the third pair of ribs still form a dense connective tissue connection with the diapophyses of the fourth vertebra at STAGE 58. At STAGE 60 the tissue connecting the first and second pairs of ribs to the corresponding vertebrae starts chondrifying. This process is completed at STAGE 61, at which the attachments of the third pair of ribs pass into procartilage. Cartilage is formed here at STAGE 62 At STAGE 66 the middle portions of the ribs are ossified; proximally and distally they are still cartilaginous.

B. THE DEVELOPMENT AND DESCRIPTION OF THE AXIAL AND ABDOMINAL MUSCULATURE

The early development up to STAGE 37/38 has been described in the division "the early development of the *axial system*" on page 35.

THE DEVELOPMENT UP TO METAMORPHOSIS

The *dorsal muscles*, derived from the *myotomes s.s.* and the *ventral muscles*, derived from the "*Urwirbelfortsätze*" develop further during the STAGES 37/38 and 39. The dorsal muscle masses, consisting of the consecutively arranged dorsal myotomes, are called the *mm. dorsales trunci*. Axial mesenchyme migrates into the intermyotomic fissures, forming the *myocommata* or *transverse myosepta*, at STAGE 43.

The most anterior somites, the *head somites*, diminish gradually. At

STAGE 42 only two pre-trunk myotomes, the head myotomes III and IV are present. Myotome III disappears at STAGE 44. The last pre-trunk myotome, myotome IV, has disappeared at STAGE 47, at which also the first trunk myotome is reduced to a few fibres joining the second trunk myotome to the posterior wall of the otic capsule. The last fibres of the first trunk myotome have disappeared at STAGE 48. Now the second trunk myotome inserts onto the cartilaginous postero-lateral wall of the otic capsule and the posterior surface of the developing occipital arch.

The "*Urwirbelfortsätze*" which first appear as double rows of cells at STAGE 40 are completely detached from the dorsal muscles at STAGE 41 and now form single rows of cells by the disintegration of the lateral lamellae. The cells of the "Urwirbelfortsätze" are transformed into muscle fibres at STAGE 44, and are split into a dorsal and a ventral portion, representing respectively the *m. obliquus abdominis* and the *m. rectus abdominis*. The two groups of muscles are not yet sharply demarcated at STAGE 46. An anterior bundle of the latter inserts onto the medio-ventral surface of the hyoid and the lateral wall of the pericardial cavity at STAGE 46. Migrant mesenchyme-like cells can be observed on the medial and lateral surfaces of the m. obliquus abdominis at STAGE 49 to 50, while at STAGE 51 the m. rectus abdominis is lying against the *linea alba*, a connective tissue band located in the mid-ventral line and derived from migrant axial mesenchyme.

At STAGE 51 the first indications are found of the formation of *secondary muscles* originating from the group of primary muscles described above. At STAGE 52 the *mm. interarcuales* develop as derivatives of the mm. dorsales trunci (from their dorso-medial aspects), connecting consecutive vertebral arches (see fig. 5, RYKE '53). At STAGE 53 ten pairs can be distinguished, the first pair between the occipital arch and the atlas, and the last pair between the 9th and the 10th vertebra. The composition of the dorsal musculature remains practically unchanged up to STAGE 56.

Meanwhile some ventral secondary muscles develop from the m. obliquus abdominis, viz. a superficial proliferation on the lateral side of the primary muscle, the *pars externus superficialis* at STAGE 52 and one on the medial side, the *m. transversus abdominis* at STAGE 53. During the STAGES 54 to 56 the two ventral secondary muscles are constantly being thickened at the expense of the primary m. obliquus abdominis. The rest of the latter represents the fused *pars externus profundus* and *pars internus* of the m. obliquus abdominis. The mm. obliqui and mm. recti become demarcated by connective tissue, to which the m. transversus abdominis is attached at STAGE 56 (see fig. 9, RYKE, '53).

THE DEVELOPMENT DURING METAMORPHOSIS

During further development many new muscle bundles arise. At STAGE 57 the mm. dorsales trunci are giving rise to the *mm. intertransversarii*, connecting consecutive "transverse processes" of the vertebrae. They run from the otico-occipital complex down to the 9th vertebra (see fig. 14, RYKE, '53). At STAGE 62 two secondary muscles are derived from the m. dorsalis trunci, viz. the *pars lateralis* of the m. *ileolumbalis*, and the m. *coccygeo-sacralis* representing a caudal continuation of the mm. intertransversarii. The latter connects the urostyle with the "transverse process" of the 9th vertebra. During the STAGES 63 to 65 two new muscles are formed from the m. dorsalis trunci, viz. the m. *coccygeo-iliacus* and the *pars medialis* of the m. *ileolumbalis*. The medio-dorsal part of the m. dorsalis trunci is known as the m. *longissimus dorsi* (see fig. 22, RYKE, '53).

At STAGE 57 secondary muscles develop on the lateral side of the m. rectus abdominis. Thus, at STAGE 58 anterior to the heart the m. *sterno-hyoideus* arises. The two mm. sterno-hyoidei fuse to a single m. sterno-hyoideus at STAGE 61. In the trunk region a m. *rectus abdominis superficialis* develops at STAGE 58. The deeper portion of the m. rectus abdominis is now called m. *rectus abdominis profundus*. During the STAGES 63 to 65 the m. transversus extends rapidly as compared with the m. obliquus abdominis externus superficialis, while the remaining portions of the primary m. obliquus abdominis disappear. During the last stages of metamorphosis the muscles become more firmly attached to the skeletal elements.

TOPOGRAPHY OF THE MUSCLES AT STAGE 66

At STAGE 66 the m. *longissimus dorsi* takes origin from the urostyle and inserts onto the prootic-occipital complex. The m. *ileolumbalis, pars medialis*, underlies the transverse processes of the 2nd to the 9th vertebrae. The m. *ileolumbalis, pars lateralis*, takes origin from the anterior tip of the iliac wing and inserts onto the myocommatal aponeurosis corresponding to the 5th vertebra, and onto the distal part of the 3rd rib. The m. *coccygeo-iliacus* takes origin from the urostyle and inserts onto the distal portion of the ilium.

The m. *rectus abdominis profundus* takes origin from the epipubis and from the m. sartorius and inserts onto the sternum and the hyoid. The m. *rectus abdominis superficialis* stretches from the level of the anterior tip of the ilium to the sternum. The m. *sterno-hyoideus* takes origin from the sternum and inserts onto the hyoid. The m. *obliquus abdominis externus superficialis* takes origin from the aponeuroses of certain muscles of the hindlimb and from the tip of the iliac wing, inserting onto several aponeuroses of the forelimb level. It is almost completely covered by the zonal muscles of the forelimb, the m. latissimus dorsi and the pars abdominalis of the m. pectoralis. The m. *transversus abdominis* takes origin from the same structures as the m. obliquus and inserts onto the lateral surfaces of the oesophagus and the pericard (see also figs. 22, 23 and 26, RYKE, '53).

* * *

THE FURTHER DEVELOPMENT OF THE SKELETON AND MUSCULATURE OF THE TAIL

The early development of the *tail* up to STAGE 37/38 has been described in the division "the early development of the *axial system*" on page 37. The development of the *urostyle region* has been included in the subdivisions "development of the *axial skeleton*", on pages 104 and 106 and "development of the *axial* and *abdominal musculature*", on pages 108 and 109 of the division "the further development of *skeleton* and *musculature* of the *trunk*".

THE DEVELOPMENT OF THE TAIL UP TO METAMORPHOSIS

At STAGE 39, at which fourty-three postotic somites have formed, the *chordal cells* are vacuolated down to the level of the fortieth somite. The yolk inclusions in the notochord have been consumed at STAGE 41. The *elastica interna* appears in the anterior region of the tail between the *chordal epithelium* and the *elastica externa* at STAGE 40. The elastica interna does not develop at the tip of the tail, where the notochord is permanently surrounded by the elastica externa only. At STAGE 48 the elastica externa begins to increase in thickness. Also the elastica interna thickens, and becomes as thick as the chordal epithelium at STAGE 50. This growth continues during the following stages, so that both layers (the elastica externa and interna) are fully grown and differentiated at STAGE 56.

The first connective tissue cells appear segmentally around the elastica externa in the tail region at STAGE 39. A continuous connective tissue sheath, the *perichordal tube*, has been formed around the notochord at STAGE 51. This sheath begins to thicken at STAGE 55 and is twice as thick as the elastica interna at STAGE 57. Except for the urostyle region the spinal cord is enveloped by connective tissue *arches* only.

At STAGE 39, at which about forty-three post-otic somites can be distinguished, the myofibrils show a clearly visible cross-striation. The differentiation of the *somites* progresses. At STAGE 40, at which about forty-five post-otic somites are present, the yolk inclusions rapidly disappear. At this stage they are still present in the dorsal and the ventral parts of the somites. This yolk material has been consumed at STAGE 42. Simultaneously the somites segregate into a *sclerotomic*, a *dermatomic* and a *myotomic* portion. The histogenesis of the myotomes is completed at STAGE 43. During the following stages, until the onset of metamorphosis, the muscle segments of the tail grow in size by an increase in number and particularly in size of the fibres, while simultaneously the number of nuclei per fibre augments.

THE REDUCTION OF THE TAIL DURING METAMORPHOSIS

During metamorphosis the *notochord* begins to degenerate posteriorly at STAGE 61, but is still normal in appearance anteriorly. The degeneration proceeds rapidly, so that also the basis of the tail is affected at STAGE 62. At this stage the *elastica externa* is beginning to disappear, while simultaneously the *connective tissue sheath* and *arches* show a degenerative hypertrophy. These processes are most active posteriorly. At STAGE 63 the *chordal epithelium* is degenerating and the whole notochord is shrinking. It loses its histological characteristics at STAGE 64. The degeneration of the notochord now also begins in the *urostyle region*. At STAGE 65 the notochord is no longer distinguishable in the tail region, but is still distinct in the urostyle region. At STAGE 66 the constriction of the notochord within the urostyle has begun.

Melanin begins to be deposited intercellularly in the swollen epidermis at STAGE 58, at which the tip of the tail begins to atrophy. At the next stage the melanin also appears intercellularly underneath the epidermis, around the walls of the blood vessels and between the fibres of the muscle segments. The melanin accumulation increases during the following stages. At STAGE 61 the skin at the end of the tail undergoes a degenerative hypertrophy, while the connective tissue of the fins becomes condensed as they atrophy. Very heavy deposits of melanin are found underneath the skin at STAGE 63.

The actual degeneration of the *muscle segments* does not start before STAGE 63, at which about forty-nine post-otic muscle segments can still be distinguished. Once started, however, it proceeds very rapidly, so that only about fourteen post-otic muscle segments can be distinguished at STAGE 64, while all the tail muscle segments have already disappeared at STAGE 65. At STAGE 66 the tail is only represented by a small dorsal swelling. It contains many globules of melanin—especially in its deeper parts—which are covered by loose connective tissue and degenerating larval skin.

* * *

THE DEVELOPMENT OF SKELETON AND MUSCULATURE OF SHOULDER GIRDLE AND FORELIMBS

A. THE DEVELOPMENT OF THE SKELETON

THE INITIAL DEVELOPMENT

At STAGE 40 a thickening of the ectoderm, the anlage of the future *forelimb atrium*, appears dorso-caudal to the third visceral cleft. This thickening grows out in a dorso-caudal direction, forming a club-shaped protrusion into the mesenchyme at STAGE 41. It reaches the anterior side of the pronephros anlage at STAGE 43. The connection with the ectoderm of the dorso-caudal edge of the last visceral cleft is maintained up to STAGE 45 as a very thin thread. At STAGE 46 the first mesenchymal cells accumulate underneath the flattened epithelial plate of the future atrium. A cavity, the *atrium*, appears in the plate at STAGE 47, dividing the plate into a thick medial and a very thin lateral layer. At this stage, at which one can speak for the first time of the actual *forelimb anlage*, the mesenchyme accumulation has markedly increased. At STAGE 48 the mesenchyme of the forelimb anlage already forms a well defined bud, capped by the ectoderm of the atrium. The spinal nerves already extend up to the anlage at this stage. The limb bud is situated ventral to the posterior portion of the pronephros and dorsal to that of the gill region.

The limb bud rapidly increases in size. At STAGE 51 it can be divided into a vertically disposed part adjacent to the somatic coelomic epithelium, the *pars zonalis*, out of which the *shoulder girdle* will develop, and a distal, protruding portion, the *pars appendicularis*, which will give rise to the *forelimb s.s.* The first nerves and blood vessels have already penetrated into the anlage at STAGE 51.

Whereas the *ectoderm* of the limb bud is still single-layered at STAGE 51, it becomes double-layered at STAGE 52, while simultaneously a *collagenous cutis* develops. The atrium has now become a spacious cavity.

THE DEVELOPMENT OF THE SHOULDER GIRDLE

At STAGE 53 the *pars zonalis* can again be divided into a dorsal *pars scapularis* and a ventral *pars coracoidea*, out of which the *scapular* and *coracoidal regions* of the *shoulder girdle* will develop.

In the pars zonalis first the *scapula*, the *procoracoid* and the *coracoid* differentiate. They are in a procartilaginous state at STAGE 55 and are chondrified at STAGE 56. At this stage the proximal part of the coracoid is slightly ossified perichondrally, and a *suprascapula*, situated ventral to the myotomes, as well as a *processus epicoracoideus*, connecting later the procoracoid with the coracoid, are now present in a procartilaginous state. They are chondrifying at STAGE 57, a process completed at STAGE 58. A small *cartilago paraglenoidalis* is discernable between the posterior part of the scapula and the coracoid, where it forms part of the borderline of the *glenoid cavity* which shows a *foramen glenoideum*, situated in between scapula, procoracoid and coracoid. In between the procoracoid and the coracoid the *incisura obturatoria* is demarcated at STAGE 56. The complete chondrification of the processus epicoracoideus at STAGE 58 leads to the formation of the *fenestra obturatoria* out of the incisura obturatoria. The *suprascapula* increases greatly in size and covers a large part of the lateral aspect of the myotomes at STAGE 59, while the *scapula* hardly increases in size. Ossification has also made progress. At STAGE 58 the scapula is ossified perichondrally, and in the coracoid ossification has proceeded rapidly, leaving only the postero-ventral end cartilaginous. In later development (STAGE 63) the two procoracoid cartilages abut against each other ventrally, rendering this part of the shoulder-girdle *firmisternal*, whereas the posterior parts of the processus epicoracoidei overlap each other midventrally, rendering this part of the shoulder-girdle *arciferous*. At STAGE 58 two membrane bones have appeared, viz. the *cleithrum*, which has ossified on the ventrolateral aspect of the suprascapula, and the *clavicula*, which is situated mainly on the ventral aspect of the procoracoid. Both bones increase in size during the following stages, the latter investing the ventro-lateral part of the procoracoid and abutting against the scapula at STAGE 59, and fusing with the scapula at STAGE 63. At STAGE 60 the cleithrum also invests the medio-ventral aspect of the suprascapula.

At STAGE 59 the two *sternal anlagen* appear as thickened patches of blastematous tissue in the ventral aponeurosis, where they are situated lateral to the processus epicoracoidei. They become partly procartilaginous at STAGE 60 and partly cartilaginous at STAGE 61. They are entirely cartilaginous at STAGE 62, approaching each other in the mid-ventral line. The medial ends are thickened and curve upwards around the processus epicoracoidei to form the roofs of the *sternal pouches*. Posteriorly each anlage forms a thin cartilaginous plate, the *ala sterni*, situated in between the m. rectus abdominis superficialis and the m. pectoralis, pars abdominalis at STAGE 63. The two sternal anlagen have fused to form a single *sternum* at STAGE 64. It consists now of a median *corpus sterni* and the two broad, flat lateral *alae sterni*. The anterior part of the corpus sterni forms the sternal pouches for the reception of the processus epicoracoidei. In the STAGES 65 and 66, the roofs of the sternal pouches are reduced in size and only *sternal fossae* remain.

THE DEVELOPMENT OF THE FORELIMB

The *pars appendicularis* differentiates in a proximo-distal direction; its regional structures will be discussed separately.

The *humerus* becomes procartilaginous at STAGE 55 and is chondrified at STAGE 56, when its shaft is even slightly ossified perichondrally. A *caput humeri* is also formed. At STAGE 57 the humerus is extensively ossified perichondrally, while the two *epicondyles* are formed at its distal end. Simultaneously the *synovial cavities* between humerus and radio-ulna are formed. A *crista ventralis* is formed on the ventral side of the proximal part of the humerus at STAGE 58, serving as a place of insertion for a large number of zonal muscles. At STAGE 66 the humerus is also ossified enchondrally.

The *radio-ulna* is already extensively chondrified and its shafts are slightly ossified perichondrally even at STAGE 56. The fused bones are extensively ossified in the middle region at STAGE 57, at which the *synovial cavities* are formed between the humerus, the radio-ulna and the proximal carpalia. The *patella ulnaris* is chondrified also at STAGE 57.

Six *carpalia* are chondrified at STAGE 56, viz. the *ulnare, radiale, centrale* and the second to fourth *distal carpalia*, while the first distal carpale is still procartilaginous. Two further carpalia are added at STAGE 58, viz. the *radiale externum*, and the *praepollex*, articulating with the first distal carpale. The *synovial cavities* are formed between the radio-ulna and the proximal carpalia at STAGE 57, and between the distal carpalia and the metacarpalia at STAGE 58, while there are indications of synovial cavities in between the carpalia at the same stage.

The first to the fourth *metacarpalia* are chondrified at STAGE 56 and are ossified perichondrally at STAGE 58.

The first to the fourth *digits* are already demarcated from the rest of the hand at STAGE 56, at which the *phalanges* are represented by patches of thickened blastematous tissue. Their chondrification proceeds also in a proximo-distal direction. The proximal and middle phalanges of the third and fourth digits and the proximal phalanges of the first and second digits are procartilaginous at STAGE 57. All the phalanges of the first to the fourth digits are completely developed and ossified perichondrally at STAGE 58. Since a fifth digit is absent, the *phalanges-formula* is 2,2,3,3,0.

B.   THE DEVELOPMENT AND DESCRIPTION OF THE MUSCU-LATURE

The development of the *zonal* and the *appendicular musculature* is treated separately. The topography of the various muscles is given after each section.

THE DEVELOPMENT OF THE ZONAL MUSCULATURE

At STAGE 55 the ventral edge of the pars zonalis tapers out into a thin strand of blastematous tissue, the anlage of the future *ventral aponeurosis*, which will later serve as the place of origin of the hypaxonic zonal muscles. On the lateral aspects of the pars zonalis mesenchymatous blastematous cells show signs of myoblast formation as the anlage of the *zonal muscles*. Their differentiation progresses very rapidly, so that at STAGE 56 it is already complete, except for the insertiones and origiones of the muscles. Several muscles can be distinguished e.g.: *m. latissimus dorsi, m. rhomboideus anterior, mm. levatores scapulae superior* and *inferior, mm. serrati superior, medius* and *inferior, mm. pectorales partes abdominalis* and *sternocoracoidea, mm. supracoracoidei medius* and *posticus, m. coraco-radialis* and *coraco-brachialis, m. deltoideus, mm. scapula-humerales profundi anterior* and *posterior, m. interscapularis, m. dorsalis scapulae, m. anconeus* and *m. mylo-pectori-humeralis*. At STAGE 57 also the origiones and insertiones are well developed. At STAGE 58 the hypaxonic trunk musculature, viz. the *m. sterno-hyoideus* and the derivatives of the *m. rectus abdominis* (see under the further development of the *axial* and *abdominal musculature*, on page 109) develops. After the appearance of the sternal anlagen at STAGE 59 they become attached to the sternum by means of strands of connective tissue at STAGE 62 to 63. At STAGE 62 the *mm. levatores scapulae* change their origin and originate now from the otico-occipital region of the neurocranium.

THE DESCRIPTION OF THE ZONAL MUSCULATURE

The above-mentioned muscles show the following topography at STAGE 66:

The *m. latissimus dorsi* originates from the suprascapula and inserts onto the crista ventralis of the humerus; the *m. rhomboideus anterior* originates anteriorly and posteriorly from the trunk musculature and inserts onto the upper part of the suprascapula; the *mm. levatores scapulae superior* and *inferior* originate from the otico-occipital region of the neurocranium and insert onto the suprascapula; the *mm. serrati superior, medius* and *inferior* originate from the second and third rib and the transverse process of the third and fourth vertebrae and insert onto the suprascapula; the *mm. pectorales partes abdominalis* and *sterno-coracoidea* originate from the m. rectus abdominis, the ventral aponeurosis and the coracoid and insert onto the crista ventralis of the humerus; the *mm. supracoracoidei medius* and *posticus* originate from the procoracoid, the processus epicoracoideus, the ventral aponeurosis and the coracoid and insert onto the crista ventralis of the humerus; the *m. coraco-radialis* and the *mm. coraco-brachiales brevis* and *longus* originate as the preceding muscles and insert respectively onto the radio-ulnare by means of a long ligament, and onto the middle and distal part of the humerus; the *m. deltoideus* originates from the scapula and inserts onto the middle and distal part of the humerus; the *mm. scapulo-humerales profundi anterior* and *posterior* originate from the scapula and insert onto the humerus; the *m. interscapularis* and *m. dorsalis scapulae*

connect the suprascapula and the scapula; the *m. anconeus* originates with
four capita (i.e. *scapulare, laterale, mediale* and *profundum*) from the scapula
and various sides of the humerus and inserts onto the proximal part of
the radio-ulna, while finally the *m. mylo-humeralis* originates from the
ventral aponeurosis and the procoracoid and inserts onto the crista ventralis of the humerus.

THE DEVELOPMENT AND DESCRIPTION OF THE APPENDICULAR MUSCULATURE

The *appendicular musculature* begins to differentiate at STAGE 55. The
large majority of the muscle anlagen are still in the form of myoblast cells
at STAGE 56. At STAGE 57 the flexor and extensor muscles, especially of the
radio-ulna section are indicated, whereas the hand and digital musculature is not yet discernible. At STAGE 58 the appendicular muscles are well
developed and are grouped into a superficial and a deeper muscle layer.
On the palmar side the *m. palmaris longus* belongs to the superficial layer,
which also includes the *mm. interphalangeales* and the *mm. lumbricales breves*,
and the *flexores teretes* to the deeper one. On the dorsal side the *mm. extensores*
are grouped into a superficial and a deeper layer. The musculature of the
forearm and hand consists only of *extensors* and *flexors*, *ab-* and *adductors*
apparently being absent.

The *m. palmaris* originates from the medial epicondyle of the humerus
and inserts onto the phalanges. The *mm. lumbricales breves* originate from
the metacarpalia and insert onto the phalanges. The *mm. extensores digitorum
communis longus* and *breves superficiales*, belonging to the superficial layer,
originate from the distal carpalia, also inserting onto the phalanges. The
mm. extensores digitorum brevis profundus and *brevis medius* of the fifth digit,
originating and inserting like the previous muscles, belong to the deeper
layer.

* *
*

THE DEVELOPMENT OF SKELETON AND MUSCULA-TURE OF PELVIC GIRDLE AND HINDLIMBS

A. THE DEVELOPMENT OF THE SKELETON

THE INITIAL DEVELOPMENT

At STAGE 43 the future *hindlimb region* is recognizable for the first time as a slight concentration of mesenchyme cells dorsal and lateral to the anal tube. The mesenchyme first becomes concentrated directly under-neath the epidermis at STAGE 44 and 45. Simultaneously the epidermis thickens slightly over the limb area. At STAGE 46 the rudiments are rep-resented by clearly defined masses of mesenchyme, while at STAGE 46 to 47 the epidermis has become double-layered and is provided with a cutis layer. The epidermis of the limb bud, which is half as thick again as the surrounding epidermis at STAGE 46, has become twice as thick as this by STAGE 47. At STAGE 48 the basal layer of the epidermis (*stratum germinativum*) shows a radial arrangement of the cells and the superficial layer (*periderm*) a tangential arrangement. The limb bud, which gradually increases in size, becomes vascularized by small capillary vessels at STAGE 49. The two rudiments are connected dorsally and medially by a strand of mesenchyme cells situated dorsal to the anal tube.

At STAGE 50 the first regional differentiation appears: the future *pelvic girdle* is indicated by an ill-defined mass of mesenchyme at the base of the limb buds. Inside the limb bud a peripheral, dense and a central, sparser mesenchyme can be distinguished. The limb bud is innervated at STAGE 51, while the epidermis has become three layers thick especially proximally on the pre-axial border. In the middle layer cells begin to be transformed into gland cells, which are fully differentiated at STAGE 52.

THE DEVELOPMENT OF THE PELVIC GIRDLE

The *pelvic girdle* is represented by two local masses of mesenchyme at the base of the limb buds, connected by a strand of mesenchyme, at STAGE 51. At STAGE 53 these masses of mesenchyme begin to chondrify centrally. The two cartilages, located one at the base of each femur (see below), are nearest together postero-ventrally and furthest apart antero-dorsally. Their chondrification proceeds during the following stages. At STAGE 54 the *iliac process* is well defined and lies superficially underneath the cutis of the skin and outside the body wall musculature, the *acetabular cavity* being still very shallow. The *ischia* are still widely separated from each other but are connected by a connective tissue band. Postero-ventrally the ischia are growing towards each other at STAGE 55. They approach each other at STAGE 56, are in contact at STAGE 57 and are fused ventrally at STAGE 58. The iliac processes have grown longer at STAGE 55 and start their perichondral ossification midway along their length at STAGE 56. At STAGE 58 the ossification of the ischia has begun. The *ilium* becomes connected to the 9th vertebra of the vertebral column by a thin connective tissue strand at STAGE 59, which has changed into a thick ligament at STAGE 60. An iliac crest appears as a bony ridge at STAGE 59. Ossification proceeds in ilia and ischia at STAGE 60, at which it begins in the pubes. The enchondral ossification begins at the base of the iliac processes at STAGE 64.

During this development the position of the *girdle* has gradually changed. This rotation takes place in a paramedian plane and begins at STAGE 55. The antero-dorsal tips of the ilia have reached the level of the dorsal aorta at STAGE 58 and are level with the notochord at STAGE 61, when the anterior tips lie further forward than the 9th vertebra.

At STAGE 60 the future *epipubis* appears as two ventral procartilaginous structures, lying anterior and ventral to the mid-ventral junction of the pubes. They have joined to form a single Y-shaped cartilage, bifurcated anteriorly, at STAGE 61. Posteriorly this cartilage lies closely apposed to the pubes. While its anterior end changes into a flat plate of cartilage, its posterior end fuses with the girdle at STAGE 65. At STAGE 66, when the adult condition of the pelvic girdle has almost been achieved, the anterior tips of the ilia are still separated more widely from the transverse processes of the 9th vertebra than will later be the case.

THE DEVELOPMENT OF THE HINDLIMB

The *hindlimb* development proceeds in a proximo-distal direction.

The *femur rudiment*, for the first time indicated at STAGE 51, is in an early procartilaginous state at STAGE 52 and is chondrifying at STAGE 53. Chondrification is completed at STAGE 54. The perichondrial ossification begins along the middle region of its shaft at STAGE 55.

The *tibia* and *fibula* are indicated as separate mesenchymal condensations at STAGE 52. They are procartilaginous at STAGE 53 and fully cartilaginous at STAGE 54, when a perichondrium has been formed. Their perichondrial ossification begins at STAGE 56. Tibia and fibula are enclosed in a single bony sheath in their middle region at STAGE 58 (formation of the *tibiofibulare* or *os cruris*). This fusion has extended considerably at STAGE 61.

The *tibiale (astragalus)* and the *fibulare (calcaneum)* form procartilaginous cell masses which are not yet well defined at STAGE 53. They are chondrifying at STAGE 54 and ossifying at STAGE 56.

The *hand* skeleton is still mesenchymatous at STAGE 53. The *metatarsalia* are procartilaginous at STAGE 54 and are ossifying at STAGE 56.

At STAGE 56 the *phalanges* are still cartilaginous. They start to ossify at STAGE 57, when the tips of the toes I to III show the first signs of cornification of the epidermis, resulting in the formation of the claws at STAGE 58. At this stage all skeletal elements, except the *tarsalia* and the *hallux* show ossification. The tarsalia show ossification at STAGE 66, when the hallux is still only cartilaginous.

B. THE DEVELOPMENT OF THE MUSCULATURE

At STAGE 52 there appear local condensations of the mesenchyme, distinct from those of the skeleton, indicating *muscle formation* in the thigh and the leg. The foot mesoderm is still not visibly differentiated. By STAGE 53 the future muscle masses are more distinct and in the thigh and leg the histogenesis of *muscle fibres* has started. In the foot the segregation of muscle and skeletal tissue is under way. From this stage onward the development of the musculature is very rapid so that by STAGE 54 *individual muscles* may, in some cases, be identified and of these many have established their connections at their insertions and a few have their origins indicated. By STAGE 55 all the major muscles of the limb are present and most have established their origins. By STAGE 56 the definitive arrangement of the musculature has been achieved and the thigh and leg muscles have well-developed fibres. By STAGE 58 only the region of the terminal phalanges shows tissue in the pre-muscular stage of development. By STAGE 60 full differentiation of the muscles has occurred and they differ from adult muscles only in the relatively small amount of connective tissue associated with them. The *perimysium* first appears around some of the muscles at STAGE 56 as a thin epithelial sheath. It remains relatively thin and insubstantial even at STAGE 66.

At STAGE 53 it is still not possible to identify individual muscles, but at STAGE 54 several are distinguishable. In general the limb muscles insertions are established before the origins. Thus at STAGE 54 some muscles (*m. iliacus externus, m. iliacus internus, m. cruralis, m. tensor latae, m. semimembranosus*) have both insertion and origin indicated, while others (*m. semitendinosus, m. plantaris longus, m. obturator internus*) have established their insertions but not their origins. In yet other cases the muscle is indicated but the insertion is not defined (*m. tibialis posticus, m. pectineus*). At STAGE 55 nearly all the muscles of the thigh and leg have established both origins and insertions, and in the foot the musculature is well defined.

Once established the muscles alter little in their relationships. The lengthening of the ilia and the rotation of the girdle are accompanied by slight shifts in the origins of some of the muscles and, necessarily, by growth in length.

THE MUSCULATURE CONNECTING PELVIC GIRDLE AND HINDLIMB

The *m. triceps femoris* (Gaupp) consists of: a) the *m. tensor fasciae latae* (small) which originates from the ventral side of the ilium, and joins the m. cruralis antero-laterally in the upper thigh; b) the *m. cruralis* (large) which originates ventral to the acetabulum and inserts as the m. glutaeus; and c) the *m. glutaeus* (large) which originates from the superior process of the ilium and inserts beneath the head of the tibia. The *m. adductor longus* (small) originates ventral to the m. cruralis and immediately dorsal to the m. pectineus and inserts with the m. adductor magnus onto the femur. The *m. pectineus* originates between the m. adductor longus and the m. sartorius from pubis and epipubis immediately above the symphysis and inserts onto the ridge of the femur. Together with the m. semitendinosus the *m. sartorius* takes origin from the epi-pubis and pubic symphysis and inserts onto the head of the tibia. The ventral head of the *m. adductor magnus* originates immediately dorsal to the m. sartorius-semitendinosus, the dorsal head arises near the posterior margin of the ischium. Both parts insert onto the distal end of the femur. The *m.m. gracilis major* and *minor* originate together at the posterior margin of the ischium, the insertions are common with the m. sartorius. The undivided *m. semimembranosus* takes origin along the dorsal margin of the posterior ilium and of the ischium and inserts onto the head of the tibia. The *m. semitendinosus* originates posterior to the m. sartorius along the ventral margin of the pubis and ischium. It inserts below the head of the tibia. The main origin of the *m. iliacus externus* (large) nearly covers the anterior two-thirds of the iliac process, the accessory head originates on the anterior dorsal margin of the ilium. The insertion is onto the trochanter of the femur. The inseparable *m.m. obturator externus* and *quadratus femoris* originate ventro-posteriorly near the margin of the ischium and insert onto the head of the femur. A separate small muscle runs from the iliac symphysis to the epipubis. The *m. gemellus* originates along the dorsal ischium and inserts onto the femoral crest. The *m. obturator internus* has its origin as a semi-circular band immediately surrounding the acetabulum. It converges to a tendon attached to the joint capsule and to the dorsal side of the trochanter. The *m. ileo-femoralis* originates at the posterior end of the iliac crest and inserts onto the femoral crest latero-dorsally.

The musculature of the rest of the hindlimb is practically identical to that of other Anurans. The reader is referred to GROBBELAAR, 1924 and GAUPP, 1896.

* * *

THE DEVELOPMENT OF HEART AND VASCULAR SYSTEM, INCLUDING LYMPHATIC SYSTEM AND SPLEEN [1]) [2])

THE INITIAL DEVELOPMENT

The two presumptive *heart anlagen,* located in the anterior, latero-ventral mesoderm at STAGE 15, begin to fuse at STAGE 16. This fusion makes slow progress, so that the single median presumptive heart anlage is not established before STAGE 19, when it is located somewhat caudal to the cement gland. The anlage becomes more sharply delimited during the following stages, but does not show any symptoms of further development before STAGE 22, at which stage the mesodermal cell material located in front of the liver diverticulum begins to loosen. Behind the liver diverticulum a similar phenomenon indicates the formation of the *blood islands.* The actual differentiation of the heart anlage only starts at about STAGE 27. At STAGE 28 the *endocardial tube* appears as a condensation of cell material with the beginning of a central dilatation. Ventrally and laterally it is bordered by a rather thick layer of mesoderm. Laterally, on both sides of the ventral midline, in this mesoderm a *pericardial cavity* begins to appear. This cavity extends rapidly, so that it is also present in the mesoderm ventral to the endocardial tube at STAGE 29/30, while the lateral cavities widen considerably. From this moment on a single pericardial cavity is present. The splanchnic layer of pericardial mesoderm, the *mesocardium,* thickens and bends dorsally around the endocardial heart anlage, which now forms a short tube closed at both ends.

THE FURTHER DEVELOPMENT UP TO METAMORPHOSIS

THE DEVELOPMENT OF THE HEART

At STAGE 31 the *endocardial tube* extends forwards beyond the *pericardial cavity* as a *ventral aorta.* The anterior portion of the latter is divided into right and left channels, which at first end blindly, but are almost immedi-

---

[1]) See diagrams in MILLARD 1945: STAGE 35/36 is midway between figs. 1 & 2 in Plate IV; STAGE 37/38 corresponds to Plate IV, fig. 2; STAGE 41 corresponds to Plate IV, fig. 3; and STAGE 45 corresponds to Plate V, fig. 1.

[2]) See diagrams in MILLARD 1949: STAGE 43 corresponds to Plate I.

ately extended as the *first pair* of *aortic arches*. In the posterior region the pericardial mesoderm begins to form the *dorsal mesocardium* at STAGE 32, so that in this region the *myocardial wall* is completed. It is completed throughout the entire heart anlage at STAGE 33/34. At STAGE 32 the posterior end of the endocardial tube broadens against the anterior surface of the liver diverticulum. At STAGE 33/34 the *heart* becomes twisted both in a sagittal and a frontal plane, so that the heart anlage shows the characteristic S shape at STAGE 35/36. At this stage the chambers are distinct, the *atrium* lying dorsally to the *ventricle*. The *sinus venosus* is already broad and lies on the antero-dorsal surface of the liver diverticulum. Red blood corpuscles are by now present in the heart. At STAGE 41 the *myocardium* of the conus and the ventricle becomes thickened and *trabeculae* develop in the ventricle. At this stage a spiral *valve* is present in the *conus*, while rudiments of other valves appear at the base of the conus and in the *atrio-ventricular aperture* as thickenings of the endocardium. The anterior and posterior valves of the atrio-ventricular aperture develop further at STAGE 42 and are large enough to close the aperture at STAGE 43. At STAGE 42 dorsal and ventral valves develop at the base of the conus and lateral valves in its distal end. All these valves are fully developed at STAGE 44. At STAGE 43 the still single atrium becomes pushed slightly anterior to the ventricle. A *partition* begins to be formed in the atrium at STAGE 44, so that the atrium is now partially divided. This process is completed at STAGE 45, with the result that the sinus venosus opens into the *right atrium* and the vena pulmonalis into the *left one*.

THE DEVELOPMENT OF THE ARTERIAL SYSTEM

The *arteries* develop somewhat earlier than the veins. The *aortic arches* develop in series from in front backwards and encircle the branchial region, connecting the *ventral aorta* with *paired dorsal aortae*. The *first aortic arch* is complete at STAGE 31. It arises from the ventral endocardial tube (ventral aorta), traverses the ventral part of the hyoid arch, and passes dorsally round the mandibular arch to join a short paired dorsal aorta. The *third aortic arch* is indicated as a cleft in the mesoderm of the first branchial arch at STAGE 32, and is completed at STAGE 33/34. The *fourth aortic arch* is indicated as a cleft in the mesoderm of the second branchial arch at STAGE 33/34, and is completed at STAGE 35/36. The *fifth aortic arch* is completed in the third branchial arch at STAGE 35/36. The dorsal part of the *sixth aortic arch* appears first, joining the paired dorsal aortae, at STAGE 35/36; it is completed at STAGE 39 and gives off a *pulmonary branch*. Its dorsal part, however, (the *ductus Botalli*) breaks down at STAGE 41.

Meanwhile the *paired dorsal aorta* has been growing backwards in a posterior direction: at STAGE 31, when first visible, it reaches to the level of the first branchial arch, at STAGE 32 to the level of the liver diverticulum, and at STAGE 33/34 to the level of the pronephros. At STAGE 35/36 the paired dorsal aortae unite to form a *median dorsal aorta*, and at STAGE 37/38 the latter penetrates into the tail as the *caudal artery*.

The *third* and *fourth aortic arches* are responsible for the circulation in the external gills of the first and second branchial arches, and both become split along part of their length into primary and secondary vessels at STAGE 35/36. The *secondary* (posterior) *vessel* of each develops into a loop in the external gill, that of the third aortic arch at STAGE 39, and that of the fourth aortic arch at STAGE 40. The circulation in the external gills reaches its maximum development at STAGE 44, and then starts to diminish—the secondary vessels are considerably reduced in size at STAGE 45, circulation through them has stopped at STAGE 46, and they have disappeared completely at STAGE 47.

The *first aortic arch* does not persist. Its anterior ventral part begins to degenerate at STAGE 37/38, the origin from the heart eventually becoming the *external carotid artery*, and its dorsal junction with the paired dorsal aorta contributing to the *posterior palatine artery* at STAGE 41. The paired dorsal aorta continues forwards beyond this junction as the *internal carotid artery*. The latter is visible at STAGE 37/38, and at STAGE 41 it enters the cranium-cavity through the fenestra hypophyseos and gives off *anterior palatine*, *ophthalmic* and *cerebral branches*. At STAGE 47 a separate carotid foramen is cut off from the fenestra hypophyseos, the ophthalmic branch leaves the cranium through the oculomotor foramen, and the anterior palatine branch through a separate foramen in the floor.

The *circulation in the filter apparatus* is supplied by branches of the fourth, fifth and sixth aortic arches, and is established at STAGE 44. The third aortic arch supplies an accessory branch to the branchial constrictor muscles. This branch arises just before the junction with the paired dorsal aorta, and is first present at STAGE 46.

The dorsal aorta gives off a *coeliaco-mesenteric artery* at STAGE 41, and an *occipitovertebral artery* arises as a branch of the paired dorsal aorta just anterior to the formation of the median dorsal aorta at STAGE 46.

At STAGE 47 and 48, thus, three continuous aortic arches remain, linking the heart with the paired dorsal aortae, namely the numbers 3, 4 and 5. Of these the third is the largest (the "*larval aorta*"), and carries the main blood supply to the paired dorsal aorta. At STAGE 49 to 50 the fifth arch loses its connection with the paired dorsal aorta, but its ventral portion persists until metamorphosis.

The *arteria subclavia* appears as a branch of the arteria occipito-vertebralis at STAGE 52. The *arteriae iliacae communes* arise at the end of a ventral branch of the aorta dorsalis. This branch is present, though very small, at STAGE 49, running caudad parallel to the dorsal aorta between the posterior ends of the mesonephroi. At STAGE 50 it gives off in its turn a small branch directed ventrad between the mesonephroi towards the rectum, the *arteria mesenterica posterior*, while its distal end branches into two short vessels, the future arteriae iliacae communes. At STAGE 51 the latter bend round the caudal vein and extend to the hindlimb anlage. The ventral branch of the larval aorta dorsalis later becomes the caudal part of the adult aorta.

THE DEVELOPMENT OF THE VENOUS SYSTEM

In the *venous system* the *pronephric sinus* appears first, at STAGE 31. This becomes connected to the broadened posterior part of the heart by the *duct of Cuvier*, which first appears at STAGE 32, but is only completed at STAGE 33/34. Each pronephric sinus receives a *medial postcardinal vein* lying medial to the Wolffian duct. This is present for the first time, though short, at STAGE 32, but rapidly elongates, and at STAGE 35/36 reaches to behind the cloaca, where it joins its fellow in the midline near the posterior end of the postanal gut. The pronephric sinus also receives an *anterior cardinal vein* at STAGE 33/34. Each duct of Cuvier receives an *omphalomesenteric vein* which is present on the ventro-lateral surface of the liver diverticulum at STAGE 33/34. The veins have now reached the condition illustrated in MILLARD 1949, text fig. 9A (which corresponds to STAGE 35/36), and from now on the systems will be described separately.

Of the two *omphalomesenteric veins*, the *left* one is reduced in diameter by STAGE 37/38, and breaks into a capillary system in the liver at STAGE 40. It now crosses the gut dorsally, due to the establishment of a dorsal anastomosis (first evident at STAGE 37/38) and the breakdown of the ventral part in that region. It receives the *sub-intestinal vein* (the *gastro-duodenal vein* of the *adult*) which lies along the left side of the intestine at STAGE 40, along the mid-ventral surface by STAGE 42, and along the right side at STAGE 43. By STAGE 43 it also receives a *gastric vein* (the *median gastric vein* of the *adult*) in the pancreatic region. The *right omphalomesenteric* breaks into a capillary system in the liver at a later stage than the left (STAGE 41). It receives the *main portal vein* from the dorsal surface of the intestine, which is first present at STAGE 42 when it reaches back to the level of the coeliaco-mesenteric artery. Right and left omphalomesenterics are usually connected by one or more anastomoses in the region of the bile duct. No further change occurs other than a lengthening of all vessels up to STAGE 48 (see MILLARD 1949, text figs. 1 and 2).

The *medial postcardinals* fuse with one another in the midline to form an *interrenal vein*, which extends caudally as *caudal vein*. This fusion has already started in the posterior region at STAGE 35/36. It extends slowly in a cranial direction: by STAGE 42 it has reached a point between the opening of the cloaca and the openings of the Wolffian ducts into the cloaca, by STAGE 46 it has reached the hindmost loop of the intestine, by STAGE 47 the posterior border of the stomach, and by STAGE 48 the junction with the *posterior vena cava*. Anterior to this point the medial postcardinals do not fuse.

At STAGE 37/38 a large vein appears in the mesenchyme of the dorsal fin of the tail. At the next stage its anterior end becomes connected with the medial postcardinal veins at the level of the middle of the cloaca by two commissural vessels, bending round the spinal ord and notochord. The left one of these vessels is the largest. At STAGE 43 the right commissural vessel disappears and the left one now opens at the level of the anal opening into the unpaired *caudal vein* as its *dorsal branch*.

Lateral postcardinals appear first as disconnected portions lateral to the Wolffian ducts at STAGE 44. These gradually become continuous, a process which proceeds in a posterior direction, the anterior part being complete at STAGE 46, and the whole vein at STAGE 48. At the latter stage each lateral postcardinal communicates anteriorly with a *sinus* surrounding the Wolffian duct which is partly formed by the anterior paired part of the medial postcardinal; and posteriorly it opens into the interrenal vein just posterior to the junction of Wolffian duct and cloaca. In the centre region a *mesonephric sinus* surrounds the mesonephric tubules and puts lateral and medial postcardinals into communication with each other.

At STAGE 51 the *interrenal vein* loses its connection with the caudal vein anterior to the point of entrance of the lateral postcardinal veins (see MILLARD, 1949, text fig. 5 F). At the same time, concurrently with the backward prolongation of the mesonephroi, the lateral postcardinal veins are lengthened by a longitudinal splitting of the caudal vein, so that the persisting ventral commissural vessel (see below) now opens directly into the lateral postcardinal vein on each side (see MILLARD, 1949, text fig. 6 D).

The *posterior vena cava* is of complex origin. Its anterior end is formed by a *hepatic vein*, which is first present at STAGE 40, and which appears independently of the omphalomesenterics. The hepatic vein grows backwards through the liver to join the *right medial postcardinal*. A narrow connection is established at STAGE 44, and the vein is equal in diameter to the medial postcardinals at STAGE 47. The posterior part of the posterior vena cava is formed by the *interrenal vein* (MILLARD 1949, text figs. 3–5.)

The ventral part of the head region is drained by a pair of *musculo-abdominal veins*, which are already well developed at STAGE 35/36, and pass posteriorly to enter the sinus venosus by two roots, one dorsal and one ventral to the rectus cervicis muscle. At STAGE 39 the dorsal root is joined by a short lateral vein on each side from the ventral abdominal wall, and by STAGE 41 the ventral root has disappeared. The lateral veins receive *opercular branches* at STAGE 43, and are fused in the midline posterior to this point. The right lateral vein is more weakly developed than the left one from the beginning, and begins to degenerate distally at STAGE 56.

The ventral abdominal wall is also drained by a *median abdominal vein* in the posterior region, first present at STAGE 42. This is connected with the caudal vein by *ventral commissures* (MILLARD 1949, text figs. 6 and 7). At STAGE 49 these commissures no longer enter the caudal vein directly except, probably, the most anterior one or two. The more caudal ones open now into a ventral branch of the caudal vein, which in its turn enters the caudal vein opposite its dorsal branch coming from the dorsal part of the tail. All ventral commissural vessels except one begin to disappear at STAGE 53. The persisting one (usually the most anterior one, MILLARD, 1949) now opens directly into the lateral postcardinal vein.

The main veins responsible for the drainage of the branchial region are the *external jugulars*, which run backwards medial to the gill chambers to open into the sinus venosus at the same level as, but dorsal to, the musculo-abdominal veins. These veins are present, though small, at STAGE 40; they receive small branches from the gill region at STAGE 42, and the drainage system is fully established at STAGE 44. An accessory drainage for the gill region is provided by a pair of *branchial veins*, which pass backwards lateral to the gill chambers and open into the pronephric sinuses. These are present for the first time at STAGE 41.

In the dorsal head region a paired *vena capitis medialis* is present internal to the cranial nerve roots at STAGE 33/34. Its posterior continuation is the *anterior cardinal vein*, which enters the pronephric sinus. At STAGE 35/36 there is in addition a *vena capitis lateralis* external to the cranial nerve roots. This is connected to the vena capitis medialis at two points: (a) between the origins of nerves v and vii, and (b) immediately anterior to the vagus ganglion (MILLARD 1949, text fig. 9 A and B). The vena capitis medialis, which at first was the larger of the two, has become narrower than the vena capitis lateralis by STAGE 37/38, is discontinuous at STAGE 40, and absent at STAGE 41 in this particular region. Meanwhile, by STAGE 37/38, the vena capitis lateralis has grown further back to connect with the vena capitis medialis at a third point, posterior to the vagus ganglion (MILLARD 1949, text fig. 9 C and D). In this region the vena capitis medialis has narrowed by STAGE 40 and disappeared by STAGE 41. The vena capitis medialis is thus replaced by a vena capitis lateralis from the origin of nerves v and vii to just behind the vagus ganglion. At STAGE 41 the vena capitis lateralis receives *median cerebral, pharyngeal* and *posterior cerebral branches* (MILLARD 1949, text fig. 10 A, Plate i). Of these the *median cerebral vein* crosses ventral to the ganglion of nerve v, but by STAGE 46 it has changed its position and crosses the prootic ganglion dorsally. The *posterior cerebral vein* has disappeared at STAGE 47. The compound head vein becomes the *internal jugular vein* of the *adult*.

The lungs are drained by a single *pulmonary vein*, which opens into the anterior part of the sinus venosus in the dorsal midline at STAGE 40. At

STAGE 41 it opens into the posterior part of the atrium and at STAGE 44 into the left half of the latter.

The *branchial vein* from the forelimb bud and the *vena cutanea magna* develop at about the same time, viz. at STAGE 50 to 51. They enter the pronephric sinus together, the vena cutanea magna passing between the peritoneum and the closely apposed inner side of the limb bud. The brachial vein is still very short but lengthens during the further development of the limb. The common proximal part of both veins becomes the *vena subclavia.*

The *ischiadic vein* from the hindlimb bud cannot be distinguished clearly before STAGE 50, when it enters the persisting ventral commissural vessel (see above). During later development it greatly increases in diameter, as does the dorsal part of the commissural vessel between the ischiadic and the lateral postcardinal vein. This part becomes the proximal section of the ischiadic vein, whereas the ventral part of the commissural vessel (the *ramus abdominalis posterior* of the *adult*) remains narrow. The *vena femoralis* can be distinguished at STAGE 51 as a narrow vein opening into the persisting ventral commissural vessel near the lateral postcardinal vein.

THE DEVELOPMENT DURING METAMORPHOSIS

THE DEVELOPMENT OF THE ARTERIAL SYSTEM

In the arterial system only small changes take place. At STAGE 62 the *ductus caroticus,* the portion of the paired dorsal aorta between the points of entrance of the third and fourth aortic arches, narrows rapidly. It has disappeared at STAGE 63. At the same time the third *aortic arch* (the "*larval aorta*") decreases and the fourth increases in diameter. At the STAGES 62 and 63 they are already about equal in diameter, while at STAGE 64 the fourth arch is larger than the third. The latter is now the proximal part of the *arteria carotis internis,* whereas the fourth arch is becoming the *aortic arch* of the *adult.*

The *arteria cutanea magna* appears at STAGE 62 as a branch of the pulmonary artery. The *arteries* of the *filter apparatus* disappear with the reduction of the branchial basket, and the *caudal portion* of the *dorsal aorta* behind the ventral branch leading to the arteriae iliacae communes disappears with the reduction of the tail. The ventral branch becomes the posterior part of the adult aorta.

THE DEVELOPMENT OF THE VENOUS SYSTEM

In the venous system more extensive changes occur. The *pronephric sinuses,* and the *paired portions* of the *median postcardinal veins* leading into them, disappear at STAGE 62 with the total reduction of the pronephroi. From that moment on the *lateral postcardinal veins* are true *renal portal veins*

(MILLARD, 1949, text fig. 5 G). The *mesonephric sinus* gradually assumes the form of a "rete mirabile" during the period of metamorphosis.

The right *lateral vein*, which has already started to degenerate distally at STAGE 56, is very much reduced at STAGE 61. At STAGE 62 both lateral veins with their opercular branches have disappeared altogether. At STAGE 58 the small *larval abdominal vein* continues forwards from the level of the cloaca into a large blood sinus, situated in the ventro-caudal end of the abdominal cavity between the tips of the lungs, and receiving a number of veins from the ventral part of the tail. During the following stages a median longitudinal connective tissue ridge develops on the inner side of the wall of the abdominal cavity. At STAGE 61 to 62 numerous small veins from the ventral body wall penetrate into this ridge, forming small blood sinuses. When at STAGE 65 these sinuses have established connection with each other, with the large posterior sinus, and anteriorly with the *subintestinal vein* (the *gastro-duodenal vein* of the *adult*), a continuous *adult abdominal vein* is present. It is clear that the direction of the blood-flow in this adult vein is opposite to that in the larval abdominal vein.

Very short *musculo-thyroid veins* can be seen entering the *ducts of Cuvier* (*venae cavae anteriores* of the *adult*) just medial to the external jugular veins at STAGE 62. They develop very slowly, and unequally on the two sides. At STAGE 66 usually one is still weakly developed, while the other one has not yet reached its maximal extension. The *branchial veins* and the *branchial branches* of the *external jugulars* disappear with the reduction of the branchial basket.

At the beginning of metamorphosis the *vena capitis lateralis* acquires a new branch just anterior to its entrance into the pronephric sinus. This branch is the *vena facialis*, which is for the first time clearly visible at STAGE 58. It comes originally from the skin, while later on it takes over the drainage of the thymus gland from the *pharyngeal vein*. The latter disappears at STAGE 62.

With the reduction of the tail the *vena caudalis* disappears, together with its dorsal branch and the portions of the *lateral postcardinal veins* posterior to the junctions of the ventral commissural vessel.

* *
*

THE DEVELOPMENT OF THE LYMPHATIC SYSTEM

The anlagen of the anterior pair of *lymph hearts* appear at STAGE 33/34 between the myotomes and the skin, dorso-caudal to the pronephroi. Each has the form of a lymphatic space, communicating dorsally with the intersegmental vein between trunk myotomes 3 and 4, and ventrally with the abdominal cavity. At STAGE 35/36 segregation takes place into the *lymph heart s.s.* and a *dorsal lymph space*. The latter begins to extend anteriorly and posteriorly at STAGE 39 to 40, leading to the formation of the main *lymph vessels* of the larva, which extend later into the head and the tail. While these vessels are extending, the communication of the lymph heart with the abdominal cavity is gradually obliterated.

In the posterior trunk region *lymph hearts* appear laterally between the myotomes and the skin at STAGE 51 in connection with the intersegmental veins. Ultimately there are four pairs of posterior lymph hearts, situated in the region of the trunk myotomes 10 to 13. During metamorphosis they come to lie more dorsally, in close contact with the dorsal *subcutaneous lymph sac*, with which they communicate (for the development of the *subcutaneous lymph sacs* see further under the division: "the development of the *skin*, etc.", on page 46). At STAGE 66 only two pairs of posterior lymph hearts remain.

The development of the *ductus thoracicus* and the *internal lymph spaces* will not be described.

* * *

THE DEVELOPMENT OF THE SPLEEN

At STAGE 43 mesenchyme begins to accumulate in the dorsal mesentery near the anterior end of the stomach, indicating the future *spleen anlage*. This anlage is well defined at STAGE 45 to 46. Blood corpuscles appear in it for the first time at STAGE 47. At STAGE 49 the spleen has entered a phase of considerable growth, while it is already extensively vascularized.

* * *

THE DEVELOPMENT OF THE NEPHRIC SYSTEM

THE PHASE OF PRIMARY SEGREGATION OF THE PRONEPHRIC SYSTEM

The *pronephric anlage* is indicated as a slight thickening of the somato-pleural portion of the lateral mesoderm at STAGE 21. This condensation of mesodermal cell material is located at the level of the head somites III and IV and the first trunk somite (for numbering of somites see page 35) at STAGE 22. At this stage its segregation from the rest of the somatopleure begins, a process which progresses slowly in a cranio-caudal direction (up to STAGE 27). At STAGE 23 an internal segregation process starts with a condensation of cell material in the anterior third of the pronephros anlage, representing the future first *nephrotome*. Simultaneously the *Wolffian duct anlage* is indicated at the level of the posterior half of the first and second trunk somite (for numbering of somites see page 35). The anlage of the second *nephrotome* is segregated at STAGE 24. At this stage the modelling of the *collecting tube* begins in the form of a segregation of an antero-posteriorly elongated solid cord in which, in two distinct regions (at the level of the centres of the first and second nephrotomes), a radial arrangement of the cells appears, without, however, any trace of a lumen. This radial arrangement extends gradually in anterior and posterior directions during the following stages, while simultaneously the anlage of the *Wolffian duct* begins to stretch caudad. It reaches the level of the third trunk somite at STAGE 26 and that of the fifth trunk somite at STAGE 27. In the anterior portion of this outgrowing Wolffian duct the cells begin to arrange themselves radially, although the anlage is markedly compressed against the entoderm mass at STAGE 27.

THE DEVELOPMENT OF THE PRONEPHRIC SYSTEM UP TO FUNCTION AT APPROXIMATELY STAGE 37/38

The following development of the *nephric system* is characterized by the further organization of the *collecting tube* and the *nephrotomes*, the appearance of the *nephrostomes* and the *external glomus*, and the outgrowth of the *Wolffian duct* and *rectal diverticula*.

The *collecting tube* shows a tubular differentiation progressing mainly in an anterior direction at STAGE 28. A minor lumen appears in this anterior portion and is indicated in the centre of the middle portion at STAGE 29/30. At STAGE 32 a central lumen has been formed over nearly the entire length of the tube, except its posterior portion. It has extended over the entire length at STAGE 33/34. At STAGE 31 the anterior portion of the tube increases in length and forms a short coil.

The posterior part of the pronephros is still unorganized at STAGE 28. At this stage the first *nephrostome* and its *canaliculus* appear in the form of a cell concentration with incipient cellular arrangement, while the second and third *nephrostomes* are only indicated as unorganized, tiny cell concentrations. The first nephrostome and its canaliculus, situated in the ventral portion of the pronephros, show a central lumen in the form of a tiny slit at STAGE 29/30. At this stage the second nephrostome, situated in the dorsal portion of the pronephros, begins to differentiate, while the third one, also situated dorsally, is still undifferentiated. The first nephrostome funnel begins to function at STAGE 31, although its walls are still thick and not yet ciliated. At this stage the second nephrostome funnel is hardly open and its canaliculus is still solid, while the collecting tube has a lumen at this level. The third nephrostome and its canaliculus are still closed, like the adjacent portion of the collecting tube. At STAGE 32 the second nephrostome funnel and canaliculus have a lumen, while the third one is just about to open. The former just begins to function at STAGE 32, the latter is functional at STAGE 33/34. The *pronephric tubules* connecting the canaliculi of the nephrostomes with the collecting tube, are formed gradually by elongation and subsequent coiling. All three *nephrons* still contain numerous yolk platelets. At STAGE 33/34 the ciliation of the nephrostome funnels is formed, the lumen of the three nephrons is greatly enlarged and the entire pronephros has become potentially functional. At STAGE 35/36 a rich blood supply has developed in the form of large blood sinuses derived from the posterior cardinal veins.

The *coelomic filter chamber* becomes visible at STAGE 28 as a slight enlargement of the coelomic cavity at the level of the first nephrotome, while the *glomus* appears at STAGE 29/30 as a small and compact aortic bud at the base of the dorsal mesentery, protruding slightly into the coelomic filter chamber. The glomus and the coelomic filter chamber slowly increase in size during the following stages from STAGE 29/30 to 33/34. At STAGE 33/34 their progress in development is considerable. They extend anteriorly and posteriorly, so that they stretch along the entire pronephros at STAGE 35/36. Blood supply starts at STAGE 35/36. At STAGE 37/38 the entire pronephric system has become functional, the Wolffian ducts being fully developed (see below).

The *Wolffian ducts* continue their growth and reach the caudal border of the sixth pair of somites at STAGE 28, that of the seventh pair of somites at STAGE 29/30, and approach the rectal diverticula (see below) at STAGE 33/34. The Wolffian ducts still form solid cellular strands at STAGE 29/30. Their cells begin to arrange themselves radially, particularly in the anterior portions of the anlagen at STAGE 31. A small lumen is formed in the anterior and middle portions at STAGE 32, at which the posterior portions are still solid and compressed against the entoderm. With the further outgrowth of the Wolffian ducts their lumina get a more definitive appearance and extend into the anlagen in a posterior direction (STAGE 33/34 and 35/36). At STAGE 35/36 the Wolffian ducts make contact with the rectal diverticula, while at STAGE 37/38 the first contact of the lumina is established. The cellular differentiation of the posterior end of the Wolffian ducts is completed at STAGE 39.

The *rectal diverticula* are indicated for the first time at STAGE 32, at which they appear as tiny slits in the dorsal wall of the proctodeum. At STAGE 33/34 these diverticula grow out in an anterior direction. They open as grooves in the dorsal wall of the proctodeal cavity. The left one is somewhat better developed than the right one. The growth of the rectal diverticula in an anterior direction continues during the following stages (up to STAGE 37/38). At STAGE 35/36 the rectal diverticula begin to widen, without showing any differentiation of their walls. They communicate with the Wolffian ducts at STAGE 37/38 (see above). The tubular differentiation of the rectal diverticula, started anteriorly at STAGE 37/38, progresses in a posterior direction up to STAGE 43, at which it is completed.

> After the establishment of the functional state the *pronephros* increases in size. Meanwhile the *mesonephros* develops in a more posterior region of the trunk. By the time the mesonephros is fully developed, the pronephros begins to atrophy. It finally disappears completely. Shortly after the beginning atrophy of the pronephros the mesonephros shows signs of degeneration, after which, however, reorganization occurs.

THE FURTHER DEVELOPMENT AND SUBSEQUENT REGRESSION OF THE PRONEPHROS

During further development a gradual regression of the posterior portion of the *glomus* occurs, so that it finally becomes shorter than the pronephros. This process begins at STAGE 41 and terminates at STAGE 51. From this stage to STAGE 60 the glomus becomes more and more compact so that at STAGE 60 it has become greatly reduced. Blood irrigation has ceased at STAGE 61. The glomus is, however, still present as a tiny globule at STAGE 66.

From STAGE 37 to 47 the *pronephros*, maintaining its absolute craniocaudal extension, becomes much shorter relative to the growing axial system. Its complexity increases gradually by a further coiling of its tubuli and collecting tube. Moreover the tubes have become thicker and their lumina have widened. Meanwhile their yolk content is gradually consumed, so that the pronephros has become yolk-free at STAGE 47. From that stage on the pronephros extends in caudal direction. It flattens medio-laterally and extends caudad up to STAGE 56, at which its maximal elongation is achieved.

From STAGE 53 on degeneration phenomena start in the *pronephros*. First a congestion takes place; the blood lacunae become markedly swollen. The first signs of atrophy appear at STAGE 54. The size of the nephrostome funnels, however, still increases. Underneath the thickened nephrostomal epithelium a thin layer of connective tissue develops at STAGE 54, into which leucocytes infiltrate at STAGE 55. This thickening connective tissue layer extends underneath the coelomic epithelium around the entire pronephros at STAGE 54 to 55, while the pronephric tubules begin to atrophy and the canaliculi become thinner. The infiltration with connective tissue and leucocytes increases gradually at STAGE 56. During following stages the degeneration of the pronephros makes further progress. The lumina of the pronephric tubules and particularly of the collecting tube are obliterated, so that the pronephros is no longer functional from STAGE 58 on. The first and second nephrostome funnels, which have gradually approached each other, have fused at STAGE 60. They are situated just in front of the newly formed *ostium tubae* (see below). The third nephrostome funnel also approaches the others. The first nephron has completely disappeared at STAGE 61. From the second and third nephrons only the nephrostomes remain at STAGE 62. At the dorsal side a portion of the second nephrostome fuses with the ostium at STAGE 62. At STAGE 63 also a portion of the third nephrostome fuses with the ostium. The pronephros has finally completely disappeared at STAGE 64.

The *Wolffian duct* between the pro- and the mesonephros (see below) decreases in diameter at STAGE 59 and shows signs of degeneration at STAGE 60. Its lumen has disappeared and its epithelium has degenerated at STAGE 61. This portion of the Wolffian duct has disappeared completely at STAGE 62.

Slightly caudal to the level of the third nephrostome funnel the coelomic epithelium thickens as a first indication of the anlage of the *ostium tubae* at STAGE 56. The thickening of the coelomic epithelium extends in dorso-ventral and anterior directions at STAGE 57. It reaches the level of the second nephrostome at STAGE 58 and that of the first at STAGE 59. Simultaneously a slit-shaped cavity is formed in the inward-growing cell material. Remnants of the second and third nephrostome fuse with the ostium respectively at STAGE 62 and 63.

THE DEVELOPMENT OF THE MESONEPHRIC SYSTEM

At STAGE 39 the first *mesonephric cells* appear on both sides of the midline between aorta, myotome, medial post cardinal vein and entoderm mass, while the Wolffian ducts are situated latero-ventral to the medial post cardinal veins. The mesonephric anlagen first appear most caudally. At

STAGE 41, at which the coelomic cavity has markedly widened and the dorsal mesentery has formed, mesonephric cells appear in a rather large number in the newly formed sub-aortic space. Meanwhile the cranio-caudal area in which they can be found has extended craniad. At STAGE 41 some mesonephric cells already migrate in the direction of the Wolffian ducts, accumulating against the medial post cardinal veins which still block their further migration. With the growing distance between entoderm mass and aorta, and the consequently increasing sub-aortic space, the medial post cardinal veins and Wolffian ducts shift mediad at STAGE 42. At STAGE 43 the migration of the mesonephric cells is in full swing in the middle and posterior region, but has not yet started in the anterior area. The first *definitive mesonephric anlagen* become distinguishable at this stage. The migration of the mesonephric cells, continuing during following stages, leads to the formation of still undivided *mesonephric cords* along the Wolffian ducts at STAGE 45. At this stage local thickenings appear in three or four areas. The number of distinct regional anlagen increases during the following stages. At STAGE 46 four to five, and at STAGE 47 six to eight units can be distinguished. Their cellular organization proceeds rapidly. At STAGE 46 the cells begin to arrange themselves radially. At STAGE 47 a central lumen appears, while at their median ends the anlagen of the first *glomeruli* become visible. The glomeruli are situated on the internal side of the tubuli. At STAGE 48 the first six to eight pairs of mesonephric tubes extend rapidly and begin to coil, while two or three pairs have already made communication with the Wolffian duct and have become functional. The first *nephrostomes* and corresponding *canaliculi* appear at the ventral surface of the mesonephros. The canaliculi, which are short and straight, are well differentiated at STAGE 49, while the nephrostomes open at STAGE 50.

During the STAGES 43 to 47 the *mesonephric area* has extended further anteriorly, and also slightly posteriorly, and has reached the posterior region of the pronephros at STAGE 47, so that no *dianephric region* exists in *Xenopus*. During the following stages the formation of the mesonephros extends mainly in anterior direction, its anterior portion being markedly retarded in differentiation. At STAGE 49 the number of functional units has increased to ten to twelve and the number of glomeruli to six to eight on each side. At this stage new tubes (*second generation*) appear in between the primary ones, so that the mesonephros becomes more massive in appearance. The new, interjacent tubules are formed at the base of the dorsal mesentery, so that they actually originate from the sub-aortic region and shift laterad. They can be distinguished from the first generation of tubules by their narrower lumina, their thicker walls and the more pronounced basophily of their cells. Simultaneously a new generation of *glomeruli* develops, situated at the dorsal side of the kidney, so that the number of glomeruli increases to about twenty on each side at STAGE 50. At STAGE 51 a *third generation* of mesonephric tubes develops, so that the number of glomeruli has now augmented to forty to fifty. These new generations push the primary generation and the Wolffian duct lateralwards. The gradual cranial extension of the mesonephros and the simultaneous caudal extension of the pronephros, causing a marked overlapping of both organs, consequently leads to an S-shaped form of the Wolffian ducts, into which also the most anterior mesonephric tubules open. At STAGE 52 the mesonephric tubes reach down to the openings of the Wolffian ducts into the rectum, but from STAGE 53 on the caudal portions of the Wolffian ducts stretch, so that these again pass beyond the kidneys. The mesonephros gradually increases further in length, particularly between STAGE 54 and 55, when its length is nearly doubled, and attains its full length at STAGE 58.

At STAGE 55, however, the first signs of degeneration appear also in the *mesonephros*, at the median side of its middle portion, where some mesonephric tubules and glomeruli have degenerated at STAGE 56. In comparison with the pronephros, the degeneration in the mesonephros is restricted, while at STAGE 59 again new tubes are formed. During the STAGES 60 and 61 the degeneration of mesonephric tubes is quite pronounced, particularly in the middle and posterior regions. The numerous mesonephric tubes are in complete disorder. Their cells have disaggregated and have greatly changed in appearance. Even the less affected tubes show lumina almost entirely obstructed by pigmented cell fragments. At STAGE 62 a large number of new mesonephric tubes and glomeruli and a number of nephrostomes is again formed. Numerous nephrostome funnels with canaliculi appear at the ventral side of the mesonephros at STAGE 64. The reorganization of the mesonephros makes further progress during the following stages, but is not yet completed at STAGE 66.

At STAGE 55 the wall of the posterior half of the *Wolffian duct* begins to alter. In the middle and posterior regions the wall thickens and becomes pluri-stratified, its terminal region caudal to the mesonephros being less affected. At STAGE 58 the wall consists of two to three layers of cells, and a thin connective tissue layer develops around the duct. This reorganization proceeds only slowly during the following stages and is also not yet completed at STAGE 66.

The *urinary bladder* develops as a ventral diverticulum of the rectum at STAGE 55 and is, in continuation with the rectum, surrounded by a thick connective tissue layer. It develops in an anterior direction. A well developed cavity is present at STAGE 56. At STAGE 58 the bladder extends laterally. Some long connective tissue strands develop at the lateral sides, attaching the bladder to the coelomic wall. The epithelium of the bladder is slightly columnar. The development of the bladder is accelerated

markedly by STAGE 59, so that it covers the posterior portion of the mesonephros at that stage. Its wall is now thin and smooth, and it communicates
with the rectum through a wide orifice. The glandular elements have
almost disappeared from the epithelium except in the stalk region, where
the epithelium remains thicker. In following stages the bladder extends
further in a cranial direction, while a thin muscular layer develops around
it at STAGE 63.

* * *
*

THE DEVELOPMENT OF THE GONADAL SYSTEM AND ADRENAL GLANDS

THE DEVELOPMENT OF THE GONADAL SYSTEM

In the region of the future genital tract the *primordial germ cells* are embedded in the dorsal entoderm at STAGE 40. They are large and laden with yolk and therefore not easily distinguishable from the entodermal cells. Their segregation from the entoderm begins at STAGE 41 along the dorsal median line. They accumulate between the two approaching splanchnopleures, which will form the dorsal mesentery at STAGE 42. At STAGE 43 the germ cells migrate towards the dorsal base of the mesentery, where they begin to form a *median, unpaired genital ridge*. At STAGE 44 a small number of them are still located within the mesentery; the majority form an unpaired, median genital ridge, while some germ cells are already beginning to migrate towards the coelomic wall along both sides of the dorsal mesentery. This lateral migration has made good progress at STAGE 45, when the two *genital ridges*, the first rudiments of the gonads, have been formed. Each genital ridge is a flat, but thickened tract of the coelomic wall in which the large, yolk-laden germ cells are embedded, surrounded by small somatic follicle cells. The lateral migration is more or less completed at STAGE 46. The genital ridges are becoming more compact and begin to protrude into the coelomic cavity at STAGE 47. They form actual folds which markedly protrude at STAGE 48. At STAGE 49 some scattered somatic cells from the mesonephric and interrenal areas begin to penetrate into the genital ridges. The formation of this *"medullary tissue"* by somatic cells, mainly originating from the interrenal blastema, descending along the lateral walls of the vena cava and penetrating into the genital ridges, has made good progress at STAGE 50. The germ cells are exclusively located in the wall of the genital ridge, the *"cortical region"*. A complete gonadal rudiment has now been formed. The amount of medullary tissue increases gradually during the following stages. The germ cells begin to lose their yolky material at STAGE 50. The anterior part of the genital ridge, which completely lacks germ cells, represents the rudiment of the *fat-body*.

The *sexual differentiation* of the gonads begins at STAGE 52. In *male* specimens the germ cells migrate from the cortical region into the medullary tissue, where they will form *spermatogonia*. In *female* specimens, on the contrary, the medullary tissue remains free of germ cells, which, embedded in the cortical region, begin to multiply as *oogonia*.

The *testes* are compact structures at STAGE 53 and 54 without clear remnants of the cortical region. In the compact medullary tissue the spermatogonia increase in number, gradually consuming their embryonic yolky material. Although the testis is still compact, small tubular cavities begin to appear at STAGE 55. These cavities, to be considered as *primordia* of the *seminiferous tubules*, are gradually formed and increase in number during the further development up to STAGE 66. Scattered nests of first meiotic prophase *spermatocytes* begin to appear from STAGE 59 on. At the stages considered here, the *efferent testicular tubules*, communicating with the Wolffian duct, are not yet developed.

The *ovaries*, which are still compact structures at STAGE 53 and 54, show, besides a central medullary tissue, a peripheral cortical territory in which the multiplying germ cells form several groups of oogonia, the yolky material of which is gradually consumed. These germ cells form nests of *oocytes* in the early prophase stages of the first meiotic division at STAGE 55. Simultaneously the medullary tissue becomes largely excavated in its middle part. These *central cavities* greatly increase in size, while the medullary tissue shows an extreme reduction at STAGE 56. Simultaneously the number of first meiotic prophase oocytes markedly increases in the cortical territory. This number increases further at STAGE 57, when several oocytes begin their *major growth period*. The central cavities have become very large at STAGE 57. The process of oocyte growth slowly continues in the period of development up to STAGE 66.

The primordia of the *Müllerian ducts* begin to develop during the STAGES 64 to 66. They arise as invaginations of the coelomic epithelium beside the Wolffian ducts, but are still poorly developed at the stages considered here.

For the development of the *Wolffian ducts* the reader is referred to the division "the development of the *nephric system*" on page 113.

The rudiments of the *fat bodies* have become very voluminous at STAGE 51, at which some of their cells begin to accumulate *adipose substances*. In both male and female specimens the fat bodies are largely provided with adipose cells at STAGE 57.

THE DEVELOPMENT OF THE ADRENAL GLANDS

The *interrenal tissue* of the *adrenal glands* (homologous with the *adrenal cortex* of the Mammals) originates at STAGES 42 and 43, in the form of a moderate cell proliferation of the dorsal root of the dorsal mesentery. This proliferation extends longitudinally from a level slightly anterior to the coeliaco-mesenteric artery, to the caudal end of the mesonephric blastemata. The embryonic cells of the *interrenal primordium* migrate dorsally from their site of origin towards the ventral wall of the aorta, where, in the space between the two postcardinal veins behind the mesenteric artery, they form a sort of *median peduncle*, ventrally joined with the mesenteric dorsal root.

At STAGES 44 and 45, when the two medial postcardinal veins unite in the median plane, forming the interrenal vein, the median peduncle of interrenal cells is split up, and the interrenal blastema becomes compressed between the dorsal wall of the interrenal vein and the ventral wall of the aorta.

From STAGE 46 to 50, the interrenal blastemata become more voluminous; they form two cell masses, ventral to the aorta and dorsal to the interrenal vein, each lying along the medial face of the mesonephric kidney. Potential *chromaffin cells* (homologous with the *adrenal medulla* of the Mammals) migrate from the sympathetic ganglia area; they invade the interrenal masses from their medial side, thus completing the formation of the *adrenal gland primordium*. From STAGE 49, in the genital area, cells of the interrenal blastemata reach the gonad primordia, by descending along the lateral walls of the posterior vena cava. For the participation of the interrenal blastemata in the formation of the gonads, the reader is referred to the first part of this division on page 137.

From STAGE 51 to 56, the cell masses of the *adrenal glands* are displaced along the medial face of the mesonephric kidneys. From STAGE 57 to 66, the adrenal glands acquire their definitive position on the medial and ventral face of the mesonephric kidneys; chromaffin cells are mixed with cords of interrenal cells.

* *
*

THE FURTHER DEVELOPMENT OF THE ORO-PHARYNGEAL CAVITY, INCLUDING THYROID GLANDS, BRANCHIAL DERIVATIVES AND LUNGS

The early development has been described in the division "the early development of the *alimentary system* and the presumptive *visceral skeleton*, etc.", on page 39, up to STAGE 28 for the *mouth* and up to STAGE 35/36 for the *pharyngeal cavity*, while the previous development of the *lungs* has been described up to STAGE 33/34.

FURTHER DEVELOPMENT OF THE LARVAL STRUCTURES

THE MOUTH ANLAGE

The *mouth anlage* develops from two sides. The pharyngeal cavity gradually penetrates into the *entodermal oral evagination*, while the ectodermal mouth plate slowly invaginates, forming the *stomodeum;* the lumina of the entodermal evagination and the ectodermal invagination gradually approach each other. At STAGE 29/30 the pharyngeal cavity reaches approximately halfway between base and tip of the entodermal oral evagination. The evagination has become a narrow funnel by the increase of the surrounding mesenchyme. Its cavity gradually deepens and widens, leading to the formation of a small *entodermal oral plate* at its tip at STAGE 35/36. Simultaneously the *ectodermal stomodeum* with *oral plate* deepens slightly. The double-layered oral plate has already become rather thin at STAGE 35/36. At STAGE 37/38 it still consists of two thin cell layers. The stomodeum has become a transverse groove at STAGE 39, at which the oral plate is about to rupture, a process which has actually taken place at STAGE 40. During the following stages the mouth opening widens. At STAGE 41 it forms a transverse slit, which extends laterad at STAGE 43 and 44. At the latter stage the anlagen of the *tentacles* become visible as pointed protrusions on both sides of the mouth slit. They develop into short tentacles at STAGE 46, when the mouth slit has extended further. At STAGE 50 the *choanae* have perforated, so that an open communication has been formed between the olfactory organ and the oro-pharyngeal cavity.

THE VISCERAL POUCHES AND GILLS

The *visceral pouches* become deeper and the corresponding furrows in the contour of the embryo change gradually. The first and second furrows flatten at STAGE 35/36, whereas the third gets deeper. Here the partition wall between pouch and furrow becomes thinner, a process which continues during the following stages. At STAGE 39 the first and second visceral furrows have flattened and the ectodermal protrusions have lost their connections with the corresponding visceral pouches. The third and fourth visceral pouches and furrows have become very deep; pouches and furrows are separated by a thin membrane only. These membranes are perforated at STAGE 40, forming the first and second *visceral clefts*. The fifth visceral pouch and furrow also deepen very much and are only separated by a thin membrane at STAGE 41. The third visceral cleft is formed at STAGE 42.

At STAGE 35/36 the third visceral arch becomes swollen by the formation of the *external gill anlage*, and a general depression is formed in the contour of the embryo in the region of the third to the sixth visceral arches, representing the *sinus cervicalis*. The latter becomes deeper at the next stage, particularly in the region of the fifth visceral furrow. At STAGE 39 the external gills are formed on the third, fourth and fifth visceral arches, the middle one being the largest. They grow out into small buds at STAGE 40, and attain their maximal length at STAGE 41, at which that on the fifth arch sometimes bifurcates. They remain small however, and are already being reduced at STAGE 43. By STAGE 44 they can only be seen on the third and fourth arches. They are largely degenerated at STAGE 47; only remnants are present at STAGE 48 and later stages up to STAGE 53.

At STAGE 40 the third visceral arch has formed an ectodermal fold, the *operculum*, which grows caudalwards. At STAGE 44 it overarches the sinus cervicalis, leading to the formation of the cavity of the filter apparatus or the *gill chambers*. After the external gills have been completely covered over, only a narrow opening is left at STAGE 46. This opening forms a ventral oblique slit at STAGE 48, while the operculum has formed a thin caudal border. At STAGE 50 a fold has grown out from the hyoid arch, forming the basal portion of the operculum, while the original outgrowth of the third arch forms its apical portion.

During later stages special organs develop as derivatives of the visceral pouches (see under the development of the *thymus gland*, the *ultimobranchial bodies* and the *epithelial bodies*, on page 145, and the *middle ear*, on page 148).

THE FILTER APPARATUS

The oro-pharyngeal cavity shows another, very characteristic development in the formation of the *filter apparatus*. Along the caudal margin of the third visceral arch epidermal folds develop at STAGE 41, while similar formations appear along the fourth and fifth arches at STAGE 42. The anlagen of the "*rakes*" are best developed along the third visceral arch, those along the following arches being retarded. At STAGE 43 the "rakes" on the third arch are club-shaped, their length being about twice their breadth. This has increased to about three times at STAGE 44. On the fourth and fifth arches there are two rows, one directed towards the preceding and one towards the following arch. The sixth arch bears only one row. The rostrally directed rows form tooth-like protrusions which fit in between the club-shaped, caudally directed "rakes" of the preceding arches. The tooth-like protrusions on the fourth arch are the most advanced. The processes on the third arch have become branched at STAGE 45. At this stage also processes appear on the inner surface of the operculum. The "rakes" on the third, fourth and fifth arches have become feather-like and double-branched at STAGE 47.

THE PRIMITIVE TONGUE AND SENSE ORGANS

In the floor of the oro-pharyngeal cavity a median furrow is present at STAGE 37/38. On both sides of this furrow folds covered with a tall epithelium are formed at STAGE 39. They develop into two longitudinal ridges at STAGE 40, representing the anlage of the *"primitive tongue"*. At STAGE 44 they protrude as tongue-shaped processes on both sides of the place of origin of the thyroid gland, and extend caudad up to a point beyond the larynx. In the region of the larynx the mid-ventral furrow becomes a narrow slit between two considerably protruding ridges at STAGE 47. More anteriorly, at the level of the third visceral arch, lateral branches run from the longitudinal ridges of the primitive tongue towards the bases of the visceral arches at STAGE 47. The columnar epithelium of the primitive tongue spreads also over these lateral ridges. At STAGE 51 folds are formed along the roof and the walls of the pharyngeal cavity. The filter processes of the branchial arches now fit exactly into the furrows present between these folds, so that the entire pharyngeal cavity has been reduced to a zig-zag-shaped slit in cross section.

Taste buds develop in the roof of the oro-pharyngeal cavity, first caudal to the future choanae, at STAGE 48. They increase in number at STAGE 49, at which they also appear in the pseudo-stratified epithelium of the primitive tongue and in the lateral regions of the pharyngeal roof, where a similar epithelium is formed. When at STAGE 51 folds and furrows are formed in the roof and walls of the pharyngeal cavity, taste buds appear also in the walls of the furrows. Separate taste buds develop in the cranial part of the oro-pharyngeal floor in front of the primitive tongue at STAGE 51. At STAGE 56 the number of taste buds considerably increases in the furrows of the caudal part of the oro-pharyngeal roof opposite the branchial arches, where they, just as on the primitive tongue, form a continuous sense epithelium.

THE LUNGS AND LARYNX

On the posterior side of the transverse ridge, separating the pharyngeal cavity from the gastro-duodenal cavity, a horizontal ridge develops, which separates the primary cavity of the liver (below) from the cavity of the *trachea* (above) at STAGE 35/36. The latter is still in open communication with the gastro-duodenal cavity dorsally and posteriorly. At STAGE 39 a horizontal ridge has closed the communication between the cavity of the trachea and that of the pharynx, so that the tracheal cavity now only communicates with the anterior part of the oesophageal and gastric cavity. At the next stage the ventral edges of the stomach and oesophagus walls fuse in the midline along their entire length, thus closing the former communication between the tracheal and the oesophageal and gastric cavities, so that from STAGE 40 on the tracheal cavity and lungs form a closed system. The rudiments of the *lungs* contain a narrow lumen at STAGE 37/38. The walls of the anlagen are still thick and contain much yolky material. The lungs stretch caudad on both sides of the midgut at STAGE 39. At STAGE 40 the tracheal cavity is wide and those of the lungs are narrow. During the following stages a common, wide tracheal tube develops, with which the cavities of the lungs communicate. At STAGE 42 the cranial, proximal part of the lungs has a thin wall, whereas in the caudal, distal third the lumen is narrow. The developing lungs reach up to the last nephrostome of the pronephros at STAGE 43. They protrude into the dorsal portion of the abdominal cavity on both sides of the oesophagus and stomach. At STAGE 45 the lungs are dorso-ventrally compressed and stretch further caudad. At STAGE 46, at which their walls consist of a flat, squamous epithelium, they reach into the caudal part of the abdominal cavity, caudal to the stomach. At STAGE 52 they even reach the caudal extremity of the abdominal cavity. The lung epithelium begins to form folds in the caudal portion of the lungs at STAGE 51. At STAGE 52 the inner surface of the lungs consists of a very thin squamous epithelium; around the epithelial sac connective tissue and blood vessels have developed.

At STAGE 50 the cranial portion of the *coelomic cavity* forms two extensions in dorsal direction just behind the ear capsules. Proximally the *lungs* form *dorsal horns*, protruding dorsally into these outpocketings of the coelomic cavity, thus coming into direct contact with the fenestra rotunda of the ear capsule. In the further development these dorsal horns are first directed craniad and then dorsad on both sides of the vertebral column, and end in the dorsal coelomic sacs, which extend up to a point near the dorsal skin at STAGE 53. At STAGE 54, at which the outpocketings have deepened, their anterior walls lie perpendicular to the longitudinal axis of the trunk, while their posterior walls decline ventro-caudad.

The *larynx* is formed in the ventro-caudal wall of the oro-pharyngeal cavity, where it borders the tracheal cavity, simultaneously with the development of the lungs. At STAGE 39 the original connection between the lumen of the lungs and the pharyngeal cavity is still open. At STAGE 40 the connection is closed and the region of the future *glottis* is indicated by regular epithelium. On the surface of the larynx at this point a long furrow develops at STAGE 42, representing the place of the future perforation. Along the lateral sides of the larynx cartilages are formed at STAGE 43, when the larynx, still closed by a thin membrane projects slightly into the branchial cavity. At STAGE 44 the glottis begins to open by a perforation of the cranial portion of the membrane (formation of *aditus laryngis*), which, however, still partially closes it. The perforation has achieved its maximal extension at STAGE 45, at which the larynx markedly protrudes from the pharyngeal floor. This protrusion fits into a corresponding hollow in the pharyngeal roof. At STAGE 52 the larynx is lined with a thick layer of squamous epithelium underlain by a thick layer of compact connective tissue.

THE THYROID ANLAGE

The *thyroid anlage* appears as a caudally directed finger-shaped protrusion of the oro-pharyngeal floor at the level of the first visceral pouch at STAGE 33/34. It grows caudad underneath the ventral skin as a thin cell strand at STAGE 35/36. Its most caudal portion becomes thickened at STAGE 37/38 and splits into two lobes (the *thyroid anlagen s.s.*) near the branching point of the truncus arteriosus at STAGE 39. The rest of the original anlage connects the thyroid anlagen with the epithelium of the oro-pharyngeal floor as a solid cell strand, the *"ductus thyreoglossus"*. The thyroid anlagen form two longitudinal strands along the branches of the truncus at STAGE 40, at which they have already lost their connection with the median "thyreoglossal duct". In each longitudinal cell strand a node-shaped thickening appears as the anlage of the *definitive thyroid gland* at STAGE 41. The proximal part of the "thyreoglossal duct" disappears at STAGE 41, the rest at STAGE 43 when the thyroid gland anlagen are situated on both sides of the hyoid crista. They increase in thickness during the following stages. At STAGE 48 the cell strands have changed into four to six diffuse lobes, which start forming *follicles* at STAGE 49 to 50. The follicle epithelium surrounding the first colloid formations is, however, still incomplete. The first follicles are completed at STAGE 51, at which each gland contains about thirteen follicles. Resorption vacuoles appear at STAGE 51 and are numerous at STAGE 52. Meanwhile the colloid masses increase in size. The number of follicles increases to about twenty at STAGE 53, the interfollicular connective tissue augments, and the gland becomes surrounded by a well-defined capsule at STAGE 54, while moreover the original flat epithelium changes through a cuboidal into a tall columnar epithelium at STAGE 56.

THE BRANCHIAL DERIVATIVES

In the dorsal region of the second visceral pouch the epithelium has thickened at STAGE 40 and has formed a bud-shaped protrusion between eye and ear vesicle, representing the anlage of the *thymus gland*. It has become pear-shaped at STAGE 41, at which it is still attached to the second visceral pouch. At STAGE 43 the thymus gland consists of a wall of cuboidal cells surrounding a number of central cells. Simultaneously a sharp boundary has been formed between gland and surrounding connective tissue. At STAGE 44 the number of cells in the interior of the anlage has increased, so that the entire anlage has become more rounded. The connection with the epithelium of the second visceral pouch is maintained up to STAGE 46. Differentiation does not start before STAGE 47, at which one or two centers are formed, around which the cells are arranged concentrically. A large number of mitoses are found. The now oval-shaped gland is surrounded by a well defined capsule of connective tissue, which thickens during the following stages. The thymus gland, which is growing gradually, shifts dorsad and comes to lie underneath the dorsal skin cranial to the ear capsule at STAGE 51. The gland consists of a cortical region and a somewhat eccentric central mass. In the cortical layer small pigment cells appear at STAGE 52. With the further growth of the animal the gland has stretched in a cranio-caudal direction at STAGE 55.

At their ventral extremities the third and fourth visceral pouches have formed club-shaped thickenings of the epithelium at STAGE 43, representing the anlagen of the *epithelial bodies*. They have become rounder and larger at STAGE 44. At STAGE 48 those derived from the third visceral pouches have become larger than those which originated from the fourth visceral pouches. They show an onion-like internal structure, and are surrounded by a flat epithelium. The former have become egg-shaped and the latter spherical at STAGE 50. Growing in size, they are still attached to the epithelium of the branchial pouches at STAGE 52.

The sixth visceral pouch forms the anlage of the *ultimobranchial body*. The pouch deepens and forms a finger-like protrusion with a narrow lumen at STAGE 40 to 41. This protrusion is directed medially and caudally. Its lumen disappears at STAGE 44. The body now forms a spherical epithelial bud, still attached to the wall of the pharyngeal cavity by a narrow stalk. At STAGE 45 this connection has disappeared and the epithelial bud has fallen apart, forming diffuse islands. These have become somewhat more compact at STAGE 47. In the cell islands diffuse "*follicle*" formation with very small lumina can be seen at STAGE 50. The cell material gradually spreads out over a larger area at STAGE 53 to 54. At the latter stage the cells form small spheres, resembling tiny thyroid follicles, without, however, signs of secretion. The follicles have become more clear at STAGE 56, at which the entire anlage has shifted caudad to the level of the caudal end of the larynx.

THE DEVELOPMENT OF THE ADULT STRUCTURES DURING METAMORPHOSIS

During the stages of metamorphosis profound changes occur; several structures disappear or are altered, and some new structures arise.

THE MOUTH

On the upper jaw *tooth germs* appear in a single row at STAGE 55. Their number is about twenty; some of them already showing enamel formation. A second row of tooth germs has formed at STAGE 59, while a third row has become visible at STAGE 60, at which the oldest and most median teeth have perforated the oral mucous epithelium. More teeth have perforated at STAGE 64.

At STAGE 57 a *lip-fold* develops along the upper jaw and a smaller one along the lower jaw. At STAGE 62 another fold develops behind the teeth of the upper jaw, the lower jaw now fitting in between the two folds of the upper jaw.

At STAGE 57 *intermaxillary glands* (FAHRENHOLZ, 1937) appear on both sides of the median line between choana and mouth opening. Each forms two small tube-like anlagen which later split up into a number of separate glandular ducts. At STAGE 58 five to six glandular ducts can be distinguished on each side, opening through a common duct. The two glands increase in size during the following stages by a gradual increase in the number of glandular ducts, and make contact with each other in the midline at STAGE 62.

During the stages of metamorphosis, the *jaw articulation* moves caudad and the *mouth slit* becomes much longer at STAGE 61. The corners of the mouth lie at the level of the caudal margin of the eyes at STAGE 63, and already caudal to the eyes at STAGE 64. At that stage the mouth slit, which has become horse-shoe-shaped, has moved to the ventral side of the head, and has attained a position caudal to the nostrils and the anterior portion of the olfactory organs which have moved forwards.

THE ORO-PHARYNGEAL CAVITY

The *oro-pharyngeal cavity* itself undergoes pronounced changes. The *branchial clefts* begin to close at STAGE 57. The first branchial cleft closes from the oro-median side. At STAGE 60 also the second and the third branchial clefts have partially closed from the median side, while the pharyngeal floor has flattened. The closure of the branchial clefts has proceeded slowly up to STAGE 62, at which the opening of the operculum has become roundish. A slit-shaped duct now leads to the *gill chamber*. The opening of the operculum has closed at STAGE 63. The branchial clefts have become flatter and form narrow, complex slits on both sides of the caudal portion of the primitive tongue. They shrivel together during following stages and are narrow, branched grooves at STAGE 64. They are still found in the latero-caudal part of the oro-pharyngeal cavity at STAGE 66.

The degeneration of the *filter apparatus* starts at STAGE 61, at which the distal branches of the filter processes have disappeared. At STAGE 62 the entire filter apparatus shows signs of cytolysis and forms a thick, folded epithelium upon the arches.

At STAGE 57 the *folds* in the *roof* of the *oro-pharyngeal cavity* in between which the filter processes fitted, have disappeared cranially and have become flatter caudally. A free cavity between roof and floor results at STAGE 58. A large number of *goblet cells* are found in the cuboidal epithelium of the lateral parts of the oro-pharyngeal cavity. The oro-pharyngeal cavity is lined with a pseudo-stratified epithelium at STAGE 62; the folds in the roof have become still flatter and the furrows have become narrower. From the choanae a paired furrow stretches caudad in the pharyngeal roof at STAGE 63. More caudally the two furrows fuse and continue as a broad median groove up to the larynx. In the lateral parts of the oro-pharyngeal roof longitudinal furrows are present at STAGE 64.

THE PRIMITIVE TONGUE

The pharyngeal floor also undergoes changes. On the surface of the *primitive tongue* a median furrow remains as a respiratory passage from the nostrils to the larynx at STAGE 61. The aditus laryngis is situated cranial to the laryngeal elevation which fits into a hollow in the oro-pharyngeal roof. At STAGE 63 the median furrow in the primitive tongue has deepened, whereas the furrow in the oro-pharyngeal floor in front of the larynx has flattened. At this stage connective tissue has accumulated cranial to the primitive tongue, perhaps representing the *real tongue anlage*. The overlying epithelium shows a large number of folds separated by crypt-like furrows. The latter also appear on the primitive tongue at STAGE 63 and 64. At STAGE 65 the cranial part of the primitive tongue forms a short, free tip. Ventro-lateral to the primitive tongue deep, narrow furrows are formed, so that the tongue becomes mushroom-shaped in cross section. Caudally these lateral furrows gradually fade out. The cranial part of the oro-pharyngeal cavity is still slit-shaped in cross section, whereas its caudal part, which has become narrower, is more rounded in cross section.

THE LUNGS

At STAGE 61 the cranial one third of the *lungs* still consists of a smooth epithelium, but the caudal two thirds shows honeycomb-like compartments by the formation of folds. By the further extension of the proximal portions of the lungs the *dorsal horns* become directed cranio-dorsally, while the ends of these horns, protruding into the dorsal coelomic sacs, direct dorsad. At STAGE 66 the dorsal coelomic sacs have been filled with lymphatic tissue, and the dorsal horns of the lungs have disappeared. Remnants of the dorsal horns are probably still present in the form of a lateral diverticulum of each lung. In the meantime the formation of folds upon the inner surface of the lung sacs has extended in a proximal direction, so that only the most proximal portion of the lung epithelium is still smooth at STAGE 66.

THE THYROID GLAND

The *thyroid gland* gradually increases in size during metamorphosis, while the distance between the two anlagen also increases. Up to STAGE 66 no further changes occur.

THE MIDDLE EAR

When the first visceral pouch at STAGE 39 has lost its connection with the ectoderm, it has the form of a solid epithelial strand, directed cranio-laterally. Its proximal part is in connection with the pharyngeal epithelium at the transverse level of the eye (the *anlage* of the *tuba Eustachii*) and its distal end (the *rudiment* of the *middle ear*) is somewhat thickened, club-like, and is situated lateral to the palatoquadratum, close to the processus muscularis. The anlage of the tuba Eustachii which is still a continuous strand of cells at STAGE 43, becomes discontinuous at STAGE 44. This situation is maintained till STAGE 53 or 54, at which the fragments fuse again, so that at STAGE 54 the anlage of the middle ear and its connection with the oro-pharyngeal cavity (tuba Eustachii or pharyngo-tympanica) look very much as they did at STAGE 39. The distal end however is some-what thicker and larger. At STAGE 56 to 57 the anlage of the middle ear has a narrow, slit-shaped lumen just like the cranial end of the anlage of the tuba pharyngo-tympanica, whereas the caudal part is still solid. Its cells already show a radial arrangement. At STAGE 60 the anlage of the middle ear has become a bilobed vesicle and is still situated lateral to the mouth slit, close to the lateral side of the palatoquadratum. From the larger portion of the vesicle, which is lens-shaped, the tuba pharyngo-tympanica now leads to the pharyngeal roof near the eye. During the caudad displacement of the jaw articulation the anlage of the middle ear moves caudad, lying still somewhat cranial to the eye at STAGE 61. The tuba pharyngo-tympanica forms a narrow duct which widens before it opens into the oro-pharyngeal cavity at the level of the caudal portion of the eye. At STAGE 62 the middle ear vesicle, which gradually widens, has moved further caudad and is now situated below the caudal margin of the eye. The pharyngo-tympanic duct opens now through a rather narrow opening into the dorso-lateral part of the oro-pharyngeal cavity caudal to the eye at a short distance from the midline. The middle ear cavity lies against the cranial side of the ear capsule at STAGE 63. The slit-like cavity lies at only a short distance from the skin. In the skin a ring-shaped depression is formed, the anlage of the "*tympanic membrane*", which is rather thick. At STAGE 64 the middle ear is still separated from the skin by a thick connective tissue layer. The skin depression, the tympanic membrane has become narrower and oval-shaped. The pharyngo-tympanic duct first runs mediad as a rather wide tube, then turns ventrad as a narrow tube and finally opens as a wide funnel into the oral cavity. At STAGE 65 the middle ear cavity has become wider and is partly surrounded by cartilage. The caudal portions of the left and right pharyngo-tympanic ducts have fused and now open through a common opening into the pharyngeal cavity just cranial to the aditus laryngis. At STAGE 66, at which the middle ear cavity has widened still further, the cartilage in the middle ear wall at the beginning of the pharyngo-tympanic duct has become thicker and is continuous around the proximal portion of the duct. A real tympanic membrane cannot yet be seen; along the ventral margin of its anlage a narrow slit-shaped depression is found in the surface of the skin, where the skin is somewhat thinner. (See the development of the *plectral apparatus* in the division "the further development of the *skeleton* and *musculature* of the *head*", on page 98.)

THE OTHER BRANCHIAL DERIVATIVES

The *thymus glands* also gradually increase in size, remaining oval-shaped. At STAGE 59 they have been displaced in a ventral direction, and are now lying at a greater distance from the dorsal surface near the oro-pharyngeal roof. Their caudal ends are pressed against the ear capsules at STAGE 62. The ventrad displacement is continued during following stages, so that at STAGE 64 the glands are situated ventro-lateral to the ear capsules, just under the body surface.

The *epithelial bodies* are still attached to the branchial pouches at STAGE 57. The first pair is pear-shaped, the second pair egg-shaped. They have lost their connection with the branchial pouches at STAGE 59. They become gradually displaced during metamorphosis, so that the first pair is situated at the level of the ventral surface of the pericard and the second pair somewhat more dorso-caudally at STAGE 61. At STAGE 63 the first pair lies in the immediate neighbourhood of the ventral branch of the truncus arteriosus at a short distance from its branching point, and the second pair somewhat more caudally on the ventral side of the dorsal branch of the truncus. At STAGE 66 they finally lie between the branches of the truncus arteriosus at the level of the dorsal surface of the pericard.

The loose cells lying between the follicles of the *ultimobranchial body* have disappeared at STAGE 59. At STAGE 60 the grape-like formation is situated caudal to the tracheal bifurcation in the lateral wall of the oro-pharyngeal cavity, dorsal to the vena cava anterior. Some follicles have become larger at STAGE 63. At STAGE 66 they still form diffuse spherical cell masses, the cells of which are grouped onion-like around a small lumen.

* * *

THE FURTHER DEVELOPMENT OF THE INTESTINAL TRACT AND GLANDS

The early development of the *alimentary system* has been described in the division "the early development up to STAGE 15" on pages 22 to 28 and subsequently in the division "the early development of the *alimentary system* and the *presumptive visceral skeleton*, etc.", on pages 39 to 42, where it has been followed up to STAGE 28, while the segregation of the *oro-pharyngeal cavity* from the *gastro-duodenal cavity* and the development of the *liver rudiment* have been followed up to STAGE 35/36. The segregation of the *tracheal cavity* from the *gastro-duodenal cavity* has been described together with the development of the *lungs* in the division "the further development of the *oro-pharyngeal cavity*, etc." on page 143. The development of the *postanal gut* has been described in the division "the early development of the *axial system*" on page 37 up to STAGE 41, at which this anlage has entirely disappeared. Finally, the development of the *urinary bladder* has been dealt with in the division "the development of the *nephric system*" on page 135.

A. THE DEVELOPMENT UP TO METAMORPHOSIS

THE PHASE OF FURTHER SEGREGATION

The further primary segregation of the *alimentary system* is briefly described before its development is treated systematically.

At STAGE 37/38 a *transverse ridge*, which has been formed between the oro-pharyngeal and the gastro-duodenal cavities, has locally reduced the lumen of the alimentary canal to a very narrow transverse slit representing the *most posterior portion* of the *pharynx* and the *beginning* of the *oesophagus*. On the contrary, the cavity of the *posterior part* of the *oesophagus* and the *stomach* is represented by a narrow vertical slit at this stage. At STAGE 40 the communication between the *gastric cavity* and the *cavity* of the *trachea* has been cut off by the fusion of the ventral edges of the stomach and oesophagus walls in the midline.

In the *midgut* the *dorsal anlage* of the *pancreas* becomes visible at STAGE 35/36 as an abrupt thickening of its antero-dorsal wall. Immediately anterior to the dorsal rudiment of the pancreas, which is clearly segregated from the adjoining parts by narrow slits at STAGE 37/38, a thick mass of entoderm has descended from the dorsal wall of the alimentary canal, forming the *posterior wall* of the *stomach*.

Ventrally the cavity of the stomach merges, still imperceptibly, into the cavity of the *duodenum*, which is also in broad communication with the *primary hepatic cavity*. At STAGE 37/38 the two *ventral pancreatic rudiments* are vaguely distinguishable as masses of entoderm to the left and right of the junction between liver and duodenum. At STAGE 40 the folding of the epithelial wall of the *liver* rudiment has resulted in the occlusion of the primary hepatic cavity, the original communication of the hepatic cavity with the duodenum persisting as the *hepato-pancreatic duct*, in which the two pancreatic ducts from the left and right ventral rudiments of the pancreas debouch. The latter are now clearly segregated from the duodenal tissue. Meanwhile the two ventral and the single dorsal pancreatic rudiments have fused. At STAGE 35/36 the cavity of the proximal part of the *cloaca* has formed a pair of *rectal diverticula* to meet the pronephric ducts. An open communication between the Wolffian ducts and these rectal diverticula is not established before STAGE 39 (cf. the development of the *nephric system*, on page 131). Caudally, the cavity of the cloaca has started to expand just in front of the anal opening at STAGE 37/38. This expansion becomes very pronounced at STAGE 42 but has disappeared again at STAGE 43.

The morphological segregation of the *alimentary canal* in *oesophagus*, *stomach*, *duodenum* [1]), *ileum* and *rectum* takes place during the *coiling process* of the alimentary canal, during which also the histological differentiation of the various parts makes further progress. After the description of the coiling process the development of the individual sections will be briefly described.

THE COILING PROCESS

The coiling of the intestinal tube leads to marked displacements of the various sections of the alimentary tract. The posterior portion of the *oesophagus* curves sharply to the left at STAGE 41, the rest remains in the median plane, slightly inclining ventrad at STAGE 43.

The *stomach* undergoes strong, rather irregular displacements. At STAGE 40 it curves to the left side of the embryo. At STAGE 41 it runs almost horizontally from right to left. At STAGE 42 it has taken up a more oblique position, slanting downwards and to the left. At STAGE 43 the stomach has been shifted away from the left part of the abdominal cavity and lies more medially, its axis being approximately vertical. At STAGE 45 the posterior end of the stomach has been displaced still further to the right; the axis of the stomach is now almost transverse.

The *duodenum* curves to the left of the embryo and then, ascending in a

---

[1]) The posterior part of the *duodenum*, viz. the part behind the entrance of the *bile duct*, cannot be distinguished histologically and morphologically from the *ileum*.

dorso-caudal direction, returns to the median plane at STAGE 39. At STAGE 40 it forms a spiral, first deviating to the left, next curving through the median plane to the right side of the embryo, and finally ascending upwards to the strictly median midgut. The ascending section of the duodenum comes to lie slightly anterior to the stomach at STAGE 43. The loop of the duodenum is almost in a paramedial plane, with the ascending section well cranial to the stomach at STAGE 45. It now consists of a ventral horizontal section, running forward, an ascending section, running dorsad, and a dorsal horizontal section, running caudad. The latter is continuous with the ileum, of which the first part forms the *marginal gut*, which runs along the right, caudal and left body wall to the point where the coils of the *helix* (see below) start.

At STAGE 41 an *oblique furrow* appears on the surface of the *intestine* at the right side, cutting deeper into the entodermal cell material at STAGE 42. Opposite the furrow, where a ventro-lateral *bulge* has appeared on the surface of the intestine at the left side at STAGE 41, the intestine has been drawn out into a long loop at STAGE 43. This loop lies along the left side of the abdominal cavity, reaching anteriorly up to the liver. Consequently, stomach and duodenum have been shifted away from the left half of the abdominal cavity. At STAGE 44 the tip of the intestinal loop bends inwards and caudalwards. At STAGE 45 it begins to be distinctly spirally coiled and has completed one revolution[1]). The inner limb of the loop leads caudad to the cloaca and will later become rectum. At STAGE 46 the spirally coiled part of the intestine has formed one and a half to two complete revolutions. At STAGE 47 it becomes evident that the intestinal coils are being arranged into two spirals, connected by a loop at the tip of the *helix*. The coils of the outer spiral consists of the anterior part of the *small intestine*, the inner spirals of the posterior part of the intestine leading into the *rectum*. At STAGE 49 to 50 there are three or four complete revolutions in the outer, and the same number in the inner spiral of the intestinal helix. The section of the gut forming the last revolution of the inner spiral is markedly thicker and wider than the preceding sections and will differentiate as the *colon*. The boundary between the small intestine and the colon has become very well defined at STAGE 52. There are five complete revolutions in the outer as well as in the inner spiral at STAGE 52, of which the colon takes up the last two revolutions. The axis of the helix has also changed its direction. Instead of slanting ventrad it comes to run almost parallel to the longitudinal axis of the body at STAGE 52 to 53. At STAGE 54 the number of inner and of outer revolutions has increased to seven. The position of the intes-

---

[1])   The revolutions of the spiral have been counted from the end of the marginal gut, in contradistinction to the data for the external criteria, where the spiralization of the entire intestinal tract has been given.

tinal spirals has become less regular at STAGE 56, since the first two or three revolutions of the outer spiral overlap the following ones dorsally.

The coiling of the intestinal tube also leads to displacements of liver and pancreas. Due to the curving of the duodenum the *dorsal pancreas* has been shifted from its original transverse position and is oriented almost longitudinally at STAGE 41. At STAGE 43 the pancreas and the *gall bladder* have shifted to the right. The gall bladder has moreover shifted forwards, so that it lies antero-ventral to the liver. The *hepato-pancreatic duct* enters the duodenum in its ascending section. At STAGE 44 the pancreas takes up a position in front of the stomach, between the latter and the ascending section of the duodenum. At STAGE 45 the point of entrance of the hepato-pancreatic duct into the duodenum has been carried further upwards.

Meanwhile also the pancreatic ducts and the adjoining parts of the pancreas have been pulled upwards. The hepato-pancreatic duct enters the duodenum in its dorsal horizontal section at the junction of the anterior and posterior part of the duodenum, which has become clearly visible at STAGE 46. The dorsal edge of the *liver* has extended far backwards, into the space between oesophagus and duodenum, at STAGE 46. The gall bladder is now at the right antero-lateral edge of the liver. The liver is pushed forwards by the coiling helix at STAGE 51, and is only located in the anterior section of the abdominal cavity, where it spreads from one side to the other, screening anteriorly both the duodenal loop and the first coil of the intestine.

THE HISTOLOGICAL DIFFERENTIATION

In the following section the histological changes taking place in the various organs will be briefly described.

A rearrangement of the cells of the *stomach* wall, which started at STAGE 41, foreshadows the formation of the main *gastric glands*. The cells situated on the periphery of the rudiment become aggregated into groups with narrow slits appearing between them. The gastric glands have become more distinct at STAGE 42, forming pocket-like structures with very narrow cavities. Some of the cells of the gastric lining have developed cilia at STAGE 43. The formation of the main gastric glands has been accomplished at STAGE 46. The differentiation of the *pyloric part* of the *stomach* is foreshadowed at STAGE 47 in a section of the epithelial lining of the alimentary canal posterior to the already developed gastric glands, where the epithelial cells have become tall and columnar. Groups of cells in the pyloric region have become markedly elongated at STAGE 51, forming the rudiments of the *pyloric crypts*. These regions have formed distinct depressions at STAGE 57, the definitive *pyloric glands*.

During the development of the gastric glands, the *duodenal wall* retains its structure of a simple columnar epithelium. At STAGE 43 some of the epithelial cells of the duodenal lining have developed cilia. At STAGE 46 the duodenum is sharply divided into a narrower and thinner anterior part and a posterior part with a broader lumen and a thicker epithelium which merges imperceptibly into the ileum. At STAGE 47 the lumen of the anterior part has expanded and the epithelial cells have become tall and columnar. A longitudinal fold—the *typhlosole*—has been formed in the ascending section of the duodenum and in the marginal gut at STAGE 51, projecting into the lumen of the intestine on the mesenterial side. It extends along the first revolution of the intestinal spiral into the beginning of the second revolution at STAGE 53, and reaches to about one fourth of the second revolution at STAGE 55.

At STAGE 35/36 the *liver rudiment* forms a thick-walled sac communicating with the gastro-duodenal cavity through a constricted canal and opening into the submesodermal space ventro-posteriorly by means of a now much shortened and broadened funnel-like extension. At STAGE 37/38 the anterior wall of the liver rudiment shows the formation of numerous folds, while at STAGE 39 the walls of the entire liver rudiment are thrown into folds which partly fill up the primary cavity. This has led to the occlusion of the primary hepatic cavity at STAGE 40, so that the communication of the liver parenchyma with the duodenum is now only by way of a system of *hepatic ducts*. At STAGE 41 the primary cavity of the liver has been filled up completely by hepatic parenchyma. Remnants of the primary hepatic cavity have become the cavities of the hepatic ducts and the *bile duct*. In some embryos of STAGE 39, but always by STAGE 40, the opening leading ventrad from the liver cavity into the sub-mesodermal space has been closed by a thin sheet of entodermal cells, delimiting a sac-like cavity, the future *gall bladder*. The cavity opens into the main cavity of the liver by means of a short, broad canal. The wall of the gall bladder consists of columnar epithelium at STAGE 41. The draining tube, the *ductus cysticus*, leads from the gall bladder upwards to join the *main hepatic duct* which, after the fusion of the two, runs ventrad as the *ductus choledochus* or *bile duct* and opens into the duodenum after receiving the two *pancreatic ducts*. At STAGE 47 the gall bladder has expanded greatly and is lined with a flat epithelium.

The rudiment of the *pancreas* consists of three parts, a dorsal and two ventral anlagen. Very narrow canals may be traced into the ventral anlagen at STAGE 37/38, starting from the opening communicating the hepatic cavity with the duodenum. At STAGE 39 the dorsal anlage is clearly segregated from the duodenal wall by narrow vertical slits which reach right to the lumen of the latter. A small diverticulum of the duodenal lumen may be traced into the dorsal anlage, but is of a very transient nature. After the withdrawal of the anlage the duodenal wall is left incomplete and the lumen of the duodenum opens into the submesodermal space. The defect later disappears, so that a direct connection between the dorsal pancreas and the duodenum is absent in later stages; the dorsal pancreas has not a duct of its own. At STAGE 40 the right and left ventral pancreatic rudiments have fused just above the gastro-duodenal vein. At about the point where the two ventral rudiments join, they have also fused with the right end of the dorsal pancreatic rudiment. The two pancreatic ducts pass the subintestinal vein on the right and left. At STAGE 41 there is a faint indication of the pancreatic cells becoming arranged in groups as a first step in the formation of *pancreatic acini*. At STAGE 42 blood vessels penetrate into the crevices between the pancreatic acini, thus making them more conspicuous. The larval pancreas finally consists of a system of closely packed acini. The *insular tissue* becomes distinguishable at STAGE 48.

The primary cavity of the *midgut* may become occluded at STAGE 41 in the part of the intestine anterior to the oblique fold formed at that stage. Posterior to the fold the cavity of the midgut is patent. Throughout the intestine the cavity has become wider at STAGE 42, at which there is a moderate amount of cell disintegration on the inner surface of the intestine. Along most of the length of the intestine, except the region directly follow-ing the duodenum, a tall columnar epithelium has developed at STAGE 44 and a cuticular border is formed at STAGE 47. The epithelium is highest in the marginal gut and the first external spiral of the helix, whereas it is much lower in the posterior parts. At STAGE 51 the intestine is very sharply and abruptly expanded at the transition from the small intestine to the colon. In later stages the thickness of the intestinal walls has become more uniform, particularly in the outer spiral at STAGE 52.

The *colon* segregates from the rest of the intestine at a rather late stage. At STAGE 49 to 50 the last revolution of the inner spiral has become marked-ly thicker and wider than the preceding sections and will form the colon, while the last straight part leading to the cloaca represents the *rectum*. The abrupt expansion of the intestine at the transition from the small intestine into the colon at STAGE 51 finally sharply demarcates the latter. The *cloaca*, the cellular differentiation of which begins at STAGE 40 forms the last section of the intestinal tract.

As a general feature, the amount of *yolk* in the entodermal cells is markedly diminished at STAGE 46, particularly in the stomach, the liver, the anterior part of the duodenum and the cloaca, while the oesophagus is already free of yolk. This is the stage at which food appears in the alimentary canal. Except for the middle part of the intestine the yolk has completely disappeared from the alimentary canal at STAGE 47. At STAGE 48 also the middle portion of the intestine has become free of yolk.

B. THE DESCRIPTION OF THE ALIMENTARY CANAL WITH GLANDS JUST BEFORE METAMORPHOSIS (STAGE 57).

The *oesophagus* is a straight tube, round in cross section. Starting from the pharynx it runs backwards along the dorsal side of the abdominal cavity. The lining of the oesophagus consists of a columnar glandular epithelium with mucus secreting cells. It is thrown into longitudinal folds. The length of the oesophagus is just under one fifth of the total length of the abdominal cavity.

The appearance of the gastric glands in the walls of the alimentary canal marks the beginning of the *stomach*. Here the alimentary canal is inflected at right angles to its previous course. The axis of the stomach is almost perpendicular to the antero-posterior axis of the body. Starting dorsally near the median plane the stomach goes down and to the right. The cavity of the stomach is rather narrow and is lined with columnar ciliated epithelium. Gastric glands form a thick layer underneath the surface epithelium. The main type of glands, viz. *ramified tubular glands* lined with cuboidal epithelium, are present in about three fourths of the length of the stomach, starting with its oesophageal end. The last one fourth of the stomach is characterized by the *pyloric glands*. These are crypt-like in shape, and consist of a very tall columnar epithelium, the cells having distinctly basophilic distal ends. There is no pyloric constriction, and the boundary between the stomach and the duodenum is indicated by the disappearance of the pyloric glands, and by the epithelium becoming lower, though still remaining columnar.

At the junction of the stomach and the duodenum the alimentary canal is again inflected. The *duodenum* first goes craniad for a short distance, then turns dorsad, and eventually caudad. It is thus a U-shaped structure, with two horizontal sections (a ventral and a dorsal one), and an interjacent vertical section. The dorsal horizontal section receives the *hepatopancreatic duct*. The initial part of the duodenum is lined by columnar ciliated epithelium. In the ascending section of the duodenum large numbers of *goblet cells* are present. In the dorsal horizontal section both the cilia and the goblet cells disappear, but *club-shaped glandular cells* appear, and the epithelium develops a cuticular border (as in the following section of the intestine). The dorsal horizontal section of the duodenum runs along the upper right side of the abdominal cavity merging imperceptibly into the ileum.

When the *ileum* reaches the posterior end of the abdominal cavity it crosses ventral to the rectum and to the posterior ends of the lungs on to the left side of the abdominal cavity and then pursues a course along the upper left side of the cavity, symmetrically to the position of the dorsal horizontal section of the duodenum but in a reversed direction, i.e. craniad, up to the point behind the antero-ventral edge of the liver. This entire initial section of the ileum is called *"marginal gut"*. From the end of the marginal gut the ileum crosses back along the edge of the liver to the right side of the abdominal cavity, this time adhering to the ventral body wall. This section of the ileum may be designated as the first coil of the *intestinal spiral* or *helix*. The remainder of the small intestine is coiled into a double spiral: an outer spiral, adhering to the body wall, and an inner spiral, contained within the first. Both spirals are clock-wise, and at the posterior end of the abdominal cavity the last coil of the outer spiral is continuous with the most posterior coil of the inner spiral, the intestine bending upon itself and forming a loop. The flow of food in the outer spiral is caudad, in the inner spiral the flow is in a craniad direction. The axes of both spirals are longitudinal, slightly slanting ventrad. The coils of both spirals are inclined, so that the ventral sections lie more anteriorly than the dorsal sections. The incline is not uniform, the first two to six coils of the outer spiral being inclined more than the others, as a result the dorsal parts of these coils overlap the dorsal parts of the next coils. There are seven to twelve coils in the outer spiral, and the same number in the inner spiral.

The first coils of the inner spiral are wound very tightly. The following coils are wider, and the intestine in these is thinner. The last two coils of the inner spiral are differentiated as the *colon*. The cavity of the intestine at this junction is gradually constricted, and then abruptly expands with the beginning of the colon. The colon at first continues the systems of coils of the inner spiral, but after two complete revolutions it turns upwards and then backwards, as the *rectum*. The colon ascends dorsad anterior to the dorsal part of the first coil of the intestine, and henceforth the rectum lies dorsal to the external spiral of the ileum. The beginning of the rectum lies very near to the posterior end of the oesophagus, and the rectum occupies in the posterior half of the abdominal cavity the same position, as the oesophagus does in the anterior part of the abdominal cavity, i.e. it runs straight backwards along the dorsal wall of the cavity. At its posterior end the rectum is continued beyond the last coil of the ileum, and then it becomes the *cloaca*. The cloaca receives the ureters which enter by two separate openings lying close to one another in its dorsal wall. On the ventral side the cloaca communicates with the *urinary bladder*. The opening of the urinary bladder is approximately opposite the openings of the ureters.

The cavity of the *intestine* is lined throughout by a columnar epithelium with cuticular border and scattered secretory cells. The anterior section of the small intestine has a longitudinal fold on its inner (mesenterial) side, which projects into the cavity of the intestine. The fold starts at the anterior border of the duodenum, but in the ascending section of the duodenum the fold becomes very low and broad, and therefore indistinct. In the dorsal horizontal section of the duodenum the fold rapidly increases in height. It is then continued all through the first coil of the external spiral of the ileum, and ends abruptly soon after the beginning of the second coil. It is named the *typhlosole*.

In the *colon* and *rectum* the cuticular border of the epithelium is less distinct than in the ileum. The epithelium of the rectum tends to be thrown into folds, and has fewer secretory cells. The cloacal epithelium is also columnar and possesses mucus-secreting cells. Posterior to the entrance of the ureters the cloaca becomes a very narrow tube wedged in between the bases of the hindlimbs and it then emerges into the ventral fin fold, at the edge of which it opens to the exterior.

The *liver* is a massive organ taking up the anterior part of the abdominal cavity from one side to the other, allowing only for the passage of the bloodvessels, the bronchi and the oesophagus. At a small distance from the anterior end of the liver the duodenum makes its appearance on the right side, followed by the pancreas on the same side. Further back the liver is encroached upon by the coils of the external spiral of the intestine. The liver parenchyma appears as a uniform meshwork of strands of liver cells and bloodvessels. It does not show a subdivision into lobules. The *gall bladder*, which is very large, lies at the antero-ventral edge of the liver, on the right side. Anteriorly it touches the pericardium. The gall bladder is lined by a simple epithelium which is columnar in some parts and flat in others and possesses a very thick cuticular border. The bladder gives off the *cystic duct* on its dorsal surface, near its anterior end. The cystic duct at first runs caudad, along the surface of the gall bladder, then ascends dorsad and to the right, passing along the surface of the liver and subsequently emerging into the space between the liver and the pancreas. In its course the duct receives about seven *hepatic ducts* and becomes the *bile duct*. The last hepatic ducts follow the bile duct into the space between the liver and the pancreas, and join it near the border of the latter organ. The bile duct approaches the dorsal horizontal limb of the duodenum and here it receives one after another the two *pancreatic ducts*, whilst already adhering to the median wall of the duodenum. The resulting *hepato-pancreatic duct* bends round the median surface of the duodenum and eventually breaks through the wall of the duodenum on its dorsal surface.

The *pancreas* is a compact organ filling the space delimited dorsally, anteriorly, and ventrally by the U-shaped duodenum, and posteriorly by the stomach. Small lobules of the pancreas penetrate beyond the stomach on either side. Laterally the pancreas reaches the right body wall over a considerable surface. The parenchyma of the pancreas consists of numerous small *acini* which are, however, packed very closely, with very little connective tissue between them. There is no longer any trace of a subdivision into a dorsal and a ventral pancreas. The secretary canals of the pancreas join the two main pancreatic ducts. These are directly derived from the original ducts of the right and left ventral pancreatic rudiments. The part of the pancreas derived from the dorsal rudiment is drained through the ducts of the ventral rudiments, mainly through the left ventral duct. The original left *ventral pancreatic duct* drains the antero-ventral part of the organ, and on leaving the pancreas it runs over a long distance upwards to join the *bile duct*, without receiving any further tributaries from the pancreatic parenchyma. The original right *ventral duct* drains the postero-dorsal part of the pancreas. This duct receives tributaries from the pancreatic parenchyma even when already nearing the *bile duct*. The left, anterior, pancreatic duct curves around the bile duct and, before fusing with the latter, comes to lie medio-dorsal to the bile duct. The right, posterior, pancreatic duct approaches the bile duct from the ventral side. Usually each of the pancreatic ducts fuses with the bile duct separately, but quite often the two pancreatic ducts first fuse with each other to form a common pancreatic duct, which is, however, very short, and soon joins the bile duct.

C. THE DEVELOPMENT DURING METAMORPHOSIS

During metamorphosis very pronounced changes occur in the *alimentary system*.

MORPHOLOGICAL CHANGES

Although the first histological signs of the beginning of metamorphosis appear at STAGE 57, the first conspicuous morphological changes do not start before STAGE 59. At this stage the lumen of the entire *intestinal tract* narrows progressively, and at the same time the abdominal cavity decreases in volume. Except for a withdrawal of the *intestinal helix* into the posterior part of the abdominal cavity and the disappearance of the perpendicular inflection between the oesophagus and the stomach at STAGE 59, the course of the intestine remains essentially unchanged until STAGE 61. Between STAGE 61 and 62 a considerable shortening of the intestinal tract takes place, so that the number of intestinal spirals is reduced to about two outer and two inner spirals at STAGE 62; the last inner spiral being formed by the *colon*. At STAGE 63 the helix makes only about one and a half irregular revolutions. This reduction continues during later stages, so that only a single loop leads from the wide half circle of the intestine into the colon at STAGE 66. It is particularly the spiralized part of the intestine that undergoes considerable shortening, since the *typhlosole*, which does not continue beyond the beginning of the first outer spiral at STAGE 62, leads into the last ascending part of the intestine up to a point at a short distance from the beginning of the colon at STAGE 63. The distance has become still shorter at STAGE 65, after which, at STAGE 66, the typhlosole disappears.

The uncoiling of the intestine leads to marked displacements of the other parts of the intestinal tract and the accessory organs. The *stomach*, the lumen of which begins to widen between STAGE 61 and 62, lies in the median plane at STAGE 63, and during the following development acquires its definitive position in the left half of the abdominal cavity, at the same time increasing in length and finally extending over about three fourths of the length of the abdominal cavity. The *duodenum* finally comes to run from the posterior end of the stomach obliquely forwards towards the median plane, then making a wide horizontal half circle to the right, after which it leads with a single loop into the medio-dorsal *colon*. During metamorphosis the *gall bladder* moves backwards through the liver until it comes to lie at the postero-ventral edge. The *cystic duct* changes its position until it finally runs mediad. After the communication with the *hepatic ducts*, the *bile duct* runs caudad between pancreas and duodenum, receives the two *pancreatic ducts* and, bending downwards, opens into the lateral wall of the duodenum. The *pancreas* changes its position together with stomach and duodenum, and ultimately becomes stretched out between the stomach and the duodenum, losing its contact with the liver.

HISTOLOGICAL CHANGES

The histological changes start at STAGE 57 in the *oesophagus* and in the *duodenum*. In the latter the changes are not conspicuous before STAGE 58. The process of metamorphosis does not start before STAGE 59 in the *stomach*, while it spreads into the *ileum* at STAGE 61 and finally into the *colon*, *rectum* and *cloaca* at STAGE 63, demonstrating the gradual spreading of the process.

In the *oesophagus* the glandular cells of the mucosa start peeling off at STAGE 57, while large parts of the oesophageal epithelium are degenerating at STAGE 58. A more or less continuous but undifferentiated epithelium has been reconstituted at STAGE 62. [1]) At STAGE 64 there is a distinct increase in glandular elements and at STAGE 65 the oesophageal epithelium is differentiated completely and contains a large number of secreting cells.

In the *stomach* histolysis is seen for the first time in the *nonpyloric part* at STAGE 59, and in the *pyloric part* at STAGE 61. In the pyloric part the cells forming the bottom of the pyloric crypts represent the cell groups engaged in the later reconstitution of the epithelium. In the nonpyloric part of the stomach, reconstitution seems to take place primarily from a zone of the submucosa situated between the glandular layer and the muscularis. In this zone of connective tissue a large number of small cell groups appear and shift towards the lumen, increasing in size, while both epithelium and

---

[1])   The exact mode of reconstitution could not be ascertained.

glands degenerate. The non-pyloric part of the stomach is more or less reconstituted at STAGE 62 and the pyloric part at STAGE 64. At STAGE 65 the stomach epithelium is completely differentiated and the pyloric glands start secreting. Finally the cells of the pyloric epithelium have become taller than those in the non-pyloric region.

In the *anterior part* of the *duodenum*, where histolysis starts at STAGE 58, (to a limited extent already at STAGE 57), reconstitution is similar to that in the pyloric part of the stomach, after local depressions have been formed in the epithelium, and has been more or less completed at STAGE 62. Glandular elements appear at STAGE 64, while the cells of the epithelium become taller.

In the *posterior part* of the *duodenum* and in the *ileum* numerous well defined cell groups lie at more or less regular intervals against the persisting basal membrane. Here they are already recognizable at STAGE 60, i.e. before histolysis sets in at STAGE 61. Reconstitution has for the most part already taken place here at STAGE 62, although often a gap still exists at STAGE 66. The epithelium of the *small intestine* is almost fully differentiated at STAGE 65. In the intestine *lymphocytes* from the lymphoid nodules, already present at very wide intervals in the submucosa at the beginning of metamorphosis, often invade the overlying differentiating or just differentiated new epithelium from STAGE 63 on, causing a certain amount of demolition. This process is similar to that observed in the skin (see page 46).

In the *colon, rectum* and *cloaca* the process of histolysis and reconstitution is similar to that in the ileum, but occurs at later stages. Here histolysis starts at STAGE 63, while a continuous reconstituted epithelium is formed at STAGE 64. At STAGE 65 the epithelium is differentiating and its cells have already become taller than those of the ileum. Many glandular cells develop in its more posterior region.

The *liver* and the *gall bladder* are hardly influenced during metamorphosis but in the region of the gall bladder some disintegration of liver tissue occurs.

The *pancreas* shows an extensive histological remodelling which starts at STAGE 59. The original structure disappears and many cells become vacuolised. This vacuolization greatly increases at STAGE 60. At STAGES 61 and 62 large lacunae filled with degenerating cells can be observed. A number of distinct cell groups, however, retain a healthy appearance, and these give rise to the new pancreatic tissue. At STAGE 63 reconstitution has already proceeded considerably, while at STAGE 64 well defined cell columns still without a lumen, are formed. At STAGE 66 the first signs of the beginning of function are seen.

The *insular tissue* of the pancreas is partially destroyed and reconstituted together with the pancreatic tissue.

* * *

CHAPTER VII

EXTERNAL AND INTERNAL STAGE CRITERIA IN THE DEVELOPMENT OF XENOPUS LAEVIS

INTRODUCTION

Besides a very restricted number of primary external and internal criteria suitable for determination, some general morphogenetical data are given, so that from this chapter the reader can also obtain a brief, but general survey of development. The primary stage criteria have been placed between ** and the various organ systems have been printed in italics. The ages mentioned in the table have been determined approximately on the basis of development at a temperature of 22–24° C under laboratory conditions.

All measurements were made on the material collected in Stellenbosch. From STAGE 22 measurements were carried out after decapsulation.

For each stage the external criteria, which are given first, are arranged topographically, viz., according to their cranio-caudal position in the embryo. From STAGE 11 on the internal criteria are arranged in the same order as the organ systems in the systematic description given in the previous chapter. Both the external and internal stage criteria have been selected by the editors, so that they are entirely responsible for this chapter.

For convenience the accurate sequence of the various organ systems given in the internal criteria is briefly repeated here:

Ectodermal derivatives

Skin, lateral line system, brain, cephalic ganglia and nerves, spinal cord, spinal ganglia and nerves, olfactory organ, eye and auditory organ (except

Mesodermal derivatives   middle ear).

Skeleton of the head (including plectral apparatus), musculature of the head, axial skeleton, axial and abdominal musculature, skeleton and musculature (and skin) of the tail, skeleton of shoulder girdle, skeleton of forelimbs, musculature of shoulder girdle, musculature of forelimbs, skeleton of pelvic girdle, skeleton of hindlimbs, musculature of pelvic girdle, musculature of hindlimbs, heart, arterial system, venous system, lymphatic system, pronephric system, mesonephric system, gonadal system and adrenals.

Entodermal derivatives

Oro-pharyngeal cavity with visceral pouches (including external gills and operculum), lungs, thyroid, branchial derivatives (including middle ear and tympanic membrane), intestinal tract and glands.

EXTERNAL AND INTERNAL STAGE CRITERIA

STAGE 1. Age 0 hours; length 1.4–1.5 mm.
Ext. Crit.: *One cell stage, shortly after fertilization*. Pigmentation darker ventrally than dorsally.
Int. Crit.: *Germinal vesicle disappeared*. Well defined cortical layer and layer of subcortical plasm, mainly located in animal half of egg. Inner plasm composed of inner animal, central and inner vegetative plasms. Special cytoplasmic inclusions near vegetative pole. Thickness of cortical and subcortical layers decreasing in animal-vegetative and dorso-ventral directions: *clear animal-vegetative and dorso-ventral polarity*.

STAGE 2⁻. Age 1¼ h.; length 1.4–1.5 mm.
Ext. Crit.: Beginning of first cleavage. *First cleavage groove has not yet reached vegetative pole*.

STAGE 2. Age 1½ h.; length 1.4–1.5 mm.
Ext. Crit.: *Advanced two cell stage*. *First cleavage groove has reached vegetative pole*. Ventral side of blastomeres darker than dorsal side.
Int. Crit.: Penetration of cortical and subcortical layers into interior of egg along cleavage furrow; *the latter cutting over about half of egg radius into animal half of egg*. *Formation of very thin partition wall between the two *blastomeres**; blastomeres not equal in size. First plane of cleavage more or less coinciding with plane of bilateral symmetry.

STAGE 3. Age 2 h.; length 1.4–1.5 mm.
Ext. Crit.: *Advanced four cell stage*. *Second cleavage groove has reached vegetative pole*. In animal view dorsal blastomeres usually smaller than ventral ones; the latter darker than the former.
Int. Crit.: Partition walls between the four blastomeres. *Cleavage cavity present for the first time*, mainly located in animal half of egg.

STAGE 4. Age 2¼ h.; length 1.4–1.5 mm.
Ext. Crit.: *Advanced eight cell stage*. Dorsal micro- and macromeres usually smaller than ventral ones; dorsal micromeres less pigmented than ventral ones.
Int. Crit.: Third plane of cleavage at about ⅓ of animal-vegetative axis. *Pigmented cortical layer and subcortical plasm penetrating over about ¼ of egg radius along third cleavage furrow*.

STAGE 5. Age 2¾ h.; length 1.4–1.5 mm.
Ext. Crit.: *Advanced sixteen cell stage*. Dorsal micromeres distinctly

smaller and less pigmented than ventral ones. *Macromeres entirely separated by cleavage grooves*.

Int. Crit.: Individual *blastomeres* varying in size and form. *Cleavage cavity* about as large as a micromere, located slightly excentrically towards dorsal side.

STAGE **6**. Age 3 h.; length 1.4–1.5 mm.

Ext. Crit.: *Advanced thirty-two cell stage*. Distinction between dorsal and ventral micromeres as in preceding stage.

Int. Crit.: *Blastomeres* roughly arranged in four rows of eight. *Cleavages still nearly synchronous*.

STAGE **6½**. Age 3½ h.; length 1.4–1.5 mm.

Ext. Crit.: *Morula stage*. About 48 blastomeres. *In animal view about 6 micromeres along meridian*.

Int. Crit.: *Blastomeres* still arranged in single layer around cleavage cavity. *Synchronism of cleavages gradually lost; animal blastomeres in advance with respect to vegetative ones*. Plasmatic organization mainly as at fertilized egg stage, but distributed now over a large number of blastomeres. Distinction possible between animal, equatorial and vegetative blastomeres; transitions between cell types more gradual at dorsal than at lateral and ventral sides.

STAGE **7**. Age 4 h.; length 1.4–1.5 mm.

Ext. Crit.: *Large-cell blastula stage*. *In animal view about 10 micromeres along meridian*.

Int. Crit.: *Tangential cleavage; formation of double-layered embryo*. Except for rather sharp boundary between outer equatorial and vegetative *blastomeres*, transition between various regions still very gradual. *Pregastrulation movements* noticeable for the first time*: slight epibolic extension of animal and equatorial areas and beginning of ascendance of plasm along cleavage furrows near vegetative pole.

STAGE **8**. Age 5 h.; length 1.4–1.5 mm.

Ext. Crit.: *Medium-cell blastula stage*. Surface not yet entirely smooth. Border of animal pigment cap more diffuse at dorsal than at lateral and ventral sides. Gradual transition in cell size from animal to vegetative pole.

Int. Crit.: Outer cell layer single, inner cell material 1–4 cells thick; * first appearance of intercellular spaces between outer and inner cell material*. *First indication of distinction between animal, marginal and vegetative areas in inner cell material*. Continuation of *pregastrulation movements*.

STAGE **9**. Age 7 h.; length 1.4–1.5 mm.

Ext. Crit.: *Fine-cell blastula stage*. Animal cells smaller at dorsal than at ventral side. *Border between marginal zone and vegetative field distinct*, particularly dorsally, owing to difference in cell size.

Int. Crit.: Sharp delimitation of outer cell layer in animal and equatorial regions; outer layer segregated into presumptive *ectodermal* and *entodermal* areas with rather abrupt transition at level below equator. Inner cell material more sharply segregated into presumptive *ectodermal* (animal), *mesodermal* (equatorial) and *entodermal* (vegetative) areas. *Blastocoel* has attained its full size; inner surface smooth.

STAGE **10**. Age 9 h.; length 1.4–1.5 mm.

Ext. Crit.: *Initial gastrula stage*. *First indication of blastopore only by pigment concentration*. No formation of groove.

Int. Crit.: *Formation of bottle-necked cells in outer cell layer in area of future *blastopore groove**. *Beginning of formation of *mesodermal mantle* at dorsal side* by rolling-in of *presumptive prechordal plate* mesoderm around *inner blastopore lip.*

STAGE **10¼**. Age 10 h.; length 1.4–1.5 mm.

Ext. Crit.: *Early gastrula stage*. *First formation of dorsal blastopore groove*; groove still straight.

Int. Crit.: *Initial invagination of *archenteron**. *Extension of *inner blastopore lip* from dorsal to lateral, and even partially to ventral side of inner marginal zone*. Gradual delimitation of *prechordal* portion of definitive mesodermal mantle from central entoderm mass.

STAGE **10½**. Age 11 h.; length 1.4–1.5 mm.

Ext. Crit.: *Crescent-shaped blastopore stage*. Blastopore groove angular. Vegetative field slightly decreased in size, *ventral border of future yolk plug indicated by pigment concentration*.

Int. Crit.: *Epibolic extension* of presumptive ectoderm, particularly of its inner, sensorial layer, and reduction of vegetative area. Marked progress in formation of *definitive mesodermal mantle*, *which has extended up to equator at dorsal side and is forming all around vegetative area*. *First contact between *prechordal part of archenteron roof* and most caudal portion of *sensorial layer of ectoderm**. *Invagination of short slit-shaped *archenteron* over 10–15°*. Clear topographical and temporal distinction between rolling-in of inner marginal zone and invagination of archenteron.

STAGE **11**. Age 11¾ h.; length 1.4–1.5 mm.

Ext. Crit.: *Horse-shoe-shaped blastopore stage*. Blastopore groove surrounding about half of future yolk plug and indicated at its ventral side; future yolk plug often rounded rectangular, slightly elongated in dorsoventral direction, *diameter more than ⅔ of diameter of egg (± 50° of circumference)*.

Int. Crit.: *Mesodermal mantle* extending to 30–40° above equator at dorsal side*, to 20–30° above equator at lateral side and approximately to equator at ventral side.

*Invagination of *archenteron* extended to lateral side; archenteron still slit-shaped, extending dorsally over 40–50°*. *Blastocoel* beginning to be displaced towards ventral side.

STAGE 11½. Age 12½ h.; length 1.4–1.5 mm.
Ext. Crit.: *Large yolk plug stage*. *Blastopore groove closed ventrally*; yolk plug not yet quite circular. At ventral side concentrated superficial pigment still visible. *Diameter of yolk plug about ⅓ of diameter of egg (± 40°)*.
Int. Crit.: *First indication of *neural anlage* in *sensorial layer of ectoderm**; *epithelial layer* now a flattened epithelium.
Mesodermal mantle extending dorsally to 20–30° from animal pole*, ventrally not beyond equator.
Archenteron extending over 80–90°, still slit-shaped*. Entodermal mass displaced towards ventral side.

STAGE 12. Age 13¼ h.; length 1.4–1.5 mm.
Ext. Crit.: *Medium yolk plug stage*. Yolk plug circular; *diameter somewhat less than ¼ of diameter of egg (± 25°)*. More and less pigmented fields radiating from yolk plug.
Int. Crit.: *Neural anlage* extended craniad; *neural and epidermal areas clearly distinguishable anteriorly*.
Mesodermal mantle reaching to some distance from animal pole at dorsal side*; beginning of segregation of *prechordal* and *chordal* areas.
Archenteron extending over more than 90°*; first indications of widening.

STAGE 12½. Age 14¼ h.; length 1.4–1.5 mm.
Ext. Crit.: *Small yolk plug stage*. Future position of neural plate and median groove indicated by darker pigment lines. Yolk plug usually ovoid, variable in size.
Int. Crit.: Clear delimitation of ecto-, meso- and entodermal germ layers.
Neural anlage in sensorial layer of ectoderm reaching nearly up to animal pole of egg*.
Mesodermal mantle has reached its definitive extension, at dorsal side close to animal pole, at lateral and ventral sides to 40–50° from animal pole*; first indication of *notochord* formation.
Archenteron extending up to animal pole and markedly widened*.

STAGE 13. Age 14¾ h.; length 1.5–1.6 mm.
Ext. Crit.: *Slit-blastopore stage*. Neural plate faintly delimited; slight elevation of its rostral part and slight flattening of its caudal part. *Caudal part of median groove formed*.

Int. Crit.: Formation of *neural anlage* still restricted to sensorial layer o ectoderm; first more intimate attachment in dorsal midline between epithelial and sensorial layer of ectoderm, and between the latter and the underlying archenteron roof.

Presumptive somite mesoderm double-layered.

Archenteron extended to about 10° ventral to animal pole; caudal archenteron ring-shaped around entodermal protuberance. *Blastocoel* rapidly decreasing in size.

STAGE 13½. Age 15½ h.; length 1.5–1.6 mm.
Ext. Crit.: *Initial neural plate stage*. *Neural plate clearly delimited*.
Int. Crit.: First indication of formation of *neural crest zone* in margin of anterior half of neural anlage.

Continuous *myocoelic slit* in presumptive somite mesoderm.

Archenteron widened and extended to definitive position at about 45° ventral to animal pole; first formation of *liver diverticulum*.

STAGE 14. Age 16¼ h.; length 1.5–1.6 mm.
Ext. Crit.: *Neural plate stage*. Cerebral part of neural plate bent downwards, with median elevation at rostral end of median groove. *Initial elevation of neural folds*, most pronounced in future nucal region. Blastopore always slit-shaped.
Int. Crit.: *Neural folds* fading out in caudal direction.
Presumptive somite area clearly demarcated against lateral mesoderm.

STAGE 15. Age 17½ h.; length 1.5–1.6 mm.
Ext. Crit.: *Early neural fold stage*. *Presumptive cement gland faintly circumscribed*. *Anterior part of neural plate roundish*. *Neural folds distinct*, except medio-rostrally; *initial formation of sharp inner ridges on neural folds* in rhombencephalic region.
Int. Crit.: **Cement gland anlage* delimited*. First symptoms of segregation of *neural crest* from thickening and narrowing neural plate (s.s.). *Presumptive *eye anlagen* beginning to sink in*.
Beginning of *dorsal convergence* and *stretching movements*. Clearly discernable *myocoel*.
Blastocoel disappeared or greatly reduced.

STAGE 16. Age 18¼ h.; length 1.5–1.6 mm.
Ext. Crit.: *Mid neural fold stage*. *Anterior part of neural plate rectangular*; darkly pigmented eye anlagen present; *neural plate sharply constricted in the middle*. Inner ridges on neural folds forming angle of about 90° with neural plate in rhombencephalic region.
Int. Crit.: First visible participation of epithelial layer of ectoderm in *neural plate*. *Distinct elevation of *neural folds* along entire length of neural

anlage*. *Eye anlagen forming deep depressions in antero-lateral edges*. Left and right *heart anlagen* beginning to fuse.

STAGE 17. Age 18¾ h.; length 1.5–1.6 mm.
Ext. Crit.: *Late neural fold stage*. *Anterior part of neural plate oblong triangular*, angles formed by eye anlagen. Neural folds approaching each other from blastopore up to anterior trunk region.
Int. Crit.: In anterior half of embryo *neural crest* material located at lateral edges of neural anlage; in caudal half sharp delimitation of neural crest material from *neural plate (s.s.)*. *Eye anlagen* showing first signs of lateral evagination*.
*First indications of *somite* segregation*, but still continuous myocoelic cavity.

STAGE 18. Age 19¾ h.; length 1.5–1.6 mm.
Ext. Crit.: *Neural groove stage*. *Anterior part of neural plate narrow, more or less club-shaped*, often narrower towards rostral end. *Parallel neural folds in trunk region very close to each other*, not yet touching.
Int. Crit.: *First segregation of mes- and rhombencephalic *neural crest**. *Eye evaginations* with slit-shaped cavities.
*Anterior 3–4 *somites* in process of segregation*.

STAGE 19. Age 20¾ h.; length 1.5–1.6 mm.
Ext. Crit.: *Initial neural tube stage*. *Neural folds touching each other*, except for inconstant openings at anterior and posterior end and behind nucal region. Considerable lateral extension of brain. Lateral outline of embryo still convex.
Int. Crit.: Embryonic pigment in *epithelial layer of ectoderm* dispersed over entire cell body. *Neural crest* segregated into 4–5 cell masses; beginning of lateral migration*.
*Segregation of anterior 4–6 *somites**. Presumptive *heart rudiment* a single median mesodermal thickening.

STAGE 20. Age 21¾ h.; length 1.7–1.8 mm.
Ext. Crit.: *Neural folds fused, suture still present*. The two eye anlagen showing through dumb-bell-shaped; eyes hardly protruding. Beginning of stretching of embryo. Lateral outline flat.
Int. Crit.: Embryonic pigment of *epithelial layer of ectoderm* concentrated at apical side of cells. *Stomodeal-hypophyseal anlage* indicated by median thickening of sensorial layer of ectoderm just in front of brain. *Brain* subdivided into arch- and deuterencephalon*. Massive *neural crest* extending anteriorly approximately to anterior border of eye anlagen. In trunk region neural crest still in contact with suture of closing *neural tube*.

*Anterior 6–7 *somites* segregated*. *Coelomic cavity* indicated in most dorsal portion of lateral plate.

Anlage of hypochorda indicated. *First indications of 1st and 2nd *visceral pouches* in pharyngeal wall*; *oral evagination* visible.

STAGE **21**. Age 22½ h.; length 1.9–2.0 mm.

Ext. Crit.: *Suture of neural tube completely closed*. *Delimitation of frontal field by pigment lines*. Primary eye vesicles showing through in the form of two separate, obliquely placed oval spots; *beginning of protrusion of eyes*. Lateral outline of embryo just becoming concave, ventral outline flat.

Int. Crit.: Middorsal suture of *neural tube* disappeared. First indication of development of *cephalic flexure* and of *rhombencephalic roof* formation. Closure of *canalis neurentericus*. *Primary eye vesicles* formed. *Ear placode* visible for the first time as thickening of sensorial layer of ectoderm.

*Segregation of *maxillary* and *mandibular portions of mesectoderm*. *Anterior 8–9 *somites* segregated*. First indication of *pronephros anlage*.

STAGE **22**. Age 24 h. (= 1 day); length 2.0–2.2 mm.

Ext. Crit.: *Distinct protrusion of eyes*. *Initial groove between jaw- and gill-areas* only at latero-dorsal side. Lateral and ventral outlines of embryo slightly concave. Anal opening displaced to ventral side. Vitelline membrane becoming wider.

Int. Crit.: First indication of formation of *dorso-lateral placodes* in sensorial layer of ectoderm. *Segregation of *brain* in *pros-*, *mes-* and *rhombencephalon*; *cephalic flexure* approximately 135°. In anterior *spinal cord* beginning of medio-lateral constriction of central canal and withdrawal of anterior *trunk neural crest*. *Primary eye vesicles* in broad contact with epidermis.

*Segregation of *hyal portion of mesectoderm*. *Anterior 9–10 *somites* distinct*. Formation of *blood islands* indicated.

STAGE **23**. Age 1 d., ¾ h.; length 2.2–2.4 mm.

Ext. Crit.: *Jaw- and gill-areas completely separated by groove*. Ventral outline of embryo more concave.

Int. Crit.: *Segregation of prosencephalon into *tel-* and *diencephalon*. *Spinal cord* completed at anterior end. First appearance of *olfactory placodes* as thickenings of sensorial layer of ectoderm. *First slight depression in *ear placode* underneath epithelial layer of ectoderm*.

*Segregation of 1st *branchial portion of mesectoderm*. *Approximately 12 *somites* segregated*. Beginning of segregation of *nephrotomes*; *Wolffian duct anlage* indicated.

*First contact of ectoderm and entoderm in 1st *visceral pouch* (formation of *mandibular arch*); 3rd visceral pouch indicated*.

STAGE **24**. Age 1 d., 2¼ h.; length 2.5–2.7 mm.

Ext. Crit.: *Eyes protruding less far laterally than gill-area*. *Gill-area more prominent than jaw-area*, gill-area not yet grooved. Ventral outline of embryo nicked. Tail bud discernible. Initial motor reactions to external stimulation.

Int. Crit.: Distinct nerve fibres in root of *trigeminus nerve*, and some fibres emerging from *gangl. profundus* and *Gasseri*. Medio-lateral constriction of central canal extended over ⅜ of *spinal cord*, and *anterior half of spinal cord completed*.

*15 *Somites* segregated*; *axial mesenchyme* becoming liberated from somites. *Tail bud* for the first time distinct. Segregation of collecting tube anlage in anterior portion of *pronephros*.

Segregation of *hypochorda* starting anteriorly.

STAGE **25**. Age 1 d., 3½ h.; length 2.8–3.0 mm.

Ext. Crit.: *Eyes protruding equally far or further laterally than gill-area*. *Gill-area grooved*. *Invagination of ear vesicle indicated by pigment spot*. *Beginning of fin formation*.

Int. Crit.: *Frontal glands* indicated in epithelial layer of ectoderm of dien-cephalic region. *Cephalic flexure* of brain about 90°. First part of *n. maxillo-mandibularis* distinct. Most anterior *neural crest* of spinal cord has begun lateral migration; *only posterior ¼ of *spinal cord* not yet completed*. *Primary eye vesicles* fully developed; medial wall decreasing, lateral wall increasing in thickness.

Head somite ı diminishing; *16 somites segregated*.

*4th *visceral pouch* indicated*.

STAGE **26**. Age 1 d., 5½ h.; length 3.0–3.3 mm.

Ext. Crit.: *Ear vesicle protruding*. *Pronephros distinctly visible*. Myo-tomes showing through for the first time. *Fin somewhat broadened at dorso-caudal end of body*. Beginning of spontaneous movements.

Int. Crit.: First indications of *olfactory lobe* formation in antero-lateral portion of telencephalic wall; beginning of evagination of *pineal body*. In posterior ⅕, *spinal cord* not yet completed and central canal still uncon-stricted. *Lateral wall of *primary eye vesicle* showing local protrusions on inner surface*.

Head somite ı disintegrated; 17 somites segregated*. *Wolffian duct* anlage extended to level of trunk somite 3*.

*First contact of ectoderm and entoderm in 2nd *visceral pouch* (formation of *hyoid arch*)*. *Postanal gut* indicated.

STAGE **27**. Age 1 d., 7¼ h.; length 3.4–3,7 mm.

Ext. Crit.: Lateral flattening of eyes. *Fin translucent*, except for region just behind anus. *Tail bud formation accentuated in lateral outline*.

Int. Crit.: First part of *n. ophthalmicus profundus* distinct. *Retinal layer of *optic vesicle* invaginating at antero-dorsal margin*; beginning of *lens* formation as thickening of sensorial layer of ectoderm. *Ear vesicle* closed, but still connected with ectoderm*.

*19 *Somites* segregated*. *Endocardial anlage* delimited. *Wolffian duct* extending to level of trunk somite 5*.

*First contact of ectoderm and entoderm in 3rd *visceral pouch* (formation of 1st *branchial arch*)*.

STAGE **28**. Age 1 d., 8½ h.; length 3.8–4.0 mm.

Ext. Crit.: Fin extending up to anus. *Fin broadened and distinctly divided into outer transparent and inner translucent band*.

Int. Crit.: Beginning of secretion of *cement gland*. *Hypophyseal anlage* penetrated beyond optic stalks; *infundibulum* rather well developed, ventral wall thinning out; *first fibres (*white matter*) developing along ventro-lateral portion of mes- and rhombencephalon and along varying portion of anterior half of spinal cord*. First *epibranchial placodes* segregated. *Marginal invagination of retinal layer extended around *eye vesicle*, central portion still convex*. *Ear vesicle* detached from epidermis*.

*20–22 *Somites* segregated*. *First formation of *endocardial tube* and beginning of formation of *pericardial cavity**. *1st *nephrostome funnel* clearly indicated*, the following two just distinguishable; *coelomic filter chamber* indicated.

STAGE **29/30**. Age 1 d., 11 h.; length 4.0–4.5 mm.

Ext. Crit.: Gray eye cup showing through for the first time. *Fin transparent up to the base over its whole length*. Tail bud distinct.

Int. Crit.: *Chiasmatic ridge* rostrally separated from *lamina terminalis* by shallow groove. *Nerve fibres extending over 50–60 % of length of *spinal cord**. *Eye vesicle* invaginated, but central portion still convex*.

Head somites II and III much reduced; 24–25 somites segregated, segregation reaching tail*. *Endocardial heart anlage* a short tube, closed at both ends; *pericardial cavity* surrounding endocardial anlage on lateral and ventral sides*. Minor lumen indicated in collecting tube of *pronephros anlage*, and first appearance of *glomus*.

Floor of *pharyngeal cavity* somewhat raised in front of liver diverticulum.

STAGE **31**. Age 1 d., 13½ h.; length 4.2–4.8 mm.

Ext. Crit.: *Tail bud equally long as broad*.

Int. Crit.: Mushroom-shaped *epiphyseal* evagination; *hypophyseal* cell plate reaching to caudal border of chiasmatic ridge, but still connected with ectoderm. Nerves beginning to grow out from *gangl.* VII; appearance of *gangl.* VIII. *Neural crest* of spinal cord entirely withdrawn from neural tube; *thin layer of nerve fibres extending over 70 % of length of *spinal cord**. *Nasal pit* indicated. Central portion of *retinal layer* concave.

*Segregation of 2nd *branchial portion of mesectoderm**. *22–23 post-otic *somites**; appearance of "*Urwirbelfortsätze*" on ventral sides of anterior myotomes. *1st pair of *aortic arches* completed, and beginning of paired *dorsal aorta* formed*; *pronephric sinus* formed. 1st *nephrostome funnel* completed.

STAGE 32. Age 1 d., 16 h.; length 4.5–5.1 mm.
Ext. Crit.: Eye cup horse-shoe-shaped, standing out distinctly. *Length of tail bud about 1½ × its breadth*.
Int. Crit.: *Hypophyseal* cell strand detached from ectoderm; roof of IVth *ventricle* very thin. *First nerve fibres passing from *nasal placode* to brain; distinct root of *n. glossopharyngeus**. Layer of nerve fibres along *spinal cord* terminating at level of cloaca. *Cavity of *primary optic vesicle* disappeared except for region near optic stalk; margins of optic cup, forming *choroid fissure*, touching each other*; first pigment in outer layer of optic cup. *First appearance of anlage of *ductus endolymphaticus**.
*About 26 post-otic *somites* formed*. Chordal epithelium and elastica externa formed at base of *tail*. Posterior end of endocardial tube broadened as anlage of *sinus venosus*; short *medial postcardinal veins* present. *2nd *nephrostome funnel* completed*; small lumen in anterior portion of *Wolffian duct*; *rectal diverticula* indicated.
*First contact of ectoderm and entoderm in 4th *visceral pouch* (formation of 2nd *branchial arch*); 5th visceral pouch indicated; paired *lung anlage* clearly visible*.

STAGE 33/34. Age 1 d., 20½ h.; length 4.7–5.3 mm.
Ext. Crit.: Stomodeal invagination a shallow vertical groove. *Dorsal part of eye more pigmented than ventral part*; distinct melanophores in dorsal part. *Melanophores appearing* dorsally on the head and laterally in a row extending from just below the pronephros backwards. *Length of tail bud about twice its breadth*. Beginning of heart beat.
Int. Crit.: First *chromatophores* differentiating in head and trunk regions. Appearance of *supra*- and *infraorbital* and *trunk lateral line placodes*. *First *optic fibres* in middle of chiasmatic ridge and along optic stalks*; appearance of *n. oculomotorius*. *Appearance of *Rohon-Beard cells* at trunk levels of spinal cord*. *Nearly spherical *lens anlage* detached from ectoderm*; distal lens epithelium being formed. Sensorial anlagen of *auditory vesicle* indicated.
Anlage of *m. interhyoideus* segregated. *About 32 post-otic *somites* formed*. Dorsal mesocardium completed and *heart* twisted; 3rd *aortic arch* completed; paired dorsal *aorta* reaching to pronephros; *omphalomesenteric veins* opening into functional *ducts of Cuvier*; anterior *cardinal vein* formed. Appearance of anlagen of anterior pair of *lymph hearts*. *3rd *nephrostome* completed*; lumen in entire collecting tube of *pronephros*.
* *Thyroid anlage* clearly discernable*. Anterior wall of *liver* rudiment bulging forward.

STAGE **35/36**. Age 2 d., 2 h.; length 5.3–6.0 mm.

Ext. Crit.: Stomodeal invagination roundish. Eye entirely black, choroid fissure nearly closed. *Formation of two gill rudiments*, anterior one nipple-shaped. *Melanophores appearing on back*. Posterior outline of proctodeum still curved. *Length of tail bud about 3 × its breadth*. Beginning of hatching.

Int. Crit.: Supra- and infraorbital lateral line anlagen extending to eye; *hyomandibular lateral line placodes* proliferating on both sides of cement gland; dorsal and middle trunk lateral lines extending beyond pronephros. Interretinal part of *n. opticus* distinct; *n. trochlearis has reached m. obliquus superior*. Cavity of *optic stalk* obliterated; *beginning of differentiation of pars optica retinae*; *lens cavity* fully developed.

*About 36 post-otic *somites* formed*. Chambers of S-shaped *heart* distinct; *4th, 5th and dorsal part of 6th *aortic arches* completed*; *secondary vessels* of 3rd and 4th aortic arches formed; *paired dorsal aortae united as median *dorsal aorta*; *medial postcardinals* united behind cloaca*; *lateral head veins* and *musculo-abdominal veins* formed. Rich blood supply of *pronephros* from cardinal veins; *coelomic filter chamber* and *glomus* extending over entire length of pronephros.

Oro-pharyngeal cavity separated from gastro-duodenal cavity by high transversal ridge; first contact of ectoderm and entoderm in 5th *visceral pouch* (formation of 3rd *branchial arch*)*. *Horizontal ridge separating *primary liver cavity* from *tracheal cavity**; the former opening distally into submesodermal space; *dorsal pancreas anlage* indicated; *canalis neurentericus* occluded*.

STAGE **37/38**. Age 2 d., 5½ h.; length 5.6–6.2 mm.; length of tail with fin about 1.8 mm.

Ext. Crit.: Stomodeal invagination much deeper, opening round. *Both gill rudiments nipple-shaped, a branch of the anterior one indicated*. *Posterior outline of proctodeum straight, forming very obtuse angle with ventral border of tail myotomes*. Melanophores spreading over tail.

Int. Crit.: Appearance of anlagen of *occipital* and *ventral trunk lateral lines*. *Sub-commissural organ* differentiating; *first fibres in *comm. posterior**. *N. branchialis* x² has reached middle of 2nd branchial arch. *Rohon-Beard cells appearing in anterior tail levels of *spinal cord**. Beginning of aggregation of *spinal ganglion* cells. *Organon vomero-nasale segregated from main olfactory organ*. Nuclei arranged in three layers in *pars optica retinae*; beginning of differentiation of visual cells.

Anlagen of *m. quadrato-hyoangularis* and *m. orbito-hyoideus* segregated. *About 40 post-otic *somites* formed*. Paired dorsal aorta extending forwards as *internal carotid artery*; median dorsal aorta extending backwards into tail as *caudal artery*. Future dorsal branch of *caudal vein* appearing in dorsal tail fin.

Wolffian ducts and *rectal diverticula* communicating*; entire pronephros functional.

Stomach and *duodenum* dorsally segregated; anterior wall of primary *liver* cavity forming numerous folds; *the two *ventral pancreas anlagen* discernable*.

STAGE **39**. Age 2 d., 8½ h.; length 5.9–6.5 mm.; length of tail with fin maximally 2.6 mm.

Ext. Crit.: Melanophores appearing around nasal pits. Opening of stomodeal invagination transversely elongated. Melanophores on back arranged in a superficial and a deeper layer. *Outlines of proctodeum and tail myotomes forming angle of about 135°*. Melanophores appearing along ventral edge of tail musculature.

Int. Crit.: Infraorbital *lateral line anlagen* segregating into sense organs. *First fibres in *comm. anterior**. *N. hyomandibularis* VII connected with m. interhyoideus; *gangl. superius* IX and *gangl. petrosum* IX separated; *roots of *n. vagus* distinct*. Thin layer of fibres on ventral surface of *spinal cord*; a few cells of central gray matter bulging into white matter in anterior levels of spinal cord. *Arteria hyaloidea has reached interior of *eye cup**; delicate layer of mesenchyme around eye cup; *eye muscle anlagen* outlined. *Sensorial epithelium of *ear vesicle* split into medio-caudal and latero-cranial anlage*.

*About 43 post-otic *somites**. *6th *aortic arch* completed with *ductus Botalli* and *pulmonary artery**. First *mesonephric cells* appearing in caudal trunk region.

*Contact between ectoderm and entoderm obliterated in 1st and 2nd *visceral pouch**; 1st visceral pouch representing anlage of *middle ear* and *tuba Eustachii*; anlage of *primitive tongue* formed; cavities of *trachea* and *pharynx* separated by horizontal ridge. Duodenum curving to left side as *first symptom of *coiling of intestinal tract**; most posterior part of *postanal gut* disintegrating.

STAGE **40**. Age 2 d., 18 h.; length 6.3–6.8 mm.

Ext. Crit.: *Mouth broken through*. Length of gills about twice their breadth, posterior one sometimes also showing a branch. *Outlines of proctodeum and tail myotomes forming angle of 90°*. Beginning of blood circulation in gills.

Int. Crit.: Supraorbital and occipital *lateral line anlagen* segregating into sense organs; dorsal and middle lateral lines of trunk extending to base of dorsal fin at base of tail, and segregating proximally into sense organs. *First fibres in *comm. habenularis**. *Spinal nerves* reaching lateral plate in trunk region. Outer and inner plexiform layers forming in *pars optica retinae*; *fasciculus opticus* formed; *nuclei of central *lens* fibres degenerating**; inner *corneal* layer formed. *Latero-cranial sensorial anlage of *ear vesicle* split into medial and lateral (*crista externa*) anlagen*.

The various parts of *cranium* and *visceral skeleton* well distinguishable as mesenchymatous condensations. Individualization of *mm. levatores mandi-*

bulae and *m. intermandibularis*. Formation of *m. genio-hyoideus*. *About 45 post-otic *somites**. Elastica interna forming at base of *tail*. *Anlage of *forelimb atrium* formed dorso-caudal to 5th visceral furrow*. *Hepatic vein* present; left *omphalo-mesenteric vein*, receiving *subintestinal vein*, has lost connection with heart and has broken into capillaries in liver*; small *external jugular* and *pulmonary veins* formed.

Oral plate ruptured and 3rd and 4th *visceral pouches* perforated*; 6th visceral pouch indicated; *tracheal cavity* separated from *gastro-duodenal cavity**; paired *thyroid anlage* detached from "thyreoglossal duct"; anlagen of *thymus gland* and *ultimobranchial bodies* indicated. Primary *liver* cavity occluded; *gall bladder* anlage formed; *the three *pancreatic rudiments* fused*.

STAGE **41**. Age 3 d., 4 h.; length 6.7–7.5 mm.
Ext. Crit.: Gills broader and flatter, more laterally directed. *Formation of a left-rostral and a right-caudal furrow in yolk mass; torsion of interjacent part about 45° [1]); formation of conical proctodeum*, forming angle of about 60° with tail myotomes. *Formation of fin rostral to proctodeum*; ventral outline of yolk mass and proctodeum a smooth concave line.
Int. Crit.: Yolk consumed in *skin* and *brain* structures. Lateral *infundibular recesses* and *optic ventricles* formed; *pia mater* discernable at level of mesencephalon. First motor neurons enlarging at edge of gray matter in tail region of *spinal cord*. *Lens cavity* disappeared. *Crista posterior* split off from medio-caudal sensorial anlage of ear vesicle*; first indications of formation of *semicircular canals*.

*Several parts of *cranium* and *visceral skeleton* procartilaginous*; *fen. basicranialis* enclosed by procartilaginous structures. Segregation of *m. tentaculi*; anlagen of *m. dilatator laryngis* and *mm. constrictores laryngis* recognizable; further individualization of *visceral muscles*; *eye muscles* differentiating. Spiral valve present in *conus arteriosus*; other *heart valves* indicated; *ductus Botalli* and 1st *aortic arch* absent*; *coeliaco-mesenteric artery* formed; *right omphalomesenteric vein* has lost connection with heart and has broken into capillaries in liver*. Beginning of segregation of *primordial germ cells* from dorsal entoderm.

*Appearance of *filter process anlagen* along caudal margin of 3rd visceral arch*. *Main gastric glands* forming; *oblique right-caudal fold in surface of *intestine*, corresponding with future tip of helix*; *postanal gut* disappeared.

STAGE **42**. Age 3 d., 8 h.; length 7.0–7.7 mm.
Ext. Crit.: Beginning of formation of opercular folds. *Torsion of intestine about 90°; proctodeum connected with yolk mass by short horizontal intestinal tube*.
Int. Crit.: First appearance of *dura-endocranial membrane* at level of mesencephalon; anlage of "*Stirnorgan*" beginning to segregate from epiphyseal

---

[1]) The torsion of the intestine is indicated by diagrams added to the plates of STAGE 41 to 45.

anlage; *first fibres in *comm. cerebellaris**. *N. ophthalmicus profundus* v and. *n. maxillomandibularis* separated by *proc. ascendens palatoquadrati*; *n. glossopharyngeus* innervating m. levator arcuum branchialium. *Rods and cones distinguishable in *pars optica retinae**.

Beginning of chondrification of *visceral skeleton* and *palatoquadratum*. *Head somites* i and ii disappeared*; left and right anlagen of *mm. constrictores laryngis* meeting dorsally and ventrally. Yolk consumed in *tail* structures. Right *omphalo-mesenteric vein* receiving main *portal vein*; *abdominal vein* present. First appearance of *interrenal cells*.

*5th *visceral pouch* perforated; *filter process anlagen* appearing along caudal margin of 4th and 5th visceral arches*.

STAGE **43**. Age 3 d., 15 h.; length 7.5–8.3 mm.

Ext. Crit.: Lateral line system becoming visible externally. Cement gland losing its pigment. *Torsion of intestine about 180°*; proctodeum narrower, arched or S-shaped.

Int. Crit.: *Ventral lateral line* has reached pre-anal fin. Beginning of development of *cerebral hemispheres*; *first signs of *choroid plexus* formation in pros- and rhombencephalic roof*. Narrow cell strand connecting lateral part of *nasal organ* with pharyngeal roof. Beginning of formation of *pectinate membrane* of eye. *Latero-cranial sensorial anlage of ear vesicle split into anlagen of *crista anterior* and *macula utriculi**.

Beginning of chondrification of *cranium*; *cartilagines Meckeli, ceratohyale* and *basihyobranchiale* chondrified*. *Hindlimb* region indicated. *Anterior and posterior *valves* capable of closing atrio-ventricular aperture of *heart**. Accumulation of mesenchyme indicating future *spleen anlage*. Continuous *mesonephric anlage* formed. *Primordial germ cells* beginning to form median, unpaired genital ridge.

Hypochorda disappeared. *Epithelial bodies* indicated on ventral side of 3rd and 4th visceral pouches. *Long *intestinal loop* formed, reaching anteriorly as far as liver*.

STAGE **44**. Age 3 d., 20 h.; length 7.8–8.5 mm.

Ext. Crit.: *Appearance of tentacle rudiments*. Opercular folds protruding further. *Coiling part of intestine showing S-shaped loop; torsion about 360°*. Blood-circulation in gills usually ceased (gills smaller).

Int. Crit.: *Caudal lateral line* appearing on pre-anal fin. *Beginning of invagination of *plexus formations**. First appearance of *dura-endocranial membrane* around anterior portion of *spinal cord*; motor neurons standing out in tail region of spinal cord. Beginning of segregation of inner *choroid* and outer *scleral coat* of eye. *Ear vesicle* divided into pars superior and pars inferior*.

Beginning of chondrification of *comm. quadrato-cranialis anterior* and *arcus subocularis*; *plana branchialia* chondrified*. *Head somite* iii disappeared*. "Urwirbelfortsätze" splitting into anlagen of *m. obliquus abdominis* and m.

rectus abdominis. Valves in *conus* fully developed; circulation in filter apparatus established; *lateral postcardinal veins* indicated; *posterior vena cava* established*.
Aditus laryngis partially formed*; anlagen of *filter apparatus* developing along cranial margins of visceral arches.

STAGE **45**. Age 4 d., 2 h.; length 8–10 mm.
Ext. Crit.: Operculum partly covering gills, edge still straight. *Intestine spiralized in ventral aspect, showing 1½ revolutions*. Beginning of feeding.
Int. Crit.: *Parietal sense organs* present. *Nasal organs* beginning to withdraw from hemispheres. Motor neurons standing out in posterior trunk region of *spinal cord. Choroid fissure* closed except for passage of hyaloid artery.
Outgrowth of *parachordalia* and formation of *hypochordal commissure.* Mesenchymal condensations of *hindlimb anlagen* connected by future anlage of *pelvic girdle* dorsal to anal tube; first thickening of ectoderm over hindlimb bud area. *Partition wall between *atria* completed*. Anlage of *spleen* well defined. *3–4 pairs of *mesonephric units* discernable in mesonephric anlage*.
Paired genital ridges formed*. *Interrenal blastema* being formed.
"Rakes" on 3rd visceral arch branched; *anlagen of *filter processes* on inner surface of operculum*.

STAGE **46**. Age 4 d., 10 h.; length 9–12 mm.
Ext. Crit.: Edge of operculum becoming convex. Xanthophores appearing on eye and abdomen. *Intestine showing 2–2½ revolutions*. *Hindlimb bud visible for the first time*.
Int. Crit.: *Dura-endocranial membrane* surrounding entire brain; *cells of Purkinje* differentiating in *corpora cerebelli*. *Some *spinal ganglion* cells enlarging and equalling Rohon-Beard cells in tail region*. *Distal end of ductus endolymphaticus enlarged as *saccus endolymphaticus**; *canalis lateralis* partially separated from utriculus by septum formation.
Beginning of chondrification of *ear capsule* with *cupula anterior* and *posterior*; *complete chondrification of *visceral skeleton*; formation of *fen. subocularis* and *for. caroticum*. Anterior tip of *notochord* degenerating. *Anlage of *forelimb atrium* detached from ectoderm; first mesenchymal accumulation of *forelimb anlage.* Double-layered ectoderm over clearly defined mesenchymal masses of *hindlimb rudiments. Occipito-vertebral artery* formed. *4–5 mesonephric units* distinguishable*.
Oesophagus free of yolk. Food appearing in *intestinal tract.*

STAGE **47**. Age 5 d., 12 h.; length 12–15 mm.
Ext. Crit.: Tentacles larger. *Edge of operculum forming quarter of a circle*. Xanthophores forming opaque layer on abdomen. *Intestine showing 2½–3½ revolutions. Hindlimb bud more distinct*.
Int. Crit.: *Cement gland* beginning to degenerate. *Prosencephalic plexus branched*; pars intermedia of *adenohypophysis* beginning to differentiate.

*2–3 *Jacobson's gland anlagen* discernable on organon vomero-nasale*. *Formation of *posterior*, and beginning of formation of *anterior semicircular canal*; formation of *maculae lagenae* and *sacculi* and of *papillae amphibiorum* and *basilaris*; *saccus endolymphaticus* strongly lobed and apposed to choroid plexus of ivth ventricle; beginning of formation of *perilymphatic system*.

Chondrification of *crista otica* and *proc. muscularis capsulae auditivae*; lateral wall of ear capsule only pierced by *for. ovale*; *for. jugulare* and *prooticum* completed by procartilaginous connections. *Head somite iv disappeared and trunk somite I reduced*. *Forelimb atrium formed*. Secondary vessels of 3rd and 4th *aortic arches* disappeared. Appearance of blood corpuscles in *spleen* anlage. *Pronephros* free of yolk; caudal end of *pronephros* and cranial end of mesonephric anlage have reached each other; *6–8 *mesonephric units* segregated, central lumen in oldest units*; anlagen of first *glomeruli* appearing. *Genital ridges* beginning to protrude into coelomic cavity.

External gills degenerated to great extent. First indication of *pyloric part of stomach*; *intestinal coils* arranged in outer and inner spiral.

STAGE **48**. Age 7½ d.; length 14–17 mm.
Ext. Crit.: Beginning of pigmentation around n. acusticus. *Forelimb bud visible for the first time*. Shining gold-coloured abdomen. *Hindlimb bud semicircular in lateral aspect*.
Int. Crit.: *Peripheral nuclear layer well differentiated in rostral part of *optic tectum*; *ecto-* and *endomeninx* separated by interjacent membrane at level of mesencephalon. *Spinal nerves* 2, 3 and 4 reaching forelimb anlage; spinal nerves 8, 9 and 10 reaching hindlimb anlage. Beginning of formation of *saccus perilymphaticus*.
*Connection between *ear capsule* and *parachordalia* cartilaginous*; anlagen of *arytenoids* well defined mesenchymatous condensations. *Basal portions of *atlas arches* and 3rd and 4th *vertebral arches* procartilaginous*. *Trunk somite 1 disappeared*. *Elastica externa* and *interna* beginning to thicken in *tail* region. *Forelimb rudiment* a well defined mesenchymal bud covered by thickened epithelium of atrium. *Medial postcardinals* fused as *interrenal vein* as far as junction with *posterior vena cava*; *lateral postcardinals* completed, opening into interrenal vein at level of cloaca*; *mesonephric sinus* formed. *First 6–8 pairs of *mesonephric tubules* coiling; 2–3 pairs communicating with Wolffian ducts*; first *nephrostomes* appearing on ventral surface of mesonephros.
First *taste buds* appearing in roof of oro-pharyngeal cavity; *thyroid gland* consisting of 4–6 diffuse lobes. *Yolk completely disappeared from *alimentary canal*.

STAGE **49**. Age ± 12 d.; length 17–23 mm.
Ext. Crit.: Melanophores usually appearing around thymus gland and nerves and blood vessels of head; xanthophores appearing on pericard.

Forelimb bud distinct. *Hindlimb bud somewhat longer, distal outline still circular, no constriction at base*. Melanophores appearing on dorsal and ventral fins.

Int. Crit.: *Appearance of bilateral *telencephalic choroid invaginations**; infundibular process of *neurohypophysis* beginning to develop; first anlage of tuberal part of *adenohypophysis*; *small tubular anlage of *paraphysis*; at least five nuclear layers in *optic tectum**; first cells of molecular layer of *cerebellum* formed; *choroid plexus* formation reaching lateral border of rhombencephalon. *N. abducens* discernable. **Semicircular canals* with *ampullae* completed; *macula lagenae* located in separate diverticulum of sacculus; *saccus perilymphaticus* extending into for. jugulare*.

*Formation of *canales semicirculares anterior* and *lateralis* of ear capsule*. Cartilaginous connection between *occipital arch* and *ear capsule.* *10–12 pairs of functional *mesonephric units* and 6–8 pairs of glomeruli formed*; second generation of mesonephric tubules and glomeruli appearing. Scattered cells from mesonephric and interrenal blastemata entering *genital ridges*.

Taste buds* appearing in epithelium of *primitive tongue and in lateral regions of oro-pharyngeal roof; *follicle formation starting in *thyroid gland**. *Colon* segregating from *intestine*.

STAGE **50**. Age ± 15 d.; length 20–27 mm.

Ext. Crit.: *Forelimb bud somewhat oval-shaped in dorsal aspect*. *Hindlimb bud longer than broad, constricted at base, distal outline somewhat conical*.

Int. Crit.: Large number of *unicellular glands* developing in *larval skin*. *Olfactory bulbs* beginning to fuse; *paraphysis* beginning to ramify; "Stirnorgan" shifted to position dorsal to paraphysis*; segregation of *tectal* and *tegmental parts of mesencephalon*. *N. abducens* has reached m. rectus posterior. *Rohon-Beard cells* beginning to decline; *cells of *lateral motor columns* beginning to differentiate at *lumbar* levels of *spinal cord**. **Brachial ganglia* with a few large cells*. *Forelimb* and *hindlimb nerve* entering limb anlagen*; *sympathetic chain ganglia* formed in trunk region. **Choanae* perforated*. **Utriculus* divided into pars anterior and pars posterior; *ductus perilymphaticus* and *recessus perilymphaticus papillae amphibiorum* formed*.

*Formation of *canalis semicircularis posterior* of ear capsule; separation of *for. endolymphaticum* and *acusticum**; *proc. ventrolateralis palatoquadrati* chondrified. Beginning of differentiation of mesenchyme in *hindlimb bud*; paired *pelvic girdle anlage* indicated. *Brachial vein* and *vena cutanea magna* being formed; *ischiadic vein* clearly distinguishable. *About 20 pairs of glomeruli, and open nephrostomes in *mesonephros**. Complete *gonadal rudiments* formed. Outpocketings of *lungs* protruding into dorsal coelomic sacs behind ear capsules. Beginning of colloid formation in *thyroid* follicles. Diffuse follicles formed in *ultimobranchial bodies*.

STAGE **51**. Age ± 17 d.; length 28–36 mm.

Ext. Crit.: Tentacles much longer. *Forelimb bud oval-shaped in lateral aspect*. *Hindlimb bud conical in shape, its length about $1\frac{1}{2}$ × its breadth*; melanophores appearing on it.

Int. Crit.: *Cement gland* entirely disappeared; formation of *unicellular glands* in epidermis of hindlimb. *Paraphysis* with many tubules. *Cells of *lateral motor columns* beginning to differentiate at brachial levels of spinal cord*. 4–5 *Jacobson's glands* on organon vomero-nasale. *Ventro-lateral lobe of *saccus endolymphaticus* extending forwards up to mesencephalon*.

Crista parotica well developed; *formation of *tectum posterius*; *fen. basicranialis* closed; separation of *for. opticum* and *oculomotorium*; beginning of formation of *parasphenoid* and *frontoparietalia*. *Chondrification of basal portions of *atlas arches* completed*. First indication of formation of secondary muscles from primary somatic musculature. *Forelimb anlage segregated into anlagen of *shoulder girdle* and *forelimb (s.s.)**. First indication of *femur anlage*. *Arteria iliaca communis* extending to hindlimb anlage. *Interrenal vein* losing connection with *caudal vein*, the latter splitting backwards; appearance of *vena femoralis*. Appearance of first posterior pairs of *lymph hearts*. Third generation of *mesonephric* tubules and glomeruli; *number of *glomeruli* augmented to 40–50*. Adipose substances appearing in cells of *fat bodies*.

Taste buds appearing a.o. anterior to primitive tongue; *filter processes* of branchial arches fitting in between folds of pharyngeal roof; *folds appearing in caudal portions of *lungs**; resorption vacuoles appearing in colloid of *thyroid gland*. *Typhlosole* formed in duodenum and marginal gut*.

STAGE **52**. Age ± 21 d.; length 42–56 mm.

Ext. Crit.: *Forelimb bud irregularly conical*. *Hindlimb bud showing first indication of ankle constriction and first sign of flattening of foot*.

Int. Crit.: *Sense organs of caudal *lateral line* appearing on postanal fin*. *"*Stirnorgan*" situated just cranial to paraphysis*. *Dorsal column* ("dorsal horn") beginning to appear at postbrachial levels of spinal cord. *Lumbar ganglia* with a few large cells*. About 10 *Jacobson's glands* on organon vomero-nasale.

*Separation of *for. acusticum anterius* and *posterius**. *Chondrification completed in 4th to 9th *vertebral arches*; chondrification in lateral wall of *perichordal tube* extending from atlas down to intervertebral region between 2nd and 3rd vertebra*; 3 pairs of *ribs* discernable as mesenchymatous condensations. *Femur* rudiment procartilaginous, *tibia* and *fibula* mesenchymatous*; local condensations of mesenchyme indicating beginning of *muscle* development. *Appearance of *arteria subclavia**. *Beginning of sexual differentiation of *gonads**.

Small pigment cells appearing in cortical region of *thymus*. *5 complete revolutions in outer and in inner spiral of *intestinal helix**.

STAGE **53**. Age ± 24 d.; length 50–60 mm.

Ext. Crit.: *Fore- and hindlimbs in paddle stage*. *Hindlimb without foot somewhat longer than broad; 4th and 5th toe indicated*.

Int. Crit.: *"*Stirnorgan*" just cranial to level of lamina terminalis*; first development of granular layer from ependyma in cerebellum; *Purkinje layers fusing in dorsal midline*. N. trochlearis passing through canal in crista trabeculae. *Dorsal column extending into prebrachial levels of spinal cord*; descendance of lumbar ganglia started.

Formation of canales olfactorii and tectum anterius; beginning of formation of goniale; *beginning of chondrification of arytenoids and annulus cricoideus*. *Chondrification of occipital arches completed; chondrification completed in 3rd pair of vertebral arches*; perichordal tube beginning to chondrify anteriorly underneath vertebral arches; *ventral perichordal tube chondrified down to region of 4th vertebra*. 10 pairs of mm. interarcuales distinguishable; pars externus superficialis of m. obliquus abdominis, and m. transversus abdominis developing. *Shoulder girdle divided into scapular and coracoidal portions*. Beginning of chondrification of pelvic girdle. *Femur chondrifying, tibia and fibula, tibiale and fibulare procartilaginous, skeleton of foot mesenchymatous*. Beginning of histogenesis of muscle fibres in thigh and leg. Congestion of pronephros.

Typhlosole extending into beginning of second revolution of outer intestinal spiral.

STAGE **54**. Age ± 26 d.; length 58–65 mm.

Ext. Crit.: *All four fingers indicated*; edge of hand slightly scalloped between fingers; melanophores appearing on forelimb. *Length of hindlimb without foot nearly twice its breadth; all five toes indicated, the 2nd only very slightly*.

Int. Crit.: *Dorsal column extending throughout trunk levels and over short distance into postsacral levels of spinal cord*. Ventro-lateral lobe of saccus endolymphaticus extending from diencephalon down to 1st spinal nerve.

Chondrification completed in 1st and 2nd pairs and caudalwards down to 10th pair of vertebral arches; chondrification of ventral perichordal tube reaching down to 6th vertebra; perichondral ossification as well as destruction and calcification of cartilage starting in vertebral arches. Shallow acetabulum formed and iliac process well defined. *Chondrification of hindlimb proceeded up to tibiale and fibulare, metatarsalia procartilaginous*. Some muscles of hindlimb with indication of both origin and insertion. *First signs of atrophy in pronephros*.

Up to 7 revolutions in outer and in inner spiral of intestinal helix.

STAGE **55**. Age ± 32 d.; length 70–80 mm.

Ext. Crit.: *Hand pronated about 90°; free parts of fingers about equally long as broad*. *Length of hindlimb without foot about 3 × its breadth; length of cartilages of 4th and 5th toe about 4 × their breadth*.

Int. Crit.: **Lumbar motor column* nuclei of spinal cord showing twice cross section area of adjacent non-motor nuclei*. *Cartilage present in medial portion of *sclera* of eye*. Ventro-lateral lobe of *saccus endolymphaticus* reaching from telencephalon to 2nd spinal nerve.

*Fusion of *cristae occipitales laterales* with *tectum posterius**. *11th pair of *vertebral arches* chondrified*; chondrification of *ventral perichordal tube* extending over entire trunk region; **dorsal perichordal tube* chondrified vertebrally down to 6th pair, and intervertebrally down to 3rd pair of vertebral arches*. Mesenchymatous perichordal tube of *tail* thickening. **Scapula, procoracoid* and *coracoid* procartilaginous*. *Humerus procartilaginous*. *Shoulder girdle musculature* indicated. *Upper arm musculature* indicated. *Perichondral ossification of *femur* started*. All major *muscles* of thigh and leg present with both origins and insertions; *foot musculature* well defined. Signs of local degeneration in middle region of *mesonephros*; anlage of *urinary bladder* discernable. Small tubular cavities appearing in compact *testis*; medullary tissue of *ovary* largely excavated; *first meiotic division in ovary*. Single row of maxillary *tooth germs* developed.

STAGE **56.** Age ± 38 d.; length 70–100 mm.
Ext. Crit.: *Elbow and wrist clearly indicated; length of free parts of fingers 3 to 4 × their breadth*. *Length of cartilages of 4th and 5th toe about 6 × their breadth*.

Int. Crit.: "*Stirnorgan*" shifted to position just caudal to fused olfactory bulbs. *In descendance of *lumbar ganglia*, root of gangl. 8 overlapping part of gangl. 7, and root of gangl. 9 all of gangl. 8*. *Olfactory organ divided into two portions*; anlagen of *glandula oralis interna* and *nasolachrymal duct* formed. Ventro-lateral lobe of *saccus endolymphaticus* extending caudad down to 3rd spinal nerve, dorso-lateral lobe extending craniad up to diencephalon.

Frontoparietalia fused. *12th pair of *vertebral arches* chondrified; chondrification of *dorsal perichordal tube* extending vertebrally down to 9th vertebra and intervertebrally down to 5th vertebra*; beginning of ossification of vertebral portions of *perichordal tube*; beginning of chondrification of *hypochord*; *2nd pair of *ribs* beginning to chondrify*. *Coracoid slightly ossified perichondrally, *scapula, procoracoid* and *coracoid* chondrified*. *Chondrification of *forelimb* up to *phalanges*, which are still blastematous; ossification extended to forearm*. *Shoulder girdle musculature* differentiated except for origins and insertions. Greater part of *forelimb musculature* still in myoblast stage. Perichondral ossification of *iliac processes*. *Skeleton of *hindlimb* completely chondrified, ossification extended distally to *tibiale* and *fibulare**. Anlage of *ostium tubae* indicated.

Cavity forming in *middle ear anlage*. *Urinary bladder* forming distinct pocket on ventral side of cloaca*.

STAGE **57**. Age ± 41 d.; length 75–105 mm.
Ext. Crit.: Pigment-free spot appearing above "Stirnorgan". *Angle of elbow more than 90°; fingers stretched out in forelimb atrium, their length about 7 × their breadth*.
Int. Crit.: *First signs of cornification of epidermis at tips of 1rst, 2nd and 3rd *toe**; *stratum spongiosum* and *stratum compactum* beginning to develop underneath *stratum germinativum* of skin; the latter beginning to form new *epidermis* and numerous *glands* under remainder of larval epidermis in metamorphosing areas of skin; glands sinking into stratum spongiosum. *Cross section area of largest *lumbar motor column* nuclei of spinal cord 3–4 × that of average adjacent non-motor nuclei*. Dorso-lateral lobe of *saccus endolymphaticus* reaching to front end of telencephalon.
*10th and 11th pairs of *vertebral arches* fused; chondrification of *dorsal perichordal tube* reaching vertebrally down to 10th and intervertebrally down to 8th vertebra*. *Mm. intertransversarii* developing. *Suprascapula chondrified*· *Epicondyles of humerus* formed; *patella ulnaris* chondrified; proximal *phalanges* of 1st and 2nd finger, and proximal and middle phalanges of 3rd and 4th finger procartilaginous*. *Musculature* of *radio-ulna* section indicated. Left and right halves of *pelvic girdle* in contact with each other. *Beginning of ossification of hindlimb *phalanges**. *Oocytes* beginning their major growth period*.
Lip folds formed along upper and lower jaw; *anlagen of *intermaxillary glands* present, each with two glandular ducts*; closure of *branchial clefts* beginning. *Beginning of metamorphosis in *alimentary canal*; *oesophagus mucosa* peeling off and first signs of histolysis in *duodenum**.

(STAGE **58⁻**. Forelimbs in process of eruption, elbow piercing skin; angle of elbow about 90° and fingers folded in forelimb atrium.)

STAGE **58**. Age ± 44 d.; length 80–110 mm. (ultimate length of larva)
Ext. Crit.: *Forelimbs broken through*. Guanophores appearing on abdomen and thighs (adult skin areas). All three claws present on hindlimb.
Int. Crit.: *Metamorphosing *skin* of forelimb area caudally continuous with metamorphosing skin of trunk*. *Saccus endolymphaticus* extending down to 4th spinal nerve.
*Appearance of *maxillae*; formation of *septum* and *tectum nasi, crista intermedia, cartilago alaris* and *nasale**. Appearance of *m. intermandibularis anterior*. *Occipito-atlantal joint* formed; 12th pair of *vertebral* arches fused with 11th and 10th pairs*; *vertebrally and intervertebrally, chondrification of *dorsal perichordal tube* extending down to 10th vertebra; *hypochord* chondrified from 9th to 12th vertebra*. Appearance of *m. rectus abdominis superficialis* and paired *m. sterno-hyoideus*. *Melanin beginning to be deposited intercellularly

in swollen epidermis of *tail*; tip of tail beginning to atrophy*. **Proc. epicoracoideus* completely chondrified; *scapula* ossified perichondrally*; *supra-scapula* already much larger; *membrane bones, *cleithrum* and *clavicula*, formed*. *Last two *carpalia* chondrified, *metacarpalia* and *phalanges* ossified perichondrally*. *Arm* and *hand musculature* differentiated. *Left and right halves of *pelvic girdle* fused ventrally; ossification of *ischia* started*. *Only *tarsalia* and *hallux* purely cartilaginous, rest of hindlimb skeleton ossified perichondrally; single bony sheath of *tibia-fibula**. In region of terminal phalanges *muscles* still in premuscular stage. Appearance of *vena facialis*. *Pronephros* no longer functional.
Goblet cells appearing in oro-pharyngeal cavity; *5–6 glandular ducts formed in *intermaxillary glands**. *Extensive histolysis in *duodenum**.

STAGE **59**. Age ± 45 d.
Ext. Crit.: Tentacles beginning to shrivel up. *Stretched forelimb reaching down to base of hindlimb*. Guanophores appearing near base of forelimb (adult skin area); anterior border of adult skin area on abdomen not yet sharp; appearance of irregular dark spots on back.
Int. Crit.: Areas of metamorphosing *skin* well circumscribed, that of trunk extending partly over telencephalon; *lymph sacs* beginning to develop underneath metamorphosing skin areas; first symptoms of degeneration in larval skin between trunk and hindlimb areas of metamorphosing skin. *Cross section area of largest *lumbar motor column* nuclei of spinal cord 5 ×, of *brachial motor column* nuclei up to 4 × that of average adjacent non-motor nuclei*.
*Appearance of *operculum fen. ovalis* and *pars interna plectri* as mesenchymatous condensations*. *Intercellular melanin deposits underneath epidermis, around blood vessels and between muscle fibres of *tail**. *Paired anlage of *sternum* blastematous*. *Ilia* connected to last vertebra by connective tissue. New generation of *mesonephric* tubules formed in reorganization process. *First meiotic division in *testis**.
*Second row of *tooth germs* formed*; *epithelial bodies* detached from visceral pouches. *Histolysis in *non-pyloric part of stomach*, and *pancreas**.

STAGE **60**. Age ± 46 d.
Ext. Crit.: N. olfactorius still longer than bulbus olfactorius. Guanophores appearing on lower jaw (adult skin area). Openings of gill chambers still wide. *Distal half of fingers of stretched forelimb extending beyond base of hindlimb*; forelimb still situated behind level of heart. Adult skin area of base of forelimb covered with guanophores; anterior border of adult skin area on abdomen sharper, reaching up to heart.
Int. Crit.: *Well defined *stratum corneum* formed on upper jaw*; appearance of *horn papillae* in the other adult skin areas; strip of larval skin between trunk and hindlimb areas of metamorphosing skin disappeared for greater

part. *Cross section area of largest *lumbar motor column* nuclei of spinal cord 6 × that of average adjacent non-motor nuclei*; *neurocoel* circular in cross section. **Naso-lachrymal duct* opening in skin cranial to eye*. Rudiment of *gland of Harder* formed anterior to eye.

*Appearance of *cartilago obliqua* and *proc. praenasalis superior*; formation of *praemaxillae* and *septomaxillae*; appearance of *dentale* and *outer lobe of goniale**; beginning of reduction of *ethmoid plate* and *flanges*; *arytenoids* chondrified. *M. genio-hyoideus* segregated into pars medialis and pars lateralis. First indications of *occipito-vertebral joint* formation; tissue connecting 1st and 2nd pair of *ribs* with corresponding vertebrae chondrifying; *cranial portion of 13th pair of *vertebral arches* fused with *urostyle**; primary marrow cavities in basal portions of vertebral arches; *chondrification of *dorsal perichordal tube* extending vertebrally down to 12th pair and intervertebrally down to 11th pair of vertebral arches*. *Paired anlage of *sternum* partially procartilaginous*. *Ossification starting in *pubis*; paired *epipubis* procartilaginous*. Complete differentiation of *hindlimb musculature*. 1st and 2nd nephrostome of *pronephros* fused; *Wolffian duct* between pro- and mesonephros showing signs of degeneration.

*Third row of *tooth germs* formed*. *Middle ear anlage* a bilobed vesicle.

STAGE **61**. Age ± 48 d.

Ext. Crit.: Head narrower. Tentacles considerably shortened, mostly curved backwards. *Length of n. olfactorius equal to diameter of bulbus olfactorius*. *4th arterial arch seen just in front of adult skin area of forelimb*. Openings of gill chambers considerably narrowed. *Forelimb at level of posterior half of heart*. Adult skin area on abdomen covering posterior half of heart. *Fins considerably reduced*.

Int. Crit.: *Metamorphosis of *adult skin* areas completed; adult skin area of trunk covering entire brain cavity and caudal portion of heart*. **Spinal gangl.* 8 descended to level well posterior to root of gangl. 9, and nearly all of gangl. 9 posterior to lumbar level of spinal cord*. Finger-like protrusions developed from lateral cavity of *olfactory organ*.

*Appearance of *pars media* and *externa plectri* and *annulus tympanicus* as mesenchymatous condensations*; erosion phenomena in floor and side walls of *cranium* and in *palatoquadratum* and *hyobranchial apparatus*; *angle between *palatoquadratum* and anterior floor of *cranium* 50°*. Reduction of pars lateralis of *m. levator mandibulae anterior*; *m. interhyoideus* attached to ceratohyale and palatoquadratum. *11th pair of *vertebral arches* partially ossified perichondrally*. *Mm. sterno-hyoidei* fused to unpaired anlage. *Degenerative hypertrophy of skin at end of *tail*; atrophy of *fins* and posterior tail *notochord**. *Paired *sternal anlage* partially cartilaginous*. **Epipubis anlagen* fused to Y-shaped cartilage*. 1st nephron of *pronephros* completely disappeared.

First signs of degeneration of *filter apparatus*; *middle ear cavity shifted to position somewhat cranial to eye*, *tuba pharyngo-tympanica* opening into oro-pharyngeal cavity at level of caudal part of eye. *Histolysis in *pyloric part of stomach* and in *ileum**.

STAGE **62**. Age ± 49 d.

Ext. Crit.: *Head still somewhat broader than cranial part of trunk*. Tentacles short, straight. *N. olfactorius shorter than diameter of bulbus olfactorius*. Corner of mouth still in front of eye. Thymus gland somewhat protruding. *3rd arterial arch ("larval aorta") seen at distance of its own diameter in front of adult skin area of forelimb*. *Opening of operculum reduced to curved slit*. *Forelimb at level of middle of heart*. Anterior border of adult skin area on abdomen entirely sharp, mostly nicked medially. Ventral fin disappeared from abdomen.

Int. Crit.: Areas of *larval skin* on head beginning to shrivel. *Tiny *"Stirn-organ"* situated in front of rostral end of brain cavity*. *Tear furrow* leading from opening of naso-lachrymal duct to eye. *Lower *eyelid* forming*.

Dorsal outlets of *for. prooticum* confluent; **proc. muscularis palatoquadrati* and *cartilago tentaculi* disappeared; angle between *palatoquadratum* and anterior floor of *cranium* 65°; appearance of *squamosum* and fusion of inner and outer lobes of *goniale**. Appearance of *m. petrohyoideus*. *Intervertebral cartilage* calcifying in pre-urostyle region; *cartilaginous connection of 3rd pair of *ribs* with vertebral column*. Pars lateralis of *m. ileolumbalis*, and *m. coccygeo-sacralis* formed. *Signs of atrophy in *notochord* over entire length of *tail*; degenerative hypertrophy of mesenchymatous *perichordal tube* and *vertebral arches* of *tail**. Medial ends of cartilaginous sternal anlagen forming *sternal pouches*. *Appearance of *arteria cutanea magna**; disappearance of *arteries of filter apparatus*. *Appearance of *musculo-thyroid veins*; complete disappearance of *lateral veins**; disappearance of *pronephric sinuses* and paired portions of *medial postcardinal veins*; disappearance of *pharyngeal vein*. Of 2nd and 3rd nephron of *pronephros* only nephrostomes remaining; *2nd nephrostome fused with *ostium tubae*; portion of *Wolffian duct* between pro- and mesonephros disappeared*.

Middle ear situated below caudal margin of eye. *Number of *intestinal revolutions* reduced to about 2 in outer and 2 in inner spiral; epithelium of *oesophagus*, *non-pyloric part of stomach*, *duodenum* and *ileum* reconstituted to more or less continuous layer*, that of non-pyloric part of stomach already differentiating.

STAGE **63**. Age ± 51 d.

Ext. Crit.: *Head narrower than trunk*. Tentacles mostly disappeared. *Corner of mouth at level of caudal border of eye*. "Larval aorta" and thymus gland no longer externally visible. *Operculum closed*. Adult skin areas on abdomen and lower jaw separated by narrow band of larval skin. *Forelimb at level of anterior half of heart*. Fin mostly perforated near anus. *Tail still slightly longer than body*.

Int. Crit.: Area of *larval skin* between upper jaw and trunk areas of adult skin disappeared for greater part. *Root of *spinal gangl.* 9 just posterior to gangl. 7*. *Naso-lachrymal duct* opening on tip of papilla.

Disappearance of *proc. muscularis capsulae auditivae*; **for.* metopticum* confluent with *for. caroticum*; disappearance of *proc. cornu-quadratus lateralis, proc. ascendens*, larval *proc. oticus* and medial portion of *comm. quadrato-cranialis anterior*; *pars interna* and *media plectri* cartilaginous; angle between *palatoquadratum* and anterior floor of *cranium* 85°; appearance of *post-palatine commissure* and *pterygoid**; formation of *proc. pterygoideus*; beginning of reduction of *branchial chambers. M. orbito-hyoideus* fused with *m. suspensorio-hyoideus*. *Cartilage of 13th pair of *vertebral arches* and cartilage of anterior *ventral perichordal tube* degenerating**; *hypochord* reaching to 13th vertebra. *In *tail*, chordal epithelium degenerating and entire *notochord* shrivelling; beginning of actual degeneration of *muscle segments* (still 49 post-otic muscle segments distinguishable)*. *Alae sterni* forming*. *Ductus caroticus* disappeared*. *3rd nephrostome of *pronephros* fused with *ostium tubae**.

Operculum a shapeless mass; *gill slits* closed*. *Middle ear* lying against ear capsule; anlage of *tympanic membrane* formed. *Intestinal helix* making only 1½ irregular revolutions*; *typhlosole* extending up to short distance from beginning of colon; *histolysis in *colon**.

STAGE **64**. Age ± 53 d.

Ext. Crit.: Corner of mouth well behind eye. *Various adult skin areas joined almost everywhere, borderlines still clearly visible*. *Length of tail (from anus) ⅓ of body length*.

Int. Crit.: *Root of *spinal gangl.* 7 anterior to gangl. 6*.

Appearance of adult *proc. oticus*; *pars media* and *externa plectri* confluent; angle between *palatoquadratum* and anterior floor of *cranium* 105°*; *branchial chambers* disappeared. Formation of adult *m. levator mandibulae posterior*; *mm. constrictores branchiales, mm. subarcuales recti* and *m. transversus ventralis* disappeared. *Entire cartilage formation of *ventral perichordal tube* converted into loose connective tissue and cartilage of *lateral perichordal tube* in process of resorption*. Degeneration of *notochord* started in *urostyle* region; *only about 14 *postotic muscle segments* distinguishable*. *Paired *sternal anlage* fused*. *4th *aortic arch* (adult aorta) larger than 3rd (larval aorta)*. Disappearance of *branchial veins* and *branchial branches of external jugulars*. *Pronephros* completely disappeared. Primordia of *Müllerian ducts* developing.

Cranial margin of *mouth slit* situated caudal to level of *nostrils*; *thymus gland* moved to position ventro-lateral to ear capsule*. *Epithelium of *pyloric portion of stomach* and of *colon* reconstituted; new glandular elements in *oesophagus, duodenum* and *ileum**.

STAGE **65**. Age ± 54 d.

Ext. Crit.: Borderlines between adult skin areas partly disappeared. *Tail oblong triangular in dorsal aspect, length about $\frac{1}{10}$ of body length*.

Int. Crit.: Only small area of strongly shrivelled larval *skin* present rostral to forelimb. *Spinal gangl.* 8 situated at posterior end of lumbar area of spinal cord*.

*Fusion of adult *proc. oticus* with ear capsule; beginning of ossification in *pars media*, and beginning of chondrification in *pars externa plectri*. *Intervertebral joints formed*; 13th and 14th pair of *vertebral arches* in process of resorption. *Notochord and somites disappeared in *tail* region*. *Posterior end of *epipubis* fused with pelvic girdle*. *Adult *abdominal vein* established*.

Primitive tongue mushroom-shaped in cross section; cranial part forming free tip; *openings and proximal portions of *pharyngo-tympanic ducts* fused*. *Epithelium of *colon* differentiating*.

STAGE **66**. Age ± 58 d.

Ext. Crit.: *Border lines between adult skin areas disappeared*. *Tail only a very small triangle, no longer visible from ventral side*.

Int. Crit.: *Dorsal root of *spinal gangl.* 9 at level of gangl. 7*. *Conjunctival sac* formed.

Plectral apparatus completed and *operculum fen. ovalis* confluent with wall of ear capsule; angle between *palatoquadratum* and anterior floor of *cranium* 115°*; formation of *hyocricoid connection*. Formation of adult *m. levator mandibulae anterior* and *m. depressor mandibulae*; *m. interhyoideus* attached to palatoquadratum only. *Middle portions of *ribs* ossified*. *Tail* only represented by small dorsal swelling of loose connective tissue, covered with degenerating larval skin*. Roofs of *sternal pouches* reduced. *Distal *tarsalia* ossifying*. Reorganization of *mesonephros* not yet completed.

*Dorsal coelomic sacs filled with lymphatic tissue; dorsal horns of *lungs* greatly reduced or absent*. *Typhlosole* disappeared; *intestine* making single loop*.

* * *

CHAPTER VIII

A COMPARATIVE TABLE OF ANURAN NORMAL TABLES

The existing Normal Tables of Anuran development—as far as they could actually be considered as such—have been compared with the Normal Table of *Xenopus laevis* in the following table. This comparison could of course be only an approximate one, since in most of the tables accurate stage criteria are missing and only figures are given.

Intermediate stages have been placed at levels in between the *Xenopus* stages concerned. Where only a rough comparison could be made, the period of development of *Xenopus laevis* with which a certain stage of the table concerned might be comparable, has been given in brace.

The following Normal Tables appear in the table:

Xenopus laevis	NIEUWKOOP & FABER, 1956.
Xenopus laevis	WEISZ, 1945.
Xenopus laevis	BRETSCHER, 1949 (roman figures).
Discoglossus pictus	GALLIEN & HOUILLON, 1951.
Bufo vulgaris	ADLER, 1901.
Bufo arenarum	DEL CONTE & SIRLIN, 1952.
Hyla regilla	EAKIN, 1946.
Rana sylvatica	POLLISTER & MOORE, 1937.
Rana pipiens	SHUMWAY, 1940.
Rana pipiens	TAYLOR & KOLLROS, 1946 (roman figures).
Rana fusca	KOPSCH, 1952.
Rana dalmatina	CAMBAR & MARROT, 1954.
Rana temporaria	MOSER, 1950.

For full bibliographical particulars concerning these Normal Tables the reader is referred to the corresponding section of the "Bibliography of Anuran development" on page 237.

COMPARATIVE TABLE OF ANURAN NORMAL TABLES

Table II

Xen. laevis	Xen. laevis	Disc. pictus	Bufo vulg.	Bufo aren.	Hyla reg.	Rana sylv.	Rana pip.	Rana fusca	Rana dalm.	Rana temp.
1	1+2	1		1+2		1+2	1+2			
2	3	2		3		3	3	1	1	
3	4	3		4		4	4		2	
4	5	4		5		5	5	2	3	
5		5		6		6	6		4	
6	6			7			7		5	
6½		6								
7	7			8		7		3		
		7								
8	} 8					8	8		6	
9				9		9	9			
10										
10¼		8		10		10	10	4	7	
									8	
10½	} 9	9							9	
11				11		11	11	5	10	
11½		10							11	
12	10			12		12	12		12	
									13	
12½	11	} 11						6	14	
13	12	12				13	13		15	
13½				} 13						
14	13	13						7	16	

Xen. laevis	Xen. laevis	Disc. pictus	Bufo vulg.	Bufo aren.	Hyla reg.	Rana sylv.	Rana pip.	Rana fusca	Rana dalm.	Rana temp.
15						14	14			
16		14						} 8	17	
17		} 15		14	15		15		18	
18			1			15				
19			2	15	16					
20						16		9		
21	14		} 3	16			16		19	
22										
23	15	16	4					10		
24	16		5		} 17		} 17			
25				17		} 17			20	
26	17		6							
27		17			} 18			} 11	21	
28										
29/30		18	7		} 19	18	18		22	
31		19		18						
32	18									1
33/34	19	20	8		} 20	19	19	12	23	
35/36		21		19						
37/38	20	22		20		20	20	13	24	2
39		23			21	} 21		} 14	25	3
40		24+25	9	} 21	22		21		26	4
41					23	22	22	15	27	5
		26							28+29	

Xen. laevis	Xen. laevis	Disc. pictus	Bufo vulg.	Bufo aren.	Hyla reg.	Rana sylv.	Rana pip.	Rana fusca	Rana dalm.	Rana temp.
42	21				24		23	16	30	
									31 + 32	
43		27	10	22		23			33	
44		28		23				17		6
									34 + 35	
45	22	29	11				24		36	
		30								7
46				24			I			
		31							37 + 38	
47		32		25			25 II		39	8
								18		9
48									40	10
49		12							41	
50	I / 23						III	19		11 + 12
51	II / III	13					IV / V	20 / 21	42	13
52							VI	22	43	
53	IV						VII / VIII	23	44 / 45	
54	V						IX / X		46	
55	VI						XI / XII	24	47	
56	VII						XIII		48	
57		14					XIV	25		
58							XV / XVI	26	49	
59							XVII		50	
60							XVIII / XIX			
61							XX	27		
62							XXI		51	
63		15					XXII	28	52	
64							XXIII	29	53	
65							XXIV			
66							XXV	30	54	

CHAPTER IX

A BIBLIOGRAPHY OF ANURAN DEVELOPMENT
(SYSTEMATICALLY ARRANGED)

In composing this bibliography we have restricted ourselves mainly to titles dealing with normal development and regeneration. Titles related to abnormal development and developmental physiology have been included only if they are of general interest or elucidate also certain points of normal morphology. In the field covered by this bibliography we have aimed at completeness. We are, however, aware of the fact that some titles may have escaped us, and will appreciate to be informed of supplementary entries.

The bibliography has been arranged in a number of sections, corresponding with the chapters and divisions of the Normal Table. These are preceded by a general section.

As far as *Xenopus* and the Pipidae are concerned, also publications on the adult morphology and anatomy have been included. The titles of publications in which is dealt with the genus *Xenopus* have been marked with an asterisk (*).

In the section on Normal Tables also a number of Urodelan Normal Tables is listed for comparative purposes.

An attempt to compile all the existing literature relating to *Xenopus laevis* has been made by H. ZWARENSTEIN, N. SAPEIKA and H. A. SHAPIRO in: "*Xenopus laevis*, a bibliography", 1946, The African Bookman, Cape Town.

(A supplement to this list has been issued in 1955 by H. ZWARENSTEIN and A. C. J. BURGERS, Medical Library, Univ. of Cape Town.)

BIBLIOGRAPHY

GENERAL SUBJECTS

Development and regeneration (general publications)

ALCOCK, TH., 1884. On the development of the common frog. Mem. Manch. Lit. Philos. Soc., 3rd Ser., **8**.

BAMBEKE, CH. VAN, 1868. Recherches sur le développement du Pélobate brun (Pelobates fuscus Wagl.). Mém. des concours et des savants étrangers de l'Acad. Roy. Belgique, **34**.

*BLES, E. J., 1905. The life-history of Xenopus laevis Daud. Trans. Roy. Soc. Edinburgh, **41**, 789 ff.

——, 1907. Notes on Anuran development: "Paludicola", "Hemisus" and "Phyllomedusa". The Budgett Memorial Volume, 443–458.

BONJOUR, A. E., 1929. Notas sobre la embryologia de algunos Batracios Sudamericanos (Fam. Leptodactylidae). Arch. Soc. Biol. Montevideo, **1**, 385–395.

BRACHET, A., 1902. Recherches sur l'ontogénèse des Amphibiens Urodèles et Anoures (Siredon pisciformis—Rana temporaria) Arch. Biol., **19**. 1–243.

BRAUS, H., 1904. Einige Ergebnisse der Transplantation von Organanlagen bei Bombinatorlarven. Verhandl. Anat. Ges. Jena, **18**, 53–66.

CAMERANO, L. and M. LESSONA, 1890. Ricerche intorno allo sviluppo ed alle cause del polimorfismo dei girini degli Anfibii anuri. Atti Accad. Sci. Torino, 1890.

*CHANG, C. Y., 1953. Parabiosis and gonad transplantation in Xenopus leavis Daudin. Jour. Exptl. Zool., **123**, 1–28.

*CRÉZÉ, J., 1950. Action de la colchicine sur les têtards de Xenopus laevis. Compt. Rend. Soc. Biol., **144**, 1317–1318.

DEAL, R. E., 1931. The development of sex characters in the tree frog. Anat. Rec., **48**, 27–45.

*DREYER, T. F., 1914. The morphology of the tadpole of Xenopus laevis. Trans. Roy. Soc. S. Africa, **4**, 241 ff.

GASSER, E., 1882. Zur Entwicklung von Alytes obstetricans. Sitzungsber. naturwiss. Ges. Marburg, **5**, 73–104.

GÖTTE, A., 1869. Untersuchungen über die Entwicklung des Bombinator igneus. Arch. Entwicklungsmech., **5**, 90–125.

——, 1875, Die Entwicklungsgeschichte der Unke, Bombinator igneus, Rösel. Leipzig, 1875.

HAMBURGER, V., 1947. A manual of experimental embryology. (2nd impr.) Chicago (contains the figures of the Normal Tables of SHUMWAY, 1940 and POLLISTER and MOORE, 1937, and figures after HARRISON's unpublished series by S. E. SCHWEICH).

HÉRON-ROYER, L. F., 1885. Note sur les amours, la ponte et le développement du Discoglosse (Discoglossus pictus Otth.). Bull. Soc. Zool. France, **10**, 565 ff.

HINCKLEY, M. H., 1882. The development of the Tree-toad. Amer. Nat., **16**.

——, 1884. Notes on the development of Rana sylvatica. Proc. Boston Soc. Nat. Hist., **22**.

HOSKINS, E. R. and M. M. HOSKINS, 1920. The inter-relation of the thyroid and hypophysis in the growth and development of frog larvae. Endocrinology, **4**, 1 ff.

IKEDA, S., 1897. Notes on the breeding habits and development of Rhacophorus Schlegeli Günther. Annot. Zool. Japan, **1**, 113 ff.

LECAMP, M., 1952. Régénération et métamorphose chez un batracien anoure, Alytes obstetricans. Compt. Rend. Acad. Sci. Paris, **235**, 1699–1700.

MORGAN, T. H., 1891. Some notes on the breeding habits and embryology of frogs. Amer. Nat., **25**.

————, 1897. The development of the frog's egg.

————, 1904. Die Entwickelung des Froscheies. Leipzig.

NOKA, H., 1930. Experimentelle Untersuchungen über die Entstehung des Situs viscerum et cordis bei Anurenlarven. Japanese Jour. Med. Sci. (Anat.), **2**, 201–216.

*PARKER, ROBBINS and LOVERIDGE, 1947. Breeding, rearing and care of the South-African clawed frog. Amer. Nat., **81**.

PISANÒ, A., 1939. Autoriparazione di giovani larve di Discoglossus pictus cui è stata asportata la regione ventrale posteriore. Giorn. Sci. Nat. ed Econ. Palermo, **40**.

PRESSLER, K., 1912. Beobachtungen und Versuche über den normalen und inversen Situs viscerum et cordis bei Anurenlarven. Arch. Entwicklungsmech., **32**.

*ROSTAND, J., 1932. De la faculté régénératrice chez Xenopus laevis. Compt. Rend. Soc. Biol., **111**, 451 ff.

RUFFINI, A., 1907. Contributo alla conoscenza della ontogenesi degli Anfibi anuri ed urodeli, nota 1a. Arch. Ital. Anat. e Embriol., **6**, 129–155.

————, 1907. Contributo alla conoscenza della ontogenesi degli Anfibi anuri ed urodeli, nota 2a. Anat. Anz., **31**, 448–472.

RUGH, R., 1948. Experimental embryology, a manual of techniques and procedures (revised ed.). Minneapolis (contains the figures of the Normal Tables of SHUMWAY, 1940, MUELLER, TAYLOR and KOLLROS, 1946, POLLISTER and MOORE, 1937, WEISZ, 1945, EAKIN, 1947, ANDERSON, 1943 and TWITTY and BODENSTEIN, and figures after HARRISON's unpublished series by N. LEAVITT).

————, 1951. The frog, its reproduction and development (Rana pipiens). Philadelphia, 1951.

RUSCONI, D. M., 1826. Développement de la grenouille commune. Tome I, Milano.

SAMPSON, L. V., 1904. A contribution to the embryology of "Hylodes Martinicensis." Amer. Jour. Anat., **3**, 473–504.

*SCHAUINSLAND, H., 1890. Die Entwicklung von Xenopus capensis. Verhandl. Ges. Deutscher Naturf. u. Ärzte, **63**, 135 ff.

SIEDLECKI, M., 1908. Ueber Bau, Lebensweise und Entwicklung des javanischen Flugfrosches (Rhacophorus reinwardti Boie). Bull. Internatl. Acad. Cracov, 1908, 682 ff.

SPENCER, B., 1885. Some notes on the development of Rana temporaria. Quart. Jour. microscopical Sci., **25**, suppl.

STEINHEIM, 1820. Die Entwicklung der Frösche. Hamburg.

————, 1846. Die Entwicklung des Froschembryos, insbesondere des Muskel- und Genitalsystems. Abhandl. Nat. Wiss., Hamburg, 1846.

STEPHENSON, N. G., 1951. Observations on the development of the Amphicoelous frogs, Leiopelma and Ascaphus. Jour. Linn. Soc. Zool., **42**, 18 ff.

STRICKER, S., 1860. Entwicklungsgeschichte von Bufo cinereus bis zum Erscheinen der äusseren Kiemen. Sitzungsber. Akad. Wiss. Wien, math.-nat. Kl.

VOGT, C., 1842. Untersuchungen über die Entwicklungsgeschichte der Geburtshelferkröte (Alytes obstetricans). Solothurn, 134 pp.

WANG, C. H., 1941. Development of swimming and righting reflexes in the frog (Rana guentheri): effects thereon of transection of central nervous system before hatching. Jour. Neurophysiol., **4**, 137–146.

*WEISZ, P. B., 1945a. The development and morphology of the larva of the South-African clawed toad, Xenopus laevis. I. Jour. Morphol., **77**, 163–192.

*————, 1945b. The development and morphology of the larva of the South-African clawed toad, Xenopus laevis. II. Jour. Morphol., **77**, 193–217.

WESTENHÖFER, 1929. Zur Morphogenese der Anuren. Sitzungsber. Ges. Naturf. Fr. Berlin.

YOUNGSTROM, K. A., 1938. Studies on the developing behavior of Anura. Jour. Comp. Neurol., **68**, 351–379.

Metamorphosis (general publications)

ADOLPH, E. F., 1931. Body size as a factor in the metamorphosis of tadpoles. Biol. Bull., **61**, 376–386.

ANGLAS, J., 1904. Observations sur les métamorphoses internes des batraciens anoures. Assoc. Franc. Avanc. Sci., **33**, 855 ff.

BARFURTH, D., 1886. Experimentelle Untersuchungen über die Verwandlung der Froschlarven. Anat. Anz., **1**.

————, 1887. Versuche über die Verwandlung der Froschlarven. Arch. microsc. Anat., **29**.

BATAILLON, E., 1891. Recherches anatomiques et expérimentales sur la métamorphose des Amphibiens anoures, Ann. Univ. Lyon, **2**, 1 ff.

DENT, J. N. and E. L. HUNT, 1952. An autoradiographic study of iodine distribution in larvae and metamorphosing specimens of Anura. Jour. Exptl. Zool., **121**, 79–97.

DUESBERG, J., 1906. Contribution à l'étude des phénomènes histologiques de la métamorphose chez les Amphibiens anoures. Arch. Biol., **22**, 163–228.

EBERTH, C. J., 1894. Die Sarkolyse (Froschlarve). Festschr. 200-jähr. Jubelf. Univ. Halle.

ETKIN, W., 1932. Growth and resorption phenomena in anuran metamorphosis. Physiol. Zool., **5**.

*GASCHE, P., 1944. Beginn und Verlauf der Metamorphose bei Xenopus laevis Daud. Festlegung von Umwandlungsstadien. Helvetica Physiol. et Pharmacol. Acta, **2**, 607–626.

GEIGY, R., 1937. Entwicklungsphysiologische Untersuchungen über die Anurenmetamorphose. I. Verhandl. Schweiz. Naturf. Gesellsch. Genf., 1937.

HELFF, O. M., 1932. Studies on amphibian metamorphosis. X. Hydrogen-ion concentration of the blood of anuran larvae during involution. Biol. Bull., **63**, 405–418.

KREMER, J., 1927. Die Metamorphose und ihre Bedeutung für die Zellforschung. II. Amphibia. Zeitschr. microsc. anat. Forsch., **9**, 99–233.

LECAMP, M., 1952. Régénération et métamorphose chez un batracien anoure, Alytes obstetricans. Compt. Rend. Acad. Sci. Paris, **235**, 1699–1700.

MERCIER, L., 1906. Les processus phagocytaires pendant la métamorphose des batraciens anoures et des insects. Arch. Zool. Exptl. et Gén., **4**, 1–151.

MORSE, W., 1918. Factors involved in the atrophy of the organs of the larval frog. Biol. Bull., **34**, 149.

SATO, K., 1924. Ueber die Metamorphose von Bufo vulgaris japonica. Zeitschr. Anat. Entwicklungsgesch., **71**.

External development and hatching

ADLER, W., 1901. Die Entwicklung der äusseren Körperform und des Mesoderms bei Bufo vulgaris. Internatl. Monatschr. Anat. Physiol., **18**.

ASSHETON, P., 1894. On the growth in length of the frog embryo. Quart. Jour. microsc. Sci., **37**, 223–243.

*BLES, E. J., 1902. Exhibition of and remarks upon some living tadpoles of Xenopus laevis. Proc. Zool. Soc. London, **2**, 79 ff.

BONJOUR, A. E., 1931. Sobre el desarollo de la forma externa de Algunos Leptodactilidos Argentinos. Bol. Acad. Nac. Cienc., **31**, 141–169.

BOULENGER, G. A., 1881. Les larves des genres Pipa et Dactylethra. Bull. Soc. Zool. France, **6**, 27 ff.

————, 1891. A synopsis of the tadpoles of the European Batrachians. Proc. Zool. Soc. London, 1891, 593–627.

CAMBAR, R., 1953. Mise en évidence d'enzymes "de l'éclosion" chez la grenouille agile: intérêt de leur utilisation en embryologie expérimentale. Compt. Rend. Acad. Sci., **237**, 355–357.

———— and R. WILLIAUME, 1954. Recherches expérimentales sur le méchanisme de l'éclosion des embryons de grenouille. Compt. Rend. Soc. Biol., **148**, 112–114.

ESPOSITO SEU MARGHERITA, M., 1950. Gli enzimi della schiusa negli Anfibi. Boll. Zool., **17**, 105–107.

FERNANDEZ–MARCINOWSKI, K., 1921. Der Mechanismus des Schlüpfens bei den Amphibienlarven. Biol. Zentralbl., **41**, 9 ff.

HÉRON–ROYER, L. F., 1878. Des nuances diverses des têtards de Batraciens anoures et des causes qui les produisent. Bull. Soc. Zool. France, 1878.

————, 1887. Observations sur le développement externe et l'état adulte des Batraciens du genre Bombinator. Bull. Soc. Zool. France, **12**, 640 ff.

JOENSCH, P. A., 1921. Beobachtungen über das Auskriechen der Larven von" Rana arvalis" und "fusca" und die Funktion des Stirndrüsenstreifens. Anat. Anz., **53**, 567–582.

LATASTE, F., 1877. Quelques observations sur les tetards des Batraciens Anoures. Bull. Soc. Zool. France, **2**, 281 ff.

NOBLE, G. K., 1926. The hatching process in Alytes, Eleutherodactylus and other Amphibians. Amer. Mus. Novitates, **249**.

PETER, K., 1929. Die Kaulquappen von Rana fusca und Rana arvalis. Anat. Anz., **68**.

*————, 1931. The development of the external features of Xenopus laevis, based on material collected by the late E. J. BLES. Jour. Linn. Soc. Zool., **37**, 515–523.

SMITH, M. A., 1924. Descriptions of Indian and Indo-Chinese tadpoles. Rec. Indian Mus., **26**.

————, 1927. The tadpoles of Tylototriton verrucosus Anderson. Rec. Indian Mus., 1927.

WRIGHT, A. H. and A. WRIGHT, 1923. The tadpoles of the frogs of Okefinokee Swamp. Anat. Rec., **24**.

Head (all organ systems)

BRACHET, A., 1907. Recherches sur l'ontogénèse de la tête chez les Amphibiens. Arch. Biol., **23**, 165–257.

CORNING, H. K., 1899. Ueber einige Entwicklungsvorgänge am Kopfe der Anuren. Morphol. Jahrbuch, **27**, 173–241.

FAHRENHOLZ, C., 1925. Ueber die Entwicklung des Gesichtes und der Nase bei der Geburtshelferkröte (Alytes obstetricans). Morphol. Jahrbuch, **54**, 421—503.

HALTER, S., 1936. La morphogénèse de la face, étudiée comparativement chez trois Anoures. Ann. Soc. Roy. Zool. Belgique, **67**, 75–86.

*PATERSON, N. F., 1939. The head of Xenopus laevis. Quart. Jour. Microsc. Sci., **81**, 161–230.

*RAMASWAMI, L. S., 1941. Some aspects of the head of Xenopus laevis. Proc. Indian Sci. Congr., **28**, 183 ff.

WEBER, A., 1909. Recherches sur la régénération de la tête chez les larves de Discoglossus pictus. Compt. Rend. Assoc. Anat., **11**, 18–20.

Miscellaneous

BEDDARD, F. E., 1895. On some points in the anatomy of Pipa americana. Proc. Zool. Soc. London, 1895, 827 ff.

*BONDE, C. VON and D. B. SWART, 1926. The structure of the Plathander (Xenopus laevis). Part I. Trans. Roy. Soc. S. Africa, **17**, XIII.

*DREYER, T. F., 1913. The "Plathander" (Xenopus laevis). Trans. Roy. Soc. S. Africa, **3**, 341–355.

*FLOWER, S. S., 1936. Further notes on the duration of life in animals. II. Amphibians. Proc. Zool. Soc. London, 369–394.

*GROBBELAAR, K., 1924. Beiträge zu einer anatomischen Monographie von Xenopus laevis (Daud.). Zeitschr. Anat. u. Entwicklungsgesch., **72**, 131–168.

LATTER, O. H., 1923. An overlooked feature in the four-legged tadpole of Rana temporaria. Nature, **111**, 151 ff.

*LEHMAN, H. E., 1953. Observations on macrophage behavior in the fin of Xenopus larvae. Biol. Bull., **105**, 490–495.

MONTALENTI, G., 1932. Sull'embriogenesi degli ibridi fra bufo vulgaris e bufo viridis. Rend. R. Acacd. Naz. Lincei, **15**, ser. 6a, 994–1000.

*MORON, Y., 1947. Observations sur Xenopus tropicalis Gray. Bull. Soc. Zool. France, **72**, 128 ff.

NOBLE, G. K., 1931. The biology of the Amphibia. New York.

TAXONOMIC POSITION AND GEOGRAPHICAL DISTRIBUTION OF XENOPUS

*BLANCHARD, R., 1885. Remarques sur la classification des Batraciens anoures. Bull. Soc. Zool. France, **10**, 584–589.

*BOULENGER, G. A., 1882. Catalogue of the Batrachia Salientia, British Museum, **2**.

*COPE, E. D., 1865. Sketch of the primary groups of Batrachia Salientia. Nat. Hist. Rev., **5**, 97–120.

*———, 1889. The Batrachia of North America. Bull. U.S. Nat. Mus., **34**.

*DAUDIN, E. M., 1802. Histoire naturelle des rainettes, des grenouilles et des crapauds.

*DUMÉRIL, A., 1859. Reptiles et poissons de l'Afrique occidentale. Arch. Mus. Paris, **10**.

*DUMÉRIL, C. and G. BIBRON, 1841. Erpétologie Général, **8**.

*DUNN, E. R., 1948. American frogs of the family Pipidae. Amer. Mus. Novitates, **1384**, 1–13.

*GRAY, J. E., 1825. A synopsis of the genera of Reptiles and Amphibia, with a description of some new species. Ann. Philos., **10**.

*GÜNTHER, A., 1858. Catalogue of the Batrachia Salientia, British Museum.

*LATASTE, F., 1879. Etude sur le Discoglosse. Act. Soc. Linn., Bordeaux, **33**, 275–342.

*MERREM, B., 1820. Versuch eines Systems der Amphibien.

*MILLARD, N., 1945. The development of the arterial system of Xenopus laevis, including experiments on the destruction of the larval aortic arches. Trans. Roy. Soc. S. Africa, **30**, 217–234.

*———, 1949. The development of the venous system of "Xenopus laevis". Trans. Roy. Soc. S. Africa, **32**, 55–97.

*MIRANDA RIBEIRO, A. DE, 1926. Notas para serviren ao estudo dos Gymnobatrachos (Anura) Brasileiros. Arch. Mus. Nac. Rio de Janeiro, **27**, 1–227.

*MIVART, ST. G., 1869. On the classification of the Anurous Batrachians. Proc. Zool. Soc. London, 1869, 280–295.

*MONARD, A., 1951. Résultats de la Mission Zoologique Suisse au Cameroun. Mém. Inst. Français d'Afrique noire, Centre de Cameroun, 1951.

*NICHOLLS, G. E., 1916. The structure of the vertebral column in the Anura Phaneroglossa and its importance as a basis of classification. Proc. Linn. Soc. London, **128**, 80–92.

*NOBLE, G. K., 1922. The phylogeny of the Salientia. I. Bull. Amer. Mus. Nat. Hist., **46**, 3–87.

*————, 1931. The biology of the Amphibia. New York.

*PARKER, H. W., 1932. Scientific results of the Cambridge Expedition to the East African lakes, 1930–1; V, Reptiles and Amphibians. Jour. Linn. Soc. London, **38**, 215 ff.

*————, 1936. Reptiles and Amphibians collected by the Lake Rudolf Rift Valley Expedition, 1934. Ann. Mag. Nat. Hist., **18**, 596–601.

*PATERSON, N. F., 1946. The skull of Hymenochirus curtipes. Proc. Zool. Soc. London, **115**, 327–354.

*————, 1949. The development of the inner ear of Xenopus laevis. Proc. Zool. Soc. London, **119**, 269–291.

*————, 1951. The nasal cavities of the toad Hemipipa carvalhoi and other Pipidae. Proc. Zool. Soc. London, **121**, 381–415.

*PLETZEN, R. VAN, 1953. Ontogenesis and morphogenesis of the breast-shoulder apparatus of Xenopus laevis. Ann. Univ. Stellenbosch, **29A**, 137–184.

*SLABBERT, G. K. and W. A. MAREE, 1945. The cranial morphology of the Discoglossidae and its bearing upon the phylogeny of the primitive Anura. Ann. Univ. Stellenbosch, **23**, 91–97.

*STEINDACHNER, A., 1867. Reise Novara, Amphibien.

*TSCHUDI, J. J. VAN, 1838. Classification der Batrachien mit Berücksichtigung der fossilen Thiere dieser Abteilung der Reptilien.

*VILLIERS, C. G. S. DE, 1924. On the anatomy of the breast-shoulder apparatus of Xenopus. Ann. Transv. Mus., **10**, 197–211.

*————, 1929. The comparative anatomy of the breast-shoulder apparatus of the three aglossal anuran genera Xenopus, Pipa and Hymenochirus. Ann. Trans. Mus., **13**, 37–69.

*————, 1934. Studies on the cranial anatomy of Ascaphus truei Stejneger, the American Liopelmid. Bull. Mus. Comp. Zool. Harvard, **77**, 1–38.

*WAGLER, 1827. (Footnote to letter from H. BOIE) Isis, **20**, 726 ff.

ECOLOGY AND METHODS OF REARING

Influence of external factors and food

DEEN, R., 1953. Der Einfluss verschiedener Belichtungszeiten auf Hypophyse, Schilddrüse und Larvenentwicklung von Rana temporaria. Zool. Jahrb. Abt. 3 Allg. Zool. u. Physiol. Tiere, **63**, 477–500.

DOMS, H., 1916. Ueber den Einfluss der Temperatur auf Wachstum und Differenzierung der Organe während der Entwicklung von Rana esculenta. Arch. mikroskopische Anat., **87**.

EECKHOUT, J. P. V. D., 1947. Sur l'influence de la lumière sur la métamorphose de Rana temporaria. Ann. Soc. Roy. Zool. Belgique, **77**, 78–82.

GUDERNATSCH, J. F., 1912. Feeding experiments on tadpoles. I. The influence of specific organs given as food on growth and differentiation. Arch. Entwicklungsmech., **35**.

————, 1914. Feeding experiments on tadpoles. II. Amer. Jour. Anat., **15**.

*TOIVONEN, S., 1952a. Ein Fall von partieller Neotenie bei Xenopus laevis Daudin und experimentelle Untersuchungen zu seiner kausalen Erklärung. Arch. Soc. Zool. Bot. Fennica "Vanamo", **6**, 107–123.

*————, 1952b. Insufficiency of light as a cause of partial neoteny in Xenopus laevis. Acta Endocrinol., **10**, 243–254.

Life history; natural and artificial breeding

*ALEXANDER, S. S. and C. W. BELLERBY, 1935. The effect of captivity upon the reproductive cycle of the South African clawed toad (Xenopus laevis). Jour. Exptl. Biol., **12**, 306–314.

*———— and ————, 1938. Experimental studies on the sexual cycle of the South African clawed toad (Xenopus laevis). I. Jour. Exptl. Biol., **15**, 74–81.

*ANDRES, G., A. BRETSCHER, F. E. LEHMANN and D. ROTH, 1948. Einige Verbesserungen in der Haltung und Aufzucht von Xenopus laevis. Experientia, **5**, 1–3.

*ARONSON, L. R., 1944. Breeding Xenopus laevis. Amer. Nat., **78**, 131–141.

*BELLERBY, C. W., 1938. Experimental studies on the sexual cycle of the South African clawed toad (Xenopus laevis). II. Jour. Exptl. Biol., **15**, 82–90.

*————, and L. HOGBEN, 1938. Experimental studies on the sexual cycle of the South African clawed toad (Xenopus laevis). III. Jour. Exptl. Biol., **15**, 91–100.

*BERK, L., 1938. Studies in the reproduction of Xenopus laevis. I. The relation of external environmental factors to the sexual cycle. S. Afr. Jour. Med. Sci., **3**, 72–77.

*————, 1939. Studies in the reproduction of Xenopus laevis. III. The secondary sex characters of the male Xenopus laevis: the pads. S. Afr. Jour. Med. Sci., **4**, 47–60.

*————, and H. A. SHAPIRO, 1939. Studies in the reproduction of Xenopus laevis. II. The histological changes in the accessory sex organs of female Xenopus induced by the administration of endocrine preparations. S. Afr. Jour. Med. Sci., **4**, 13–17.

*BLES, E. J., 1901. On the breeding habits of Xenopus laevis, Daud. Proc. Camb. Phil. Soc., **11**, 220–222.

*————, 1905. The life-history of Xenopus laevis Daud. Trans. Roy. Soc. Edinburgh, **41**, 789 ff.

*CAMERON, S. B., 1947. Successful breeding of Xenopus laevis, the South African clawed toad. Amer. Jour. Med. Techn., May 1947.

*DEANESLY, R. and A. S. PARKES, 1945. The preparation and biological effects of iodinated proteins. VIII. Use of Xenopus tadpoles for the assay of thyroidal activity. Jour. Endocrinol, **4**, 324–355.

*DREYER, T. F., 1913. The "Plathander" (Xenopus laevis). Trans. Roy. Soc. S. Africa, **3**, 341–355.

*ELKAN, E. R., 1939. Xenopus laevis, Daudin. Problems and observations. Aquarist and Pondkpr., **9**, 95–100.

*GASCHE, P., 1943. Die Zucht von Xenopus laevis Daudin und ihre Bedeutung für die biologische Forschung. Rev. Suisse Zool., **50**, 262–269.

HÉRON-ROYER, L. F., 1885. Note sur les amours, la ponte et le développement du Discoglosse (Discoglossus pictus Otth.). Bull. Soc. Zool. France, **10**, 565 ff.

HONIKMAN, S., H. A. SHAPIRO and H. ZWARENSTEIN, 1935. Variations in the ovarian response of Xenopus to the gonadokinetic principle of the anterior pituitary. Trans. Roy. Soc. S. Africa, **23**, XVIII.

IKEDA, S., 1897. Notes on the breeding habits and development of Rhacophorus Schlegeli Günther. Annot. Zool. Japan, **1**, 113 ff.

*LANDGREBE, F. W. and G. L. PURSER, 1941. Breeding of Xenopus in the laboratory. Nature, London, **148**, 115.

*LESLIE, J. M., 1890. Notes on the habits and oviposition of Xenopus laevis. Proc. Zool. Soc. London, 1890, 69–71.

MORGAN, T. H., 1891. Some notes on the breeding habits and embryology of frogs. Amer. Nat., **25**.

*OCHSÉ, W., 1948. Die Zucht des südafrikanischen Krallenfrosches Xenopus laevis Daudin. Gynaecologia, **126**, 57–77.

*PARKER, ROBBINS and LOVERIDGE, 1947. Breeding, rearing and care of the South-African clawed frog. Amer. Nat., **81**.

SAVAGE, R. M., 1938. The ecology of tadpoles etc. and the function of the envelope. Proc. Zool. Soc. London, **108**, 465 ff.

*SHAPIRO, B. G. and H. A. SHAPIRO, 1934. Histological changes in the ovaries and ovarian blood vessels of Xenopus laevis associated with hypophysectomy, captivity and the normal reproductive cycle. Jour. Exptl. Biol., **11**, 73–80.

*SHAPIRO, H. A., 1935. Experimental induction of coupling in Xenopus laevis, with the production of fertilized eggs. Nature, London, **135**, 510 ff.

SIEDLECKI, M., 1908. Ueber Bau, Lebensweise und Entwicklung des javanischen Flugfrosches (Rhacophorus reinwardti Boie). Bull. Internatl. Acad. Cracov, 1908, 682 ff.

*VAILLANT, L., 1908. La réproduction des Xenopus laevis Daudin à la ménagerie du Muséum d'Histoire naturelle. Bull. Mus. Hist. Nat., Paris, 1908, 203–204.

*VANDERPLANK, F. L., 1935. The effects of pH on breeding. Aquarist and Pondkpr., **6**, 135–136.

*———, 1939. Xenopus laevis: light as stimulus to breeding. Aquarist and Pondkpr., **9**, 183–184.

*WEISMAN, A. J., and C. W. COATES, 1944a. Ovarian activity of Xenopus laevis while in captivity. Amer. Nat., **78**, 383–384.

———— and ————, 1944b. The South-African frog in pregnancy diagnosis. A research bulletin. New York, 1944.

EARLY DEVELOPMENT

General publications

ADLER, W., 1901. Die Entwicklung der äussere Körperform und des Mesoderms bei Bufo vulgaris. Internatl. Monatschr. Anat. Physiol., **18**.

BAUTZMANN, H., 1932. Experimentelle Analyse des organisatorischen Geschehens in der Primitiventwicklung von Amphibien: Determinationszustand und Aufgabenverteilung der Randzonenanlagen im Organisationsprozess. Verhandl. Anat. Ges., Vers. 41, Lund.; Erg. H. zum Anat. Anz., **75**.

———, 1933. Ueber Determinationsgrad und Wirkungsbeziehungen der Randzonenteilanlagen (Chorda, Ursegmente, Seitenplatten und Kopfdarmanlage) bei Urodelen und Anuren. Arch. Entwicklungsmech., **128**, 665–765.

BECCARI, N., 1924. Studi sulla prima origine delle cellule genitali nei Vertebrati. III: Ricerche nel Bufo viridis. Arch. Ital. Anat. e Embriol., **21**, 352–374.

BOUNOURE, L., 1929. Sur un caractère cytologique essentiel des gonocytes primaires chez la Grenouille rousse. Compt. Rend. Soc. Biol., **101**.

———, 1939. L'origine des cellules reproductrices et le problème de la lignée germinale. Paris.

BYTINSKI-SALZ, H., 1929. Untersuchungen über die Determination und die Induktionsfähigkeit einiger Keimbezirke der Anuren. Arch. Entwicklungsmech. **118**, 121–163.

HÉRON-ROYER, L. F., 1879. Note sur l'oeuf et la première periode embryonnaire du Pélodyte ponctué (Pelodytes punctatus, Dugès). Bull. Soc. Zool. France, 1879.

HOLTFRETER, J., 1943. Properties and functions of the surface coat in amphibian embryos. Jour. Exptl. Zool., **93**, 251–323.

KING, H. D., 1902. Experimental studies on the formation of the embryo of Bufo lentiginosus. Arch. Entwicklungsmech., **13**, 545–564.

MARCHETTI, L., 1918. Sui primi momenti dello sviluppo di alcuni organi primitivi nel germe di Bufo vulgaris. Arch. Ital. Anat. e Embriol., **16**, 175–254.

————, 1914. Sui primi momenti dello sviluppo di alcuni organi primitivi nel germe di Bufo vulgaris. Seconda nota preventiva. Anat. Anz., **47**, 496–508, 524–539.

MONROY, A., 1939. Sulla localizzazione delle cellule genitali primordiali in fasi precoci di sviluppo. Ricerche sperimentali in Anfibi anuri. Arch. Ital. Anat. e Embriol., **41**, 368–389.

MOQUIN, T., 1876. Recherches sur les premières phases du développement des Batraciens Anoures. Ann. Sci. Nat. Zool., 1876.

MORGAN, T. H., 1894. The formation of the embryo of the frog. Anat. Anz., 1894, 697–705.

RONDININI, R., 1928. Particolarità formative in alcuni organi primitivi e sviluppo della coda nelle larve di Bufo vulgaris. Arch. Ital. Anat. e Embriol., **25**, 98–130.

ROUX, W., 1902. Bemerkungen über die Achsenbestimmung des Froschembryos und die Gastrulation des Froscheies. Arch. Entwicklungsmech., **14**, 600–624.

SCHOTTÉ, O., 1931. Transplantationsversuche über die Determination der Organanlagen von Anurenkeimen. I. Allgemeines und Technik der Transplantation. Arch. Entwicklungsmech., **123**, 179–205.

TAKASHIMA, R., 1922. Ein Beitrag zur Abstammung der primordialen Keimzellen bei den anuren Amphibien. Folia Anat. Japonica, 1922 (in Japanese).

WINTREBERT, P., 1931a. Analyse du développement de Discoglossus pictus Otth. par le procédé des marques colorées. Mode d'extension de la face ventrale. Origine des feuillets de la lèvre caudale. Compt. Rend. Soc. Biol., **107**, 1214–1218.

————, 1931b. Analyse du développement de Discoglossus pictus Otth. par le procédé des marques colorées. La destinée de bandes colorées placées sur le méridian frontal. Compt. Rend. Soc. Biol., **107**, 1443–1446.

Initial development

BAMBEKE, M. VAN, 1896. Sur une groupement de granules pigmentaires dans l'oeuf en segmentation d'Aphibiens anoures. Bull. Acad. Roy. Belgique. Cl. Sci., **31**.

BOUNOURE, L., 1930. Sur la présence de cellules germinales distinctes dans la blastula de la Grenouille rousse. Compt. Rend. Acad. Sci., **190**, 1143–1145.

————, 1931a. A quel moment peut-on parler de cellules germinales initiales dans le développement de la Grenouille rousse? Compt. Rend. Soc. Biol., **107**, 988–991.

————, 1931b. Sur l'existence et la localisation dans les premiers blastomères de la Grenouille d'un matériel cytoplasmique distinct destiné au germen. Compt. Rend. Soc. Biol., **107**, 991–993.

————, 1931c. Sur la nature golgienne d'un élément cytoplasmique caractéristique du germen dans les premiers stades du développement de la Grenouille. Compt. Rend. Acad. Sci., **193**, 297–299.

————, 1931d. Sur l'existence d'un déterminant germinal dans l'oeuf indivi de la Grenouille rousse. Compt. Rend. Acad. Sci., **192**, 402–404.

————, 1934. Recherches sur la lignée germinale chez la Grenouille rousse aux premiers stades du développement. Ann. Sci. Nat. Zool., **17**, 69–245.

————, 1935. Sur la possibilité de réaliser une castration dans l'oeuf de la Grenouille rousse; résultats anatomiques. Compt. Rend. Soc. Biol., **120**, 1316–1319.

BRACHET, A., 1923. Recherches sur les localisations germinales et leur propriétés ontogénétiques dans l'oeuf de Rana fusca. Arch. Biol., **33**, 343–430.

————, 1927. Etude comparative des localisations germinales dans l'oeuf des Amphibiens urodèles et anoures. Arch. Entwicklungsmech., **111**, 250–291.

DALCQ, A., 1937. Sur l'existence de foyers pronéphretiques et d'autres "prédispositions" organogènes dans la blastula et la jeune gastrula du Discoglosse. Compt. Rend. Soc. Biol., **124**, 833 ff.

————, 1939. Exploration du plancher du blastocèle et de la zone marginale chez le Discoglosse. Evolution de ces ébauches en situation anormale. Compt. Rend. Soc. Biol. Belg., **131**, 781 ff.

GIERSBERG, H., 1924. Beiträge zur Entwicklungsphysiologie der Amphibien. I. Furchung und Gastrulation bei Rana und Triton. Arch. Entwicklungsmech., **103**, 368–386.

HÉRON-ROYER, L. F., 1885. Note sur les amours, la ponte et le développement du Discoglosse (Discoglossus pictus Otth.). Bull. Soc. Zool. France, **10**, 565 ff.

KING, H. D., 1908. The ovogenesis of Bufo lentiginosus. Jour. Morphol. **19**, 369–438.

PASTEELS, J., 1932. Etude des localisations germinales de l'oeuf insegmenté des Amphibiens anoures. Arch. Biol., **43**, 521–574.

ROUX, W., 1888. Ueber die Lagerung des Materials des Medullarrohres im gefurchten Froschei. Verhandl. Anat. Ges., 1888.

SCHECHTMAN, A. M., 1938. Localization of the neural inductor and tail mesoderm in a frog egg (Hyla regilla). Proc. Soc. Exptl. Biol. and Med., **39**, 236–239.

SMRECZYNSKI, ST., 1928. La position de l'embryon virtuel dans l'oeuf de Rana fusca. Compt. Rend. Assoc. Anat., **23**, 426–430.

VINTEMBERGER, P., 1934. Sur la situation de l'ébauche chordale présomptive avant le creusement du blastocèle, dans l'oeuf de Rana fusca. Compt. Rend. Soc. Biol., **115**, 852 ff.

WARING, H., F. W. LANDGREBE and R. M. NEILL, 1942. Ovulation and oviposition in Anura. Jour. Exptl. Biol., **28**, 11–25.

WINTREBERT, P., 1929. Les voiles gris, premiers territoires ectodermiques de grande multiplication, et leur mode d'extension sur l'oeuf de Discoglossus pictus Otth. Compt. Rend. Soc. Biol., **102**, 997–999.

————, 1931. Les plans d'ébauches de Discoglossus pictus Otth. à la fin de la blastula et à la fin de la neurula. Compt. Rend. Acad. Sci., **193**.

*WITTEK, M., 1952. La vitellogénèse chez les amphibiens. Arch. Biol., **63**, 133–198.

Gastrulation and neurulation

*BÄCKSTRÖM, S., 1954. Morphogenetic effects of lithium on the embryonic development of Xenopus. Arkiv Zool., **6**, 527–536.

BERGEL, A., 1927. Ueber natürlich entstandene Spinae bifidae bei Rana fusca, nebst Bemerkungen über die Gastrulationsvorgänge. Arch. Entwicklungsmech., **109**, 253–282.

BERTACCHINI, P., 1899a. Morfogenesi e teratogenesi negli Anfibi anuri. Ia Serie. Blastoporo e doccia midolare. Internatl. Monatschr. Anat. Physiol., **16**, 140–154.

————, 1899b. Morfogenesi e teratogenesi negli Anfibi anuri. IIa Serie. Blastoporo e organi assili dorsali dell'embrioni. Internatl. Monatschr. Anat. Physiol., **16**, 269–300.

BOUNOURE, L., 1929. Sur l'existence des gonocytes primaires dans l'embryon de la Grenouille rousse à partir du début de la gastrulation; localisation et migration de ces gonocytes aux différents stades. Compt. Rend. Soc. Biol., **101**, 706–708.

DALCQ, A., 1937. Sur l'existence de foyers pronéphretiques et d'autres "prédispositions" organogènes dans la blastula et la jeune gastrula du Discoglosse. Compt. Rend. Soc. Biol., **124**, 833 ff.

————, 1946. Remarques sur la cinématique et la dynamique de l'ébauche préchordale chez les Anoures. Compt. Rend. Soc. Biol., **140**, 1152–1154.

DELSMAN, H. C., 1917. De gastrulatie van Rana esculenta en van Rana fusca. Versl. Gew. Verg. Wis-Nat. Afd. Kon. Ned. Akad. Wet., **25**, 780–794.

EKMAN, G., 1919/20. Experimentelle Untersuchungen über die Gastrulation und das erste Längenwachstum des Embryos bei Rana esculenta. Overs. Finska Vet.-Soc. Förh., **62**.

————, 1926. Einige Bemerkungen über die Gastrulation bei Rana esculenta. Ann. Zool. Soc. Zool. Bot. Fennicae "Vanamo", **4**, 238–243.

ERLANGER, R. v., 1890. Ueber Blastoporus der anuren Amphibien, sein Schicksal und seine Beziehungen zum bleibenden After. Zool. Jahrb., **4**.

————, 1891. Zur Blastoporusfrage bei den anuren Amphibien. Anat. Anz., **6**.

FAUTREZ, J., 1942. Critique et contrôle expérimentale de la conception de T. Dettlaff sur la formation de la plaque neurale chez les Amphibiens. Bull. Acad. Roy. Belgique, Classe Sci., **28**, 594–604.

FUMAGALLI, Z., 1951. Sulla morfogenesi della cresta neurale deglia nfibi. Biol. Latina, **3**, 97–111.

GIERSBERG, H., 1924. Beiträge zur Entwicklungsphysiologie der Amphibien. I. Furchung und Gastrulation bei Rana und Triton. Arch. Entwicklungsmech., **103**, 368–386.

————, 1926. Beiträge zur Entwicklungsphysiologie der Amphibien. III. Neue Untersuchungen zur Neurulation bei Rana und Triton. Arch. Entwicklungsmech., **108**, 283–321.

HAMECHER, H., 1904. Ueber die Lage des kopfbildenden Teiles und der Wachstumszone für Rumpf und Schwanz (FR. KOPSCH) zum Blastoporusrande bei Rana fusca. Internatl. Monatschr. Anat. Physiol., **21**.

HOLTFRETER, J., 1938. Differenzierungspotenzen isolierter Teile der Anurengastrula. Arch. Entwicklungsmech., **138**, 657–738.

IKEDA, S., 1902. Contributions to the embryology of Amphibia: The mode of blastopore closure and the position of the embryonic body. Jour. Coll. Sci., Tokyo Imp. Univ., **17**.

JOLLY, J. and G. LIEURE, 1942. La culture de la plaque médullaire des Batraciens et la formation des vaisseaux de l'embryon. Compt. Rend. Soc. Biol., **135**, 352 ff.

KING, H. D., 1902. The gastrulation of the egg of Bufo lentiginosus. Amer. Nat., **36**.

KOPSCH, FR., 1895a. Beiträge zur Gastrulation beim Axolotl und Froschei. Verhandl. Anat. Ges. Basel, 1895, 181–189.

————, 1895b. Ueber die Zellenbewegungen während des Gastrulationsprozesses an den Eiern vom Axolotl und vom braunen Grasfrosch. Sitzungsber. Ges. Naturf. Freunde, 1895, 21–30.

————, 1952. Bildung und Längenwachstum des Embryos von Rana fusca und von Bufo vulgaris nebst Bemerkungen zur Konkreszenzlehre und zur Coelomtheorie. Acta Anat., **16**, 122–147.

LASELL, J. and P. WINTREBERT, 1930. Analyse du développement de Discoglossus pictus Otth. par le procédé des marques colorées. Le développement de l'ectoderme à la surface de l'oeuf. Compt. Rend. Soc. Biol., **104**, 1229–1234.

MAYER, C., 1931. Ueber die Sonderungsvorgänge im Urdarmdach, die Bedeutung und das Schicksal der hypochordalen Platte bei anuren Amphibien. Arch. Entwicklungsmech., **124**, 469–521.

MOSZKOWSKI, M., 1902. Zur Frage des Urmundschlusses bei Rana fusca. Arch. microsc. Anat., **60**.

*NIEUWKOOP, P. D. and P. FLORSCHÜTZ, 1950. Quelques caractères spéciaux de la gastrulation et de la neurulation de l'oeuf de Xenopus laevis, Daud. et de quelques autres Anoures. I. Etude descriptive. Arch. Biol., **61**, 113–150.

PASTEELS, J., 1936. Critique et controle expérimental des conceptions de P. WINTREBERT sur la gastrulation du Discoglosse. Arch. Biol., **47**, 631–677.

————, 1939. Une version nouvelle du plan des ébauches de la jeune gastrula du Discoglosse. Compt. Rend. Soc. Biol., **131**, 779–781.

————, 1942. New observations concerning the maps of presumptive areas of the young amphibian gastrula (Amblystoma and Discoglossus). Jour. Exptl. Zool., **89**, 255–281.

*————, 1943. Fermeture du blastopore, anus et intestin caudal chez les Amphibiens anoures. Acta Neerl. Morph., **5**, 11–25.

*————, 1949. Observations sur la localisation de la plaque préchordale et de l'ento-blaste présumptifs au cours de la gastrulation chez Xenopus laevis. Arch. Biol., **60**, 235–250.

*PETER, K., 1934. Die Gastrulation von Xenopus laevis (Unter Benutzung des Werkstoffes, gesammelt von Dr. E. J. BLES). Zeitschr. microsc. anat. Forsch., **35**, 181–194.

*————, 1940. Untersuchungen über die Entwicklung des Dotterentoderms. V. Die Entwicklung des Entoderms bei Amphibien. Zeitschr. microsc.-anat. Forsch., **47**, 322–350.

PISANÒ, A., 1941. Territori presuntivi caudali di neurule di Rana esculenta allevati in espianto. Ricerca Sci., **12**.

————, 1942. Espianti di territori presuntivi caudali di neurule di Rana esculenta. Ricerche Morfol., 63 pp.

ROBINSON, A. and R. ASSHETON, 1891. The formation and fate of the primitive streak with observations on the archenteron and germinal layers of Rana temporaria. Quart. Jour. Microsc. Sci., **32**, 451–499.

ROUX, W., 1902. Bemerkungen über die Achsenbestimmung des Froschembryos und die Gastrulation des Froscheies. Arch. Entwicklungsmech., **14**, 600–624.

RUGH, R., 1943. The neurenteric canal in the frog, Rana pipiens. Proc. Soc. Exptl. Biol. and Med., **52**, 304–307.

SCHOTTÉ, O., 1930. Der Determinationszustand der Anurengastrula im Transplantations-experiment. Arch. Entwicklungsmech., **122**, 663–664.

SCHULTZE, O., 1888. Die Entwicklung der Keimblätter und der Chorda dorsalis von Rana fusca. Zeitschr. wiss. Zool., **47**, 325–352.

————, 1889. Ueber die Entwicklung der Medullarplatte des Froscheies. Verhandl. Phys. Med. Ges. Würzburg, 1889–N.F., **23**.

SEEMANN, J., 1907. Ueber die Entwicklung des Blastoporus bei Alytes obstetricans. Anat. Hefte, **33**, 315 ff.

SKREB, N., 1952. Etude des rapports topographiques de la chorde céphalique et de la plaque préchordale chez les Amphibiens. Arch. Biol., **63**, 85–108.

VOGT, W., 1929. Gestaltungsanalyse am Amphibienkeim mit örtlicher Vitalfärbung. II. Gastrulation und Mesodermbildung bei Urodelen und Anuren. Arch. Entwick-lungsmech., **120**, 384–706.

WILSON, H. V., 1900. Formation of the blastopore in the frog egg. Anat. Anz., **18**, 209–239.

————, 1901. Closure of blastopore in the normally placed frog egg. Anat. Anz., **20**, 123–128.

WINTREBERT, P., 1931. Les plans d'ébauches de Discoglossus pictus Otth. à la fin de la blastula et à la fin de la neurula. Compt. Rend. Acad. Sci., **193**.

Early larval development

*BÄCKSTRÖM, S., 1954. Morphogenetic effects of lithium on the embryonic development of Xenopus. Arkiv. Zool., **6**, 527–536.

BERTACCHINI, P., 1899a. Morfogenesi e teratogenesi negli Anfibi anuri. Ia Serie. Blas-toporo e doccia midolare. Internatl. Monatschr. Anat. Physiol., **16**, 140–154.

————, 1899b. Morfogenesi e teratogenesi negli Anfibi anuri. IIa Serie. Blastoporo e organi assili dorsali dell'embrioni. Internatl. Monatschr. Anat. Physiol., **16**, 269–300.

BOUNOURE, L., 1929. Sur l'existence des gonocytes primaires dans l'embryon de la
     Grenouille rousse à partir du début de la gastrulation; localisation et migration
     de ces gonocytes aux différents stades. Compt. Rend. Soc. Biol., **101**, 706–708.

EKMAN, G., 1919/20. Experimentelle Untersuchungen über die Gastrulation und das
     erste Längenwachstum des Embryos bei Rana esculenta. Overs. Finska Vet.-Soc.
     Förh., **62**.

ERLANGER, R. v., 1890. Ueber Blastoporus der anuren Amphibien, sein Schicksal und
     seine Beziehungen zum bleibenden After. Zool. Jahrb., **4**.

HAYCK, H. v., 1930. Die Einschaltung der hypochordalen Platte in das Darmdach bei
     Bufo. Anat. Anz., **69**, 243–247.

KNOUFF, R. A., 1935. The developmental pattern of ectodermal placodes in Rana pipiens.
     Jour. Comp. Neurol., **62**, 17 ff.

KOPSCH, FR., 1952. Bildung und Längenwachstum des Embryos von Rana fusca und
     von Bufo vulgaris nebst Bemerkungen zur Konkreszenzlehre und zur Coelom-
     theorie. Acta Anat., **16**, 122–147.

LASELL, J. and P. WINTREBERT, 1930. Analyse du développement de Discoglossus
     pictus Otth. par le procédé des marques colorées. Le développement de l'ecto-
     derme à la surface de l'oeuf. Compt. Rend. Soc. Biol., **104**, 1229–1234.

MAYER, C., 1931. Ueber die Sonderungsvorgänge im Urdarmdach, die Bedeutung und
     das Schicksal der hypochordalen Platte bei anuren Amphibien. Arch. Entwick-
     lungsmech., **124**, 469–521.

*PASTEELS, J., 1943. Fermeture du blastopore, anus et intestin caudal chez les Amphibiens
     anoures. Acta Neerl. Morph., **5**, 11–25.

*PETER, K., 1940. Untersuchungen über die Entwicklung des Dotterentoderms. V. Die
     Entwicklung des Entoderms bei Amphibien. Zeitschr. microsc.-anat. Forsch., **47**,
     322–350.

RONDININI, R., 1928. Particolarità formative in alcuni organi primitivi e sviluppo della
     coda nelle larve di Bufo vulgaris. Arch. Ital. Anat. e Embriol., **25**, 98–130.

RUGH, R., 1943. The neurenteric canal in the frog, Rana pipiens. Proc. Soc. Exptl. Biol.
     and Med., **52**, 304–307.

SCHULZE, O., 1888. Die Entwicklung der Keimblätter und der Chorda dorsalis von
     Rana fusca. Zeitschr. wiss. Zool., **47**.

VOGT, W., 1929. Chorda, Hypochorda und Darmentoderm bei anuren Amphibien.
     Verhandl. Anat. Ges., **38**; Anat. Anz., **67**, Erg. H.

SKIN, LATERAL LINE SYSTEM AND PIGMENTATION PATTERN

Skin

ASSHETON, R., 1895. Notes on the ciliation of the ectoderm of the amphibian embryo.
     Quart. Jour. Microsc. Sci., **38**, 465–484.

BAUMANN, A. and G. DERUAZ, 1932a. Recherches chez les larves de Batraciens sur la
     topographie des cellules ciliées épidermiques et sur les courants qui elle déterminent.
     Compt. Rend. Soc. Biol., **111**, 772–774.

———— and ————, 1932b. Evolution de l'appareil ciliaire dans l'épiderme d'un
     Batracien anoure. Compt. Rend. Soc. Biol., **111**, 860–862.

———— and ————, 1933. Recherches sur l'histologie et la physiologie de l'appa-
     reil ciliaire épidermique chez les larves de quelques Batraciens. Arch. Anat. Histol.
     et Embryol., **16**, 231–338.

BECCARI, N., 1914. L'organo tegumentale frontale delle larve di Anfibi. Arch. Ital. Anat.
     e Embriol., **13**, 379–400.

BHADURI, J. L., 1935. The anatomy of the adhesive apparatus in the tadpoles of Rana afghana, Günther. With special reference to the adaptive modifications. Trans. Roy. Soc. Edinburgh., **58**.

*CALABRESI, E., 1924. A proposito di speciali appendici sensoriali presenti nella pelle di Xenopus laevis (Daud.). Monitore Zool. Ital., **35**, 90–102.

CLAUSEN, H. J., 1930. Rate of histolysis of anuran skin and muscle during metamorphosis. Biol. Bull., **59**, 199–210.

DRZEWICKI, S., 1924. La métamorphose partielle de la peau des têtards de Pelobates fuscus Lanz, sous l'influence de quantités minimales de substance de la glande thyroide. Compt. Rend. Soc. Biol., **110**.

————, 1925. L'influence de la substance de la glande thyroide sur la métamorphose de la peau chez les têtards de Pelobates fuscus. Ksiega Pamiatkowa XII Zjardu Lekarzyi Przyrodnikow Polskich, 1925.

*FAHRENHOLZ, C., 1927. Die Flaschenzellen der Amphibienepidermis und ihre Beziehung zum Häutungsvorgang. Zeitschr. microsc.-anat. Forsch., **10**, 297–312.

*————, 1929. Die Tastsinnesorgane in der Haut des afrikanischen Krallenfrosches (Xenopus calcaratus Peters). Gegenbaurs morphol. Jahrb., **63**, 454–479.

FANO, L., 1903. Sull'origine, lo sviluppo e la funzione delle ghiandole cutanee degli Anfibi. Arch. Ital. Anat. e Embriol., **2**, 405 ff.

GUARDABASSI, A. and M. SACERDOTE, 1951. La lamina calcificata del derma cutaneo di Anfibi anuri nostrani ed esotici. Arch. Ital. Anat. e Embriol., **56**, 247–272.

HELFF, O. M., 1931. Studies on amphibian metamorphosis. IX. Integumentary specificity and dermal plicae formation in the anuran, Rana pipiens. Biol. Bull., **60**, 11–22.

———— and W. STARK, 1941. Studies on amphibian metamorphosis. XVIII. The development of structures in the dermal plicae of Rana sylvatica. Jour. Morphol., **68**, 303–322.

HORA, S. L., 1934. Development and probable evolution of the suctorial disc in the tadpoles of Rana afghana. Trans. Roy. Soc. Edinburgh, **57**.

HOWES, N. H., 1947. The skin of the tadpole of the common toad Bufo bufo bufo (L.) during metamorphosis. Proc. Zool. Soc. London, **116**, 602 ff.

KÖLLIKER, A. v., 1885. Stiftchenzellen in der Epidermis von Froschlarven. Zool. Anz., **200**, 439–441.

*LEHMAN, H. E., 1953. Observations on macrophage behavior in the fin of Xenopus larvae. Biol. Bull., **105**, 490–495.

LEYDIG, F., 1885. Stiftchenzellen in der Oberhaut von Batrachierlarven. Zool. Anz., **212**, 749–751.

LIEBERKIND, I., 1937. Vergleichende Studien über die Morphologie und Histogenese der larvalen Haftorgane bei den Amphibien. Thesis, Kopenhagen.

MacCALLUM, A. B., 1885. The nerve terminations in the cutaneous epithelium of the tadpole. Quart. Jour. Microsc. Sci., **26**, 53–70.

MARCHETTI, L., 1914. Sui primi momenti dello sviluppo di alcuni organi primitivi nel germe di Bufo vulgaris. Sviluppo delle ventose. Prima nota preventiva. Anat. Anz., **45**, 321–347.

MAURER, F., 1898. Die Vaskularisierung der Epidermis bei Anuren Amphibien zur Zeit der Metamorphose. Morphol. Jahrb., **26**.

MEYER, M., 1953. Einige histologische Bemerkungen über die larvale Epidermis und die äussere Cornea von Pelobates fuscus Laur. (Knoblauchkröte). Anat. Anz., **99**, 312–320.

NAVILLE, A., 1924. Recherches sur l'histogénèse et la régénération chez les Batraciens Anoures (corde dorsale et téguments). Arch. Biol., **34**, 235–344.

POSKA-TEISS, L., 1930. Ueber die larvale Amphibienepidermis. Zeitschr. Zellforsch. microsc. Anat., **11**, 445 ff.

RAPPINI, M., 1923. Sul distacimento autolitico delle ghiandole adesive (ventose) nella larva di Bufo vulgaris. Arch. Ital. Anat. e Embriol., **20**.

RIJDER, J. A., 1888. "Ventral suckers" or "Sucking disks" of the tadpoles of different genera of frogs and toads. Amer. Nat., **22**.

ROSIN, S., 1944. Ueber die orthogonale Struktur der Grenzlamelle der Cutis bei Amphibienlarven und ihre Verbreitung bei andern Chordaten. Rev. Suisse Zool., **51**, 376–382.

————, 1946. Ueber Bau und Wachstum der Grenzlamelle der Epidermis bei Amphibienlarven; Analyse einer orthogonalen Fibrillärstruktur. Rev. Suisse Zool., **53**, 133–201.

SAGUCHI, S., 1913. Ueber Mitochondrien und mitochondriale Stränge in den Epidermiszellen der Anurenlarve nebst Bemerkungen über die Frage der Epidermis-Cutisgrenze. Arch. microsc. Anat. u. Entwicklungsgesch., **83**, 177–246.

SCHUBERT, M., 1924. Untersuchungen über die Rückbildungserscheinungen der Epidermis des Froschlarvenschwanzes. Anat. Anz., **58**.

SCHULTZE, O., 1907. Ueber den Bau und die Bedeutung der Aussencuticula der Amphibienlarve. Arch. microsc. Anat. u. Entwicklungsgesch., **69**.

*SPANNHOF, L., 1954. Zur Genese, Morphologie und Physiologie der Hautdrüsen bei Xenopus laevis Daud. Wiss. Z. Humboldt-Univ., Berlin, Math.-nat. Reihe, **3**, 295–305.

STÖHR, PH., 1881. Ueber die Haftorgane der Anurenlarven. Sitzungsber. phys. med. Ges. Würzburg, **8**, 118 ff.

THIELE, J., 1888. Der Haftapparat der Batrachierlarven. Zeitschr. wiss. Zool., **46**, 67–79.

WEISS, O., 1908. Ueber die Entwicklung der Giftdrüsen in der Anurenhaut. Anat. Anz., **33**, 124 ff.

WOERDEMAN, M. W., 1925. Entwicklungsmechanische Untersuchungen über die Wimperbewegungen des Ektoderms von Amphibienlarven. Arch. Entwicklungsmech., **106**, 41–61.

YANAI, T., 1953. Structure and development of the frontal glands of the frogs, Rhacophorus schlegelii arborea and Rh. schl. schlegelii. Annot. zool. Jap., **26**, 86–90.

————, M. OUJI and K. OMURA, 1953. On the origin of the frontal glands of amphibians. Annot. zool. Jap., **26**, 193–201.

Lateral line system

*ESCHER, K., 1925. Das Verhalten der Seitenorgane der Wirbeltiere und ihrer Nerven beim Uebergang zum Landleben. Acta Zool., **6**, 307–414.

HARRISON, R. G., 1903. Experimentelle Untersuchungen über die Entwicklung der Sinnesorgane der Seitenlinie bei den Amphibien. Arch. Microsc. Anat., **63**, 35–149.

*HORST, C. J. V. D., 1934. The lateral-line nerves of Xenopus. Psychiat. Neurol. Bl. Amst., **3/4**, 426–435.

KNOUFF, R. A., 1935. The developmental pattern of ectodermal placodes in Rana pipiens. Jour. Comp. Neurol., **62**, 17 ff.

STONE, L. S., 1933a. The development of lateral-line sense organs in amphibians observed in living and vital-stained preparations. Jour. Comp. Neurol., **57**, 507–540.

————, 1933b. Developmental changes in primary lateral-line organs studied in living larvae of Anurans and Urodeles. Proc. Soc. Exptl. Biol. and Med., **30**, 1258–1259.

————, 1937. Further experimental studies of the development of lateral-line sense organs in amphibians observed in living preparations. Jour. Comp. Neurol., **68**, 83–115.

WINTREBERT, P., 1911. La distribution cutanée et l'innervation des organes lateraux chez la larve d'Alytes obstetricans. Compt. Rend. Soc. Biol., **70**, 1050 ff.

Pigmentation pattern

BYTINSKI-SALZ, H., 1938. Chromatophorenstudien. II. Struktur und Determination des adepidermalen Melanophorennetzes bei Bombina. Arch. Exptl. Zellforsch., **22**, 132–170.

COLOMBO, G., 1947. Alcune osservazioni preliminari sull'origine dei cromatofori dermali negli anfibi anuri. Boll. Soc. Ital. Biol. Sperim., **23**, 2pp.

ELIAS, H., 1931. Die Entwicklung des Farbkleides des Wasserfrosches (Rana esculenta). Zeitschr. Zellforsch. microsc. Anat., **14**, 55–72.

————, 1934. Ueber die Entwicklung der Chromatophoren und anderer Zellen in der Haut von Bufo viridis. Zeitschr. Zellforsch. microsc. Anat., **21**, 529–544.

————, 1936. Die Hautchromatophoren von Bombinator pachypus und ihre Entwicklung. Zeitschr. Zellforsch. microsc. Anat., **24**.

————, 1939. Die adepidermalen Melanophoren der Discoglossiden, ein Beispiel für den phylogenetischen Funktionswechsel eines organs, seinen Ersatz in der frühen Funktion durch ein neues Organ und sein schliessliches Verschwinden. Zeitschr. Zellforsch. microsc. Anat., **29**.

HERRICK, E. H., 1933. The structure of epidermal melanophores in frog tadpoles. Biol. Bull., **64**, 304–308.

LINDEMAN, V. F., 1929. Integumentary pigmentation in the frog, Rana pipiens, during metamorphosis, with especial reference to tailskin histolysis. Physiol. Zool., **2**, 255 ff.

*STEVENS, L. C., 1954. The origin and development of chromatophores of Xenopus laevis and other Anurans. Jour. Exptl. Zool., **125**, 221–246.

BRAIN

Brain (s. s.)

*ADAM, H., 1954. Freie kugelförmige Pigmentzellen in den Gehirnventrikeln von Krallenfroschlarven Xenopus laevis (Daudin). Zeitschr. mikrosk.-anat. Forsch., **60**, 6–32.

BAFFONI, G. M. and G. CATTE, 1951. La citomorfosi della cellula di Mauthner in Hyla arborea Savigni. Riv. Biol. N.S., **43**, 373–397.

FREY, E., 1938. Studien über die hypothalamische Optikuswurzel der Amphibien. I. Rana mugiens, Rana esculenta, Bombinator pachypus und Pipa pipa. Proc. Kon. Ned. Akad. Wet., **9**, 1004–1014.

GRAAF, H. W. DE, 1886a. Bijdrage tot de kennis van de bouw en de ontwikkeling der epiphyse bij Amphibiën en Reptielen. Thesis, Leiden.

————, 1886b. Zur Anatomie und Entwicklung der Epiphyse bei Amphibien und Reptilien. Zool. Anz., **9**, 219 ff.

*HAFFNER, K. VON, 1951. Die Pincalblase (Stirnorgan, Parietalorgan) von Xenopus laevis Daud. und ihre Entwicklung, Verlagerung und Degeneration. Zool. Jahrb. Abt. 2. Anat. u. Ontog. Tiere, **71**, 375–412.

*KAMER, J. C. V. D., 1949. Over de ontwikkeling, de determinatie en de betekenis van de epiphyse en de paraphyse van de Amphibiën. Thesis, Utrecht.

KOLLROS, J. J., J. C. KALTENBACH, R. HILL and V. PEPERNIK, 1950. The growth of mesencephalic V nucleus cells as a metamorphic event in Anurans. Anat. Rec., **108**, 1p.

LARSELL, O., 1925. The development of the cerebellum in the frog (Hyla regilla) in relation to the vestibular and lateral-line systems. Jour. Comp. Neurol., **39**, 249–289.

MANFREDONIA, M., 1937. Particolare disposizione nella chiusura del neuroporo anteriore in Discoglossus pictus. Arch. Ital. Anat. e Embriol., **38**, 143–152.

MAZZI, V., 1954. Alcune osservazioni intorno al sistema neurosecretorio ipotalamo-ipofisario e all'organo sottocommissurale nell'ontogenesi di Rana agilis. Monitore Zool. Ital., **62**, 78–82.

RIECH, F., 1925. Epiphyse und Paraphyse im Lebenszyklus der Anuren. Zeitschr. vergl. Physiol., **2**, 524–570.

SZEPSENWOL, J., 1935a. l'Existence de la cellule de Mauthner chez les Amphibiens anoures. Compt. Rend. Soc. Biol., **118**, 944–946.

————, 1935b. La différenciation primitive et la nature des cellules de Mauthner chez les amphibiens anoures et urodèles. Compt. Rend. Soc. Biol., 1935.

TENSEN, J., 1927. Einige Bemerkungen über das Nervensystem von Pipa pipa. Acta Zool., **8**.

*WEISS, P. and F. ROSETTI, 1951. Growth responses of opposite sign among different neuron types exposed to thyroid hormone. Proc. Natl. Acad. Sci., **37**, 540–556.

*WINTERHALTER, W. P., 1931. Untersuchungen über das Stirnorgan der Anuren. Acta Zool., **12**, 1–67; Viertelj. schr. Naturwiss. Ges. Zürich, **76**.

*ZACCHEI, A. M., 1953a. Ulteriori ricerche sui centri tegmentali rombencefalici degli anfibi anuri. Atti Accad. Naz. Lincei. Rend. Cl. Sci. Fis. Mat. e Nat., **15**, 107–111.

*————, 1953b. La struttura cerebellare degli anfibi anuri in rapporto alle condizioni statiche e di locomozione. Atti Accad. Naz. Lincei. Rend. Cl. Sci. Fis. Mat. e Nat. **15**, 120–125.

Hypophysis

ALLEN, B. M., E. D. TORREBLANCA and J. A. BENJAMIN, 1929. A study upon the histogenesis of the pars anterior of the hypophysis of Bufo during metamorphosis. Anat. Rec., **44**, 208 (Abstr.).

ATWELL, W. J., 1918. The development of the hypophysis of the Anura. Anat. Rec., **15**, 73–92.

*————, 1938. The pars tuberalis of the hypophysis in toads. Anat. Rec., **72** (Suppl.) 38.

*————, 1941. The morphology of the hypophysis cerebri of toads. Amer. Jour. Anat., **68**, 191–206.

BURCH, A. B., 1945. An experimental study of the histological and functional differentiation of the epithelial hypophysis in Hyla regilla. Univ. California Publ. Zool., **51**, 185–214.

*CHARIPPER, M. and J. MARTORANO, 1948. The morphology of the pituitary gland of the South African toad Xenopus laevis. Zoologia, Sci. Contr. N.Y. Zool. Soc., **33**, 157–162.

*CORDIER, R., 1948. Sur l'aspect histologique et cytologique de l'hypophyse pendant la métamorphose chez Xenopus laevis. Compt. Rend. Soc. Biol., **142**, 845–847.

*————, 1949. La réaction hypophysaire de la métamorphose chez Xenopus laevis. Compt. Rend. Assoc. Anat., **54**, 143–150.

*————, 1953. L'hypophyse de Xenopus. Interprétation histophysiologique. Ann. Soc. Roy. Zool. Belgique, **84**, 5–16.

*————, 1954. Cytologie hypophysaire et sa signification fonctionnelle chez l'amphibien Xenopus. Compt. Rend. Assoc. Anat., **79**, 484–490.

D'ANGELO, S. A., 1941. An analysis of the morphology of the pituitary and thyroid glands in amphibian metamorphosis. Amer. Jour. Anat., **69**, 407 ff.

DEEN, R., 1953. Der Einfluss verschiedener Belichtungszeiten auf Hypophyse, Schilddrüse und Larvenentwicklung von Rana temporaria. Zool. Jahrb. Abt. 3 Allg. Zool. u. Physiol. Tiere, **63**, 477–500.

*ETKIN, W., 1941. The first appearance of functional activity in the pars intermedia in the frog Xenopus. Proc. Soc. Exptl. Biol., **47**, 425–428.

*HOGBEN, L., 1942. The amphibian pituitary. Nature, London., **149**, 695–696.

HOSKINS, E. R. and M. M. HOSKINS, 1920. The inter-relation of the thyroid and hypophysis in the growth and development of frog larvae. Endocrinology, **4**, 1 ff.

KERR, F., 1939. On the histology of the developing pituitary in the frog (Rana t. temporaria) and in the toad (Bufo bufo). Proc. Zool. Soc. London B, **109**, 167–180.

*LEVENSTEIN, I. and H. A. CHARIPPER, 1939. The pituitary gland of the South African clawed toad (Xenopus laevis Daud.). Anat. Rec., **75** (suppl.), 123 ff.

MAZZI, V., 1954. Alcune osservazioni intorno al sistema neurosecretorio ipotalamo-ipofisario e all'organo sottocommissurale nell'ontogenesi di Rana agilis. Monitore Zool. Ital., **62**, 78–82.

*RIMER, G. B. G., 1931a. Histology and morphological studies of the endocrine glands of Xenopus laevis. Thesis, Cape Town, 106 pp.

*————, 1931b. The pituitary gland of Xenopus laevis. Trans. Roy. Soc. S. Africa, **19**, 341–354.

SATO, K., 1935. Studien über die Entwicklung der Hypophysenanlage. I Mitteilung: Bei den Amphibien, besonders den Embryonen von B. vulgaris japonicus. Okayama-Igakkai-Zasshi, **46**, 184–202.

SCHLIEFER, W., 1935. Die Entwicklung der Hypophyse bei Larven von Bufo vulgaris bis zur Metamorphose. Zool. Jahrb. Abt. 2. Anat. u. Ontog. Tiere, **59**, 383–454.

*SPANNHOF, L., 1954. Über das Einheilungsvermögen von Hypophysenimplantaten bei Xenopus laevis Daudin. Verhandl. Deutsch. Zool. Gesellsch. Tübingen, 1954, 122–124.

TAKASIMA, R. and K. TERATO, 1936. Eine kurze Mitteilung über die Entwicklung des Nervenlappens vom Hypophysis cerebri bei einigen japanischen Anurenarten. Folia Anat. Japonica, **14**, 421–433.

———— and S. YUBA, 1936. Ein Beitrag zur Morphogenese der parstu beralis vom Hypophysis cerebri bei den verschiedenen japanischen Amphibien. Folia Anat. Japonica, **14**, 665–674.

TERATO, K., 1935a. Embryologische Studien über die Hypophysis cerebri bei der japanischen Kröte (R. formosus); I Mitteilung: Ueber die larvale Entwicklung der Hypophyse. Jour. Kumamoto Med. Soc., **11**.

————, 1935b. Embryologische Studien über die Hypophysis cerebri bei der japanischen Kröte (R. formosus); II Mitteilung: Ueber die postlarvale Entwicklung der Hypophysis cerebri. Jour. Kumamoto Med. Soc., **11**.

———— and S. IMAMURA, 1935. Ueber die Entwicklung der Hypophyse von Rana nigromaculata. Osaka Ijishinshi, **6**.

———— and K. YAMADA, 1935. Ueber die Entwicklung der Hypophys von Rhacophorus schlegelii arborea. Osaka Ijishinshi, **6**.

YUBA, S., 1935. Ueber die Entwicklung der Hypophysis cerebri bei in Süd-Korea spezifischen Cacopoides tornieri. Mitteil. Med. Gesellsch. Osaka, **34**.

CEPHALIC GANGLIA AND NERVES

BAUMANN, A., 1933. Développement de l'innervation cardiaque primitive chez un Batracien anoure, Bombinator pachypus. Compt. Rend. Soc. Biol., **113**, 1381–1382.

————, 1934. Observations sur les premiers stades de l'innervation du cœur chez un Batracien anoure (Bombinator pachypus Bonap.). Rev. Suisse Zool., **41**, 235–261.

CAMPENHOUT, E. VAN, 1930. Intraocular optic nerves in embryos of Rana pipiens. Anat. Rec., **61**, 351–358.

————, 1935. Experimental researches on the origin of the acoustic ganglion in amphibian embryos. Jour. Exptl. Zool., **72**, 175–193.

CHIARUGI, G., 1890. Sui miotomi e sui nervi della testa posteriore e della regione prossi-
    male del tronco negli embrioni degli Anfibi Anuri. Monitore Zool. Ital., **1**, 22–59.
————, 1891. Sur les myotomes et sur les nerfs de la tête postérieure et de la région
    proximale du tronc dans les embryons des Amphibiens Anoures. Arch. Ital. Biol., **15**.
*ESCHER, K., 1925. Das Verhalten der Seitenorgane der Wirbeltiere und ihrer Nerven
    beim Uebergang zum Landleben. Acta Zool., **6**, 307–414.
*HORST, C. J. V. D., 1934. The lateral-line nerves of Xenopus. Psychiat. Neurol. Bl. Amst.,
    3/4, 426–435.
KNOUFF, R. A., 1927. The origin of the cranial ganglia of Rana. Jour. Comp. Neurol.,
    44, 259–361.
KRUIJTZER, E. M., 1931. De ontwikkeling van het chondrocranium en enkele kopzenuwen
    van Megalophrys montana. Thesis, Leiden.
LANDACRE, F. L. and M. F. McLEAN, 1912. The cerebral ganglia of the embryo of
    Rana pipiens. Jour. Comp. Neurol., **22**, 461 ff.
TENSEN, J., 1927. Einige Bemerkungen über das Nervensystem von Pipa. Acta Zool., **8**.
TOKURA, R., 1925. Entwicklungsmechanische Untersuchungen über das Hörbläschen
    und das akustische sowie faciale Ganglion bei den Anuren. Folia Anat. Japonica,
    3, 173–208.
WINTREBERT, P., 1911. La distribution cutanée et l'innervation des organes lateraux
    chez la larve d'Alytes obstetricans. Compt. Rend. Soc. Biol., **70**, 1050 ff.

SPINAL CORD, GANGLIA AND NERVES

BAMBEKE, M. VAN, 1907. Considérations sur la génèse du névraxe, spécialement sur celle
    observée chez le Pélobate brun (Pelobate fuscus Wagl.). Arch. Biol., **23**, 523–539.
BAUMANN, A., 1933. Développement de l'innervation cardiaque primitive chez un
    Batracien anoure, Bombinator pachypus. Compt. Rend. Soc. Biol., **113**, 1381–1382.
————, 1934. Observations sur les premiers stades de l'innervation du cœur chez un
    Batracien anoure (Bombinator pachypus Bonap.). Rev. Suisse Zool., **41**, 235–261.
BROWN, M. E., 1946. The histology of the tadpole tail duringmeta morphosis, with
    special reference to the nervous system. Am. Jour. Anat., **78**, 79 ff.
CAMPENHOUT, E. VAN, 1930. Contribution to the problem of the development of the
    sympathetic nervous system. Jour. Exptl. Zool., **56**, 295–320.
CHIARUGI, G., 1890. Sui miotomi e sui nervi della testa posteriore e della regione prossi-
    male del tronco negli embrioni degli Anfibi Anuri. Monitore Zool. Ital., **1**, 22–59.
————, 1891. Sur les myotomes et sur les nerfs de la tête postérieure et de la région
    proximale du tronc dans les embryons des Amphibiens Anoures. Arch. Ital. Biol., **15**.
GREEVEN, R., 1941. Sympathetic nervous system of Bombinator. Thesis, Innsbrück.
HENSEN, V., 1868. Ueber die Nerven im Schwanz der Frosch-larva. Arch. microsc.
    Anat., **4**, 115 ff.
TAYLOR, A. C., 1943. Development of the innervation pattern in the limb bud of the
    frog. Anat. Rec., **87**, 379–409.
————, 1944. Selectivity of nerve fibres from the dorsal and ventral roots in the
    development of the frog limb. Jour. Exptl. Zool., **96**, 159–185.
TENSEN, J., 1927. Einige Bemerkungen über das Nervensystem von Pipa pipa. Acta
    Zool., **8**.
WEBER, A., 1925. Différenciation des cellules nerveuses dans les ganglions rachidiens.
    Compt. Rend. Soc. Biol., **92**, 1281–1283.
————, 1932. Pénétration des fibres nerveuses dans les bourgeons des membres
    antérieurs chez un Batrachien anoure, "Bombinator pachypus". Compt. Rend.
    Assoc. Anat. **27**, 610–612.

OLFACTORY ORGAN

BANCROFT, I. R., 1895. The nasal organs of Pipa americana. Bull. Essex. Inst., **27**, 101–107.

BELL, E. T., 1907. Some experiments on the development and regeneration of the eye and the nasal organ in frog embryos. Arch. Entwicklungsmech., **23**, 457–478.

BORN, G., 1876. Ueber die Nasenhöhlen und den Tränennasengang der Amphibien. Gegenbaurs Morphol. Jahrb., **2**, 577–646.

COOPER, R. S., 1943. An experimental study of the development of the larval olfactory organ of Rana pipiens Schreber. Jour. Exptl. Zool., **93**, 415–452.

EKMAN, C., 1924. Ueber die Regeneration der Nasenanlage bei Rana fusca. Comm. Biol. Soc. Sci. Fenn., **1**, 6 ff.

FAHRENHOLZ, C., 1925. Ueber die Entwicklung des Gesichtes und der Nase bei der Geburtshelferkröte (Alytes obstetricans). Morphol. Jahrbuch, **54**, 421–503.

*FÖSKE, H., 1934. Das Geruchsorgan von Xenopus laevis. Zeitschr. Anat. u. Entwicklungsgesch., **103**, 519–550.

HINSBERG, V., 1901. Die Entwicklung der Nasenhöhle bei Amphibien. I, II. Arch. microsc. Anat., **58**.

KUREPINA, M., 1931. Die Entwicklung des Geruchsorgans der Amphibien. Zool. Jahrb. Abt. 2. Anat. u. Ontog. Tiere., **54**, 1–54.

MANFREDONIA, M., 1938. Localizzazione della zona presuntiva del placode olfattivo negli Discoglossus pictus. Arch. Ital. Anat. e Embriol., **40**, 356 ff.

*PATERSON, N. F., 1939. The olfactory organs and tentacles of Xenopus laevis. S. African Jour. Sci., **36**, 390–404.

———, 1951. The nasal cavities of the toad Hemipipa Carvalhoi Mir.-Rib. and other Pipidae. Proc. Zool. Soc. London, **121**, 381–415.

ROWEDDER, W., 1937. Die Entwicklung des Geruchsorgans bei Alytes obstetricans und Bufo vulgaris. Zeitschr. Anat. u. Entwicklungsgesch., **107**, 91–123.

THRAMS, O. K., 1936. Das Geruchsorgan von Pipa americana. Zeitschr. Anat. u. Entwicklungsgesch., **105**, 678–693.

TSUI, C. L., 1946a. Development of olfactory organ in Rana nigromaculata. Quart. Jour. Microsc. Sci., **87**, 61–90.

———, 1946b. Morphological observations on the fate of the lateral appendix in the embryonic olfactory organ of Rana nigromaculata. Quart. Jour. Microsc. Sci., **87**, 91–101.

——— and T. H. PAN, 1946. The development of the olfactory organ of Kaloula borealis (Barbour) as compared with that of Rana nigromaculata Hallowell. Quart. Jour. Microsc. Sci., **87**, 299–316.

WATANABE, M., 1936. Ueber die Entwicklung des Geruchsorgans von Rhacophorus schlegelii. Zeitschr. Anat. u. Entwicklungsgesch., **105**.

ZWILLING, E., 1940. An experimental analysis of the development of the anuran olfactory organ. Jour. Exptl. Zool., **84**, 291–318.

EYE

ALBERTI, W., 1922. Zur Frage der Linsenregeneration bei den Anuren. Arch. Entwicklungsmech., **50**, 355–374.

BELL, E. T., 1907. Some experiments on the development and regeneration of the eye and the nasal organ in frog embryos. Arch. Entwicklungsmech., **23**, 457–478.

CAMPENHOUT, E. VAN, 1930. Intraocular optic nerves in embryos of Rana pipiens. Anat. Rec., **61**, 351–358.

COLE,, W. H. and C. F. DEAN, 1987. The photokinetic reactions of frog tadpoles. Jour. Exptl. Zool., **23**, 361–370.

EAKIN, R. M., 1946. Determination and regulation of polarity in the retina of Hyla Univ. California Publ., Zool., **51**, 245 ff.

EKMAN, G., 1914. Experimentelle Beiträge zum Linsenbildungsproblem bei den Anuren mit besonderer Berücksichtigung von Hyla arborea. Arch. Entwicklungsmech., **39**, 328–352.

FUCHS, F., 1924. Ueber Augenregeneration nach Entfernung des Bulbus bei Alytes und Bufo. Zool. Jahrb. Abt. 3. Allg. Zool. u. Physiol. Tiere, **41**, 121 ff.

FUJITA, H., 1912. Regenerationsprozess der Netzhaut des Tritons und des Frosches. Arch. Vergl. Ophtalmol., **3**, 357 ff.

GIESBRECHT, E., 1925. Beiträge zur Entwicklung der Cornea und zur Gestaltung der Orbitalhöhle bei den einheimischen Amphibien. Zeitschr. wiss. Zool., **124**, 304–359.

GUARDABASSI, A., 1947. Processi di differenziamento della abbozzo oculare e di induzione della lente in condizione di espianto. Ricerche sperimentali su Bufo vulgaris e Rana dalmatina Fitzing (= Rana agilis Thom.). Arch. Ital. Anat. e Embriol., **54**, 134–143.

HARMS, W., 1923. Brillen bei Amphibienlarven. Zool. Anz., **56**, 136–142.

JOKL, A., 1921. Zur Entwicklung des Anurenauges. Anat. Heften, **59**, (Heft 178) 211–276.

————, 1922. Zur Entwicklung des Anurenauges. Arch. Entwicklungsmech., **50**, 623–624.

JOLLY, J., 1950. Recherches sur les ébauches optiques et sur la régénération de l'œil chez les Batraciens. Arch. Anat. Microsc. et Morphol. Exptl., **39**.

KALTENBACH, J. C., 1953. Local action of thyroxin on amphibian metamorphosis. II. Development of the eyelids, nictitating membrane, cornea and extrinsic ocular muscles in Rana pipiens larvae effected by thyroxin-cholesterol implants. Jour. Exptl. Zool., **122**, 41–52.

KING, H. D., 1905. Experimental studies on the eye of the frog embryo. Arch. Entwicklungsmech., **19**, 85–107.

LINDEMAN, V. E., 1929. Development of the nictitating membrane of the frog (Rana pipiens). Anat. Rec., **44**, 217 ff.

MEYER, M., 1953. Einige histologische Bemerkungen über die larvale Epidermis und die äussere Cornea von Pelobates fuscus Laur (Knoblauchkröte). Anat. Anz., **99**, 312–320.

OKADA, Y. K., 1939. Studies on the lens-regeneration in anuran Amphibia. Mem. Coll. Sci. Kyôto Imp. Univ. Ser. B, **15**, 159–166.

PASQUINI, P., 1928. Sulla presunta rigenerazione dell'occhio negli embrioni di Rana esculenta. Monitore Zool. Ital., **39**, 78–83.

————, 1929. Fenomeni di regolazione e di riparazione nello sviluppo dell'occhio degli Anfibi. Rend. R. Accad. Naz. Lincei, **9**, 99–104.

———— and A. DELLA MONICA, 1929. Rigenerazione del cristallino nelle larve di Anfibi anuri. R. Accad. Naz. Lincei, Cl. Sci. fis., math. e nat., **10**, 218–224.

PIANTI, E. DEL, 1942. Ricerche sulla ricostituzione dell' abbozzo dell'occhio di Rana esculenta dissociato nei suoi elementi. Arch. Zool. Ital., **30**, 229 ff.

POPOFF, W. W., 1934. Ueber die Morphogenese der Hornhaut bei Anura. I. Die Bildung und Entwicklung der eigentlichen Cornea bei Bufo bufo, Rana temporaria und Rana esculenta. Zool. Jahrb. Abt. 2. Anat. u. Ontog. Tiere, **58**, 661–696.

ROMANO, M., 1936. Le modificazioni dell'occhio degli Anuri durante la metamorfosi Arch. Ital. Anat. e Embriol., **36**, 433–465.

*SAXÉN, L., 1953. The development of the visual cells; embryological and physiological investigations on Amphibia. Ann. Acad. Sci. Fennica, Ser. A, **23**, 1-93.

Törö, E., 1932. Zur Frage der Entwicklung und Regeneration der Hornhaut. Zeitschr. Anat. u. Entwicklungsgesch., **98**, 97–125.

Ubisch, L. von, 1927. Beiträge zur Erforschung des Linsensystems. Zeitschr. wiss. Zool., **129**, 213–252.

*Vilter, V., 1946. Interversion dorso-ventrale des champs rétiniens chez le Xenopus laevis et répercussions sur le réflex pigmentaire rétino-cutané. Compt. Rend. Soc Biol., **140**, 760–763.

Yano, K., 1927. Die Entwicklung und Morphologie des Skleralknorpels bei den Anuren. Folia Anat. Japonica, **5**, 169–290.

AUDITORY ORGAN

Birkmann, K., 1940. Morphologisch-anatomische Untersuchungen zur Entwicklung des häutigen Labarynthes der Amphibien. Zeitschr. Anat. u. Entwicklungsgesch., **110**, 443–448.

Eisinger, K. and H. Sternberg, 1924. Beiträge zur Entwicklungsmechanik des inneren Ohres. Arch. Microsc. Anat. Entwicklungsgesch., **100**.

Guardabassi, A., 1952a. L'organo endolinfatico degli anfibi anuri. Arch. Ital. Anat. e Embriol., **57**, 241–294.

————, 1952b. I sali di calcio del sacco endolinfatico ed i processi di calcificazione delle osse durante la metamorfosi normale e sperimentale nelle larve di Bufo vulgaris, Rana dalmatina e Rana esculenta. Boll. Soc. Ital. Biol. Sperim., **28**, 355–357.

————, 1953. Les sels de Ca du sac endolymphatique et les processus de calcification des os pendant la metamorphose normale et expérimentale chez les têtards de Bufo vulgaris, Rana dalmatina et Rana esculenta. Arch. Anat. Microsc. et Morphol. Exptl., **42**, 143–167.

Guareschi, C., 1930. Studi sullo sviluppo dell'otocisti degli anfibi anuri. Arch. Entwicklungsmech., **122**, 179–203.

Herter, K., 1921. Untersuchungen über die nicht-akustischen Labyrinthfunktionen bei Anurenlarven. Zeitschr. allg. Physiol., **19**, 335–414.

Michl, R., 1924. Sur le développement du labyrinthe auditif chez Rana temporaria. Fac. de Méd. Brno, **3**, 139–165 (French summary).

*Paterson, N. F., 1949. The development of the inner ear of Xenopus laevis. Proc. Zool. Soc. London, **119**, 269–291.

Streeter, G. L., 1921. Migration of the ear vesicle in the tadpole during normal development. Anat. Rec., **21**, 115–126.

Tokura, R., 1925. Entwicklungsmechanische Untersuchungen über das Hörbläschen und das akustische sowie faciale Ganglion bei den Anuren. Folia Anat. Japonica, **3**, 173–208.

Villy, F., 1890. The development of the ear and the accessory organs in the common frog. Quart. Jour. Microsc. Sci., **30**, 523–550.

Whiteside, B., 1922. The development of the saccus endolymphaticus in Rana temporaria Linné. Amer. Jour. Anat., **30**, 231–266.

Witschi, E., 1947. Development and metamorphosis of the auxiliary apparatus of the ear of the frog. Anat. Rec., **99**, 568 (Abstr.).

————, 1949. The larval ear of the frog and its transformation during metamorphosis. Zeitschr. Naturforsch., **4**, 230–242.

*————, J. A. Bruner and W. A. van Bergeijk, 1953. The ear of the adult Xenopus. Anat. Rec., **117**, 602–603.

SKELETON AND MUSCULATURE OF THE HEAD

General publications

*BEDDARD, F. E., 1895. On the diaphragm and on the muscular anatomy of Xenopus, with remarks on its affinities. Proc. Zool. Soc. London, 1895, 841–850.

GAUPP, E., 1906. Die Entwickelung des Kopfskelettes. Hertwig, Handbuch der Vergleichende und experimentelle Entwicklung der Wirbeltiere, Jena, **3**, 573 ff.

*GROBBELAAR, C. S., 1935. The musculature of the Pipid genus Xenopus. S. African Jour. Sci., **32**, 395 ff (Abstr.).

*PUSEY, H. K., 1943. On the head of the Liopelmid frog, Ascaphus truei. I. The chondrocranium, jaws, arches and muscles of a partly-grown larva. Quart. Jour. Microsc. Sci., **84**, 105–185.

REINBACH, W., 1939. Untersuchungen über die Entwicklung des Kopfskeletts von Calyptocephalus gayi (mit einem Anhang über das Os suprarostrale der anuren Amphibien). Jena Zeitschr. Zool. Naturwiss., **72**, 211 ff.

SPEMANN, H., 1898. Ueber die erste Entwicklung der Tuba Eustachii und des Kopfskeletts von Rana temporaria. Zool. Jahrb. Abt. 2. Anat. u. Ontog. Tiere, **11**, 389–416.

Cranium

BAUSENHARDT, D., 1939. Die Bildung der Choane und das Problem der Gesichtsfortsätze bei Amphibien. Zool. Anz., **128**, 24–35.

*BEER G,. R. DE, 1937. Development of the chondrocranium of Xenopus. In: "The development of the Vertebrate skull", Oxford.

EEDEN, J. A. VAN, 1951. The development of the chondrocranium of Ascaphus truei Stejneger with special reference to the relations of the palatoquadrate to the neurocranium. Acta Zool., **32**, 41–176.

ELLIOT, A., 1907. Some facts in the later development of the frog, Rana temporaria. I. The segments of the occipital region of the skull. Quart. Jour. Microsc. Sci., **51**, 647–657.

GAUPP, E., 1892. Grundzüge der Bildung und Umbildung des Primordialcraniums von Rana fusca. Verhandl. Anat. Geselsch., Wien, **6**, 183–190.

GIESBRECHT, E., 1925. Beiträge zur Entwicklung der Cornea und zur Gestaltung der Orbitalhöhle bei den einheimischen Amphibien. Zeitschr. wiss. Zool., **124**, 304–359.

*KOTTHAUS, A., 1933. Die Entwicklung des Primordialcraniums von Xenopus laevis bis zur Metamorphose. Zeitschr. wiss. Zool., Abt. A., **144**, 510–572.

KRUIJTZER, E. M., 1931. De ontwikkeling van het chondrocranium en enkele kopzenuwen van Megalophrys montana. Thesis, Leiden.

OKUTOMI, K., 1937. Die Entwicklung des Chondrocraniums von Polypedates buergeni schlegelii. Zeitschr. Anat. Entwicklungsgesch., **107**, 28–64.

PARKER, W. K., 1871. The development of the skull of the common frog. Phil. Trans. Roy. Soc. London, Ser. B. Biol. Sci., **161**, 137 ff.

*————, 1876. On the structure and development of the skull in the Batrachia. II. Phil. Trans. Roy. Soc. London, Ser. B. Biol. Sci., **166**, 601–669.

*————, 1881. On the structure and development of the skull in the Batrachia. III. Phil. Trans. Roy. Soc. London, Ser. B. Biol. Sci., **172**, 1–266.

PEETERS, J. L. E., 1910. Over de ontwikkeling van het chondrocranium en de kraakbenige wervelkolom van eenige Urodela en Anura. Thesis, Leiden.

*PETER, K., 1895. Über die Bedeutung des Atlas der Amphibien. Anat. Anz., **10**, 565–574.

RAMASWAMI, L. S., 1940. Some aspects of the chondrocranium in the tadpoles of South Indian frogs. Half-Yearly Jour. Mysore Univ. Sect. B., **15**.

————, 1943. An account of the chondrocranium of Rana afghana and Megophrys, with a description of the masticatory musculature of some tadpoles. Proc. Natl. Inst. Sci. India, **9**, 43 ff.

————, 1944. The chondrocranium of two torrent-dwelling anuran tadpoles. Jour. Morphol., **74**, 347–374.

SETERS, W. H. VAN, 1921. De ontwikkeling van het chondrocranium van Alytes obstetricans voor de metamorphose. Thesis, Leiden.

————, 1922. Développement du chondrocrane d'Alytes obstetricans avant la métamorphose. Arch. Biol., **32**, 373–491.

SEWERTZOW, S. A., 1891. Ueber einige Eigenthümlichkeiten in der Entwicklung und im Bau des Schädels von Pelobates fuscus. Bull. Soc. Imp. Nat. Moscou, **1**.

STEEN, J. C. V. D., 1930. De ontwikkeling van de occipito-vertebrale streek van Microhyla. Thesis, Leiden.

STEPHENSON, N. G., 1951. On the development of the chondrocranium and visceral arches of Leiopelma archeyi. Trans. Zool. Soc. London, **27**, 203–253.

STÖHR, PH., 1881. Zur Entwicklungsgeschichte des Anuren-Schädels. Zeitschr. wiss. Zool., **36**, 68–103.

STOKELY, P. S., 1954. The progress of ossification in the skull of the cricketfrog Pseudacris nigrita triseriata. Copeia, 1954, 211–217.

Visceral skeleton and musculature of the head

CHIARUGI, G., 1890. Sui miotomi e sui nervi della testa posteriore e della regione prossimale del tronco negli embrioni degli Anfibi Anuri. Monitore Zool. Ital., **1**, 22–59.

————, 1891. Sur les myotomes et sur les nerfs de la tête postérieure et de la région proximale du tronc dans les embryons des Amphibiens Anoures. Arch. Ital. Biol., **15**.

EDGEWORTH, F. H., 1920. On the development of the hypobranchial and laryngeal muscles in Amphibia. Jour. Anat., **54**, 125–162.

*————, 1930. On the masticatory and hyoid muscles of larvae of Xenopus laevis. Jour. Anat., **64**, 184–188.

GAUPP, E., 1894. Das Hyo-branchial-Skelett der Anuren und seine Umwandlung. Morphol. Arb., **3**, 399 ff.

HELFF, O. M., 1928. Studies on amphibian metamorphosis. III. The influence of the annular tympanic cartilage on the formation of the tympanic membrane. Physiol. Zool., **1**, 463 ff.

KALTENBACH, J. C., 1953. Local action of thyroxin on amphibian metamorphosis. II. Development of the eyelids, nictitating membrane, cornea and extrinsic ocular muscles in Rana pipiens larvae effected by thyroxin-cholesterol implants. Jour. Exptl. Zool., **122**, 41–52.

KOTHE, K., 1910. Entwicklungsgeschichtliche Untersuchungen über das Zungenbein und die Ohrknöchelchen der Anuren. Arch. Naturgesch., **76**, 29–66.

*LUTHER, A., 1914. Ueber die vom N. trigeminus versorgte Muskulatur der Amphibien. Acta Soc. Sci. Fennica, **44**, 1–151.

MÄRTENS, M., 1895. Die Entwicklung des Knorpelgerüstes im Kehlkopf von Rana temporaria. Thesis, Göttingen.

————, 1897. Die Entwicklung der Kehlkopfknorpel bei einigen unserer einheimischen Anuren Amphibien. Anat. Hefte, I Abt. 28–30. **9**, 391–416.

MIYAWAKI, S., 1929. Die Entwicklung der Ohrkapsel und des schallleitenden Apparates von Hyla arborea japonica. Folia Anat. Japonica, **8**, 1–37.

NIESSING, C., 1933. Ueber das Zungenbein und der Kehlkopf von Pipa americana. Gegenbaurs morphol. Jahrb., **71**, 545–570.

PUSEY, H. K., 1938. Structural changes in the anuran mandibular arch during meta-morphosis, with reference to Rana temporaria. Quart. Jour. Microsc. Sci., **80**, 479–522.

RAMASWAMI, L. S., 1943. An account of the chondrocranium of Rana afghana and Megophrys, with a description of the masticatory musculature of some tadpoles. Proc. Natl. Inst. Sci. India, **9**, 43 ff.

RIDEWOOD, W. G., 1897. On the structure and development of the hyobranchial skeleton of the Parsely-frog (Pelodytes punctatus). Proc. Zool. Soc. London, 1897, 577–595.

————, 1898a. On the development of the hyobranchial skeleton of the Midwife-toad (Alytes obstetricans). Proc. Zool. Soc. London, 1898, 4–12.

*————, 1898b. On the structure and development of the hyobranchial skeleton and larynx in Xenopus and Pipa, with remarks on the affinities of the Aglossa. Jour. Linn. Soc. Zool., **26**, 53–128.

————, 1899. On the hyobranchial skeleton and larynx of the new aglossal toad Hymenochirus Boettgeri. Jour. Linn. Soc. Zool., **27**, 454 ff.

SALVADORI, G., 1928. Origine e sviluppo della columella auris negli Anfibi Anuri. Ric. Morf. Biol. anim., **1**.

SCHMALHAUSEN, I. I., 1954. Die Entwicklung der Kiemen und ihrer Blutgefässe und Muskulatur bei den Amphibien. Zoologicheskii Zhurnal (U.S.S.R.), **33**, 848–868 (in Russian).

SCHACHUNJAN, R., 1926. Die Entwicklung des visceralen Skeletts bei Amphibia Anura im Moment der Metamorphose. Rev. Zool. Russe, Moscow, **6**, 109–111.

SCHULZE, F. E., 1892. Ueber die inneren Kiemen der Batrachierlarven. II Mitteilung: Skelett, Muskulatur, Blutgefässe, Filterapparat, respiratorische Anhänge und Atmungsbewegungen erwachsener Larven von Pelobates fuscus. Abhandl. Preuss. Akad. Wiss. Berlin, 1892, XIII, 205 ff.

SEDRA, S. N., 1949. The metamorphosis of the jaws and their muscles in the toad, Bufo regularis Reuss, correlated with the changes in the animals feeding habits. Proc. Zool. Soc. London, **120**, 405 ff.

*SPANNHOF, L., 1954. Die Entwicklung des Mittelohres und des schallleitenden Apparates bei Xenopus laevis Daud. Zeitschr. wiss. Zool., **158**, 1–30.

STEINHEIM, 1846. Die Entwicklung des Froschembryos, insbesondere des Muskel-und Genitalsystems. Abhandl. Nat. wiss., Hamburg, 1846.

STEPHENSON, N. G., 1951. On the development of the chondrocranium and visceral arches of Leiopelma archeyi. Trans. Zool. Soc. London, **27**, 203–253.

TAKISAWA, A., Y. OHARA and K. KANO, 1952. Die Kaumuskulatur der Anuren (Bufo vulgaris japonicus) während der Metamorphose. Folia Anat. Japonica, **24**, 1–28.

————, Y. OHARA and Y. SUNAGA, 1952. Über die Umgestaltung der Mm. intermandi-bulares der Anuren während der Metamorphose. Folia Anat. Japonica, **24**, 217–241.

———— and Y. SUNAGA, 1951. Über die Entwicklung des M. depressor mandibulae bei Anuren im Laufe der Metamorphose. Folia Anat. Japonica, **23**, 273–293.

*VILLIERS, C. G. S. DE, 1931/32. Über das Gehörskelett der aglossen Anuren. Anat. Anz., **74**, 33–35.

VILLY, F., 1890. The development of the ear and the accessory organs in the common frog. Quart. Jour. Microsc. Sci., **30**, 523–550.

VIOLETTE, H. N., 1930. Origin of columella auris of Anura. Anat. Rec., **45**, 280 (Abstr.).

WITSCHI, E., 1947. Development and metamorphosis of the auxiliary apparatus of the ear of the frog. Anat. Rec., **99**, 568 (Abstr.).

SKELETON AND MUSCULATURE OF THE TRUNK

*BEDDARD, F. E., 1895. On the diaphragm and on the muscular anatomy of Xenopus, with remarks on its affinities. Proc. Zool. Soc. London, 1895, 841–850.

BERGFELDT, A., 1896. Chordascheiden und Hypochorda bei Alytes obstetricans. Anat. Hefte, **7**.

*BERNASCONI, A. F , 1951. Über den Ossifikationsmodus bei Xenopus laevis Daud. Thesis, Freiburg (Schweiz); Zürich.

CHIARUGI, G., 1890. Sui miotomi e sui nervi della testa posteriore e della regione prossimale del tronco negli embrioni degli Anfibi Anuri. Monitore Zool. Ital., **1**, 22–59.

————, 1891. Sur les myotomes et sur les nerfs de la tête postérieure et de la région proximale du tronc dans les embryons des Amphibien Anoures. Arch. Ital. Biol., **15**.

CLAUSEN, H. J., 1930. Rate of histolysis of anuran skin and muscle during metamorphosis. Biol. Bull., **59**, 199–210.

EMELIANOV, S., 1925. Die Entwicklung der Rippen und ihr Verhältnis zur Wirbelsäule. I. Die Entwicklung der Amphibienrippen. Rev. Zool. Russe, **5**, 1 ff.

ERDMANN, K., 1933. Zur Entwicklung des knöchernen Skeletts von Triton und Rana unter besonderer Berücksichtigung der Zeitfolge der Ossifikationen. Zeitschr. Anat., **101**, 566–651.

GEGENBAUR, C., 1861. Bau und Entwickelung der Wirbelsäule bei Amphibien überhaupt, und beim Frosche insbesondere. Abhandl. naturf. Ges. Halle, **6**, 1 ff.

*GROBBELAAR, C. S., 1935. The musculature of the Pipid genus Xenopus. S. African Jour. Sci., **32**, 395 ff (Abstr.).

GUARDABASSI, A., 1952. I sali di calcio del sacco endolinfatico ed i processi de calcificazione delle osse durante la metamorfosi normale e sperimentale nelle larve di Bufo vulgaris, Rana dalmatina e Rana esculenta. Boll. Soc. Ital. Biol. Sperim., **28**, 355–357.

————, 1953. Les sels de Ca du sac endolymphatique et les processus de calcification des os pendant la métamorphose normale et expérimentale chez les têtards de Bufo vulgaris, Rana dalmatina et Rana esculenta. Arch. Anat. Microsc. et Morphol. Exptl., **42**, 143–167.

HASSE, C., 1892. Die Entwicklung der Wirbelsäule der ungeschwänzten Amphibien. Zeitschr. wiss. Zool., **55**, 252–264.

HODLER, F., 1949a. Zur Entwicklung der Anurenwirbelsäule. Eine morphologisch-entwicklungsphysiologische Studie. Rev. Suisse Zool., **56**, 327–330.

————, 1949b. Untersuchungen über die Entwicklung von Sacralwirbel und Urostyl bei den Anuren. Ein Beitrag zur Deutung des anuren Amphibientypus. Rev. Suisse, Zool., **56**, 747–790. *

HOWES, G. B., 1893. Notes on the variation and development of the vertebral and limb-skeleton of the Amphibia. Proc. Zool. Soc. London, 1893, 268 ff.

*HERING, H. VON, 1880. Ueber die Wirbelsäule von Pipa. Morphol. Jahrb., **6**, 297–314.

*KÄLIN, J. and A. BERNASCONI, 1949. Ueber den Ossifikationsmodus bei Xenopus laevis Daud. Rev. Suisse Zool., **56**, 359–364.

KAPELKIN, W., 1900. Zur Frage über die Entwicklung des axialen Skelettes der Amphibien. Bull. Soc. Nat. Moscou, **14**.

KLAATSCH, H., 1897. Ueber die Chorda und die Chordascheiden der Amphibien. Verhandl. Anat. Ges., **11**.

*MOOKERJEE, H. K., 1931. On the development of the vertebral column of Anura. Phil. Trans. Roy. Soc. London, Ser. B. Biol. Sci., **219**, 165–196.

————, 1938. On the development of the transverse process and the rib of Salientia (Anura). Anat. Anz., **87**, 239 ff.

*———— and S. K. Das, 1939. Further investigations on the development of the vertebral column in Salientia (Anura). Jour. Morphol., **54**, 167–209.

Peeters, J. L. E., 1910. Over de ontwikkeling van het chondrocranium en de kraakbenige wervelkolom van eenige Urodela en Anura. Thesis, Leiden.

*Ridewood, W. G., 1897. On the development of the vertebral column in Pipa and Xenopus. Anat. Anz., **13**, 359–376.

*Ryke, P. A. J., 1953. The ontogenetic development of the somatic musculature of the trunk of the Aglossal Anuran Xenopus laevis (Daudin). Acta Zool., **34**, 1–70.

Steinheim, 1846. Die Entwicklung des Froschembryos, insbesondere des Muskel- und Genitalsystems. Abhandl. Nat. wiss., Hamburg, 1846.

TAIL (all organ systems)

Barfurth, D., 1887. Die Rückbildung des Froschlarvenschwanzes und die sogenannten Sarcoplasten. Arch. microsc. Anat., **29**.

Bergel, A., 1928. Ueber homioplastische Transplantation des Schwanzblastems bei Rana fusca. Arch. Entwicklungsmech., **113**, 172–209.

Bijtel, J. H., 1929. Development of amphibian tail. Thesis, Groningen.

————, 1931. Ueber die Entwicklung des Schwanzes bei Amphibien. Arch. Entwicklungsmech., **125**, 448–486.

Brown, M. E., 1946. The histology of the tadpole tail during metamorphosis with special reference to the nervous system. Am. Jour. Anat., **78**, 79 ff.

Clausen, H. J., 1930. Rate of histolysis of anuran skin and muscle during metamorphosis. Biol. Bull., **59**, 199–210.

Eberth, C. J., 1866. Zur Entwicklung der Gewebe im Schwanze der Froschlarven. Arch. microsc. Anat., **2**, 490–503.

————, 1894. Die Sarkolyse (Froschlarve). Festschr. 200-jähr. Jubelf. Univ. Halle.

Etkin, W., 1932. Growth and resorption phenomena in anuran metamorphosis. Physiol. Zool., **5**.

Guieysse, A., 1905. Etude de la régression de la queue chez les têtards des amphibiens anoures. Arch. Anat. Microsc. et Morphol. Exptl., **7**.

Harrison, R. G., 1898. The growth and regeneration of the tail of the frog larva (studied with the aid of Born's method of grafting). John's Hopkins Hosp. Bull., **103**; Arch. Entwicklungsmech., **7**, 430, 61 pp.

Helff, O. M., 1930. Studies in amphibian metamorphosis. viii. The rôle of the urostyle in the atrophy of the tail. Anat. Rec., **47**, 177 ff.

———— and H. J. Clausen, 1929. Studies on amphibian metamorphosis. v. The atrophy of anuran tail muscle during metamorphosis. Physiol. Zool., **2**, 575 ff.

Hensen, V., 1868. Ueber die Nerven im Schwanz der Frosch-larva. Arch. microsc. Anat., **4**, 115 ff.

*Jurand, A., 1954. The influence of certain drugs of the nervous system on the regeneration of the tail in Xenopus laevis tadpoles. i. Sodium bromide. Folia Biol. Krakow, **2**, 201–214.

Kuipers, H., 1931. Microscopisch onderzoek naar de chordaregeneratie bij kikvorslarven. Nederlands Tijdschr. Geneesk., **75**, 840–842.

Lindeman, V. F., 1929. Integumentary pigmentation in the frog, Rana pipiens, during metamorphosis, with especial reference to tailskin histolysis. Physiol. Zool., **2**, 255 ff.

Loos, 1889a. Ueber die Betheiligung der Leukocyten an dem Zerfall der Gewebe im Froschlarvenschwanz während der Reduktion desselben. Leipzig.

————, 1889b. Ueber Degenerationserscheinungen im Thierreich, besonders über die Reduktion des Froschlarvenschwanzes und die im Verlauf desselben auftretenden histologischen Prozesse. Jablon. Ges. Leipzig, 1889.

*Lüscher, M., 1946a. Die Entstehung polyploider Zellen durch Colchicinbehandlung im Schwanz der Xenopus-larve. Arch. Julius Klaus-Stiftung, **21**, 303–305.

*————, 1946b. Hemmt oder fordert Colchicin die Zellteilung im regenerierenden Schwanz der Xenopus-Larve? Rev. Suisse Zool., **53**, 481–486.

*————, 1946c. Die Wirkung des Colchicins auf die an der Regeneration beteiligten Gewebe im Schwanz der Xenopus-Larve. Rev. Suisse Zool., **53**, 683–732.

*————, 1946d. Die Hemmung der Regeneration durch Colchicin beim Schwanz der Xenopus-Larve und ihre entwicklungsphysiologische Wirkungsanalyse. Helvetica Physiol. et Pharmacol. Acta, **4**, 465–494.

Manicastri, 1903. La rigenerazione di parti laterali delle code di larve di Anuri. Monitore Zool. Ital., 1903.

Morgan, T. H. and S. E. Davis, 1902. The internal factors in the regeneration of the tail of the tadpole. Arch. Entwicklungsmech., **15**.

Morse, W., 1918. Factors involved in the atrophy of the organs of the larval frog. Biol. Bull., **34**, 149 ff.

Nakamura, O., 1936. Relation between limb development and tail resorption in Anura. Bot. and Zool. Jap., **6**.

Naville, A., 1924. Recherches sur l'histogénèse et la régénération chez les Batraciens Anoures (corde dorsale et téguments). Arch. Biol., **34**, 235–344.

Noetzel, W., 1895. Die Rückbildung der Gewebe im Schwanz der Froschlarve. Arch. microsc. Anat., **45**, 475–510.

Pisanò, A., 1942. Espianti di territori presuntivi caudali da neurule di Rana esculenta. Ricerche Morfol., 63 pp.

*Roguski, H., 1953. Die Regeneration des Schwanzes bei Kaulquappen von Xenopus laevis. Folia Biol. Warszawa, **1**, 7–22 (in Polish).

*————, 1954. Influence of the medulla spinalis on the regeneration of the tail in tadpoles (Xenopus laevis). Part i. Folia Biol. Warszawa, **2**, 189–200 (in Polish).

Rondinini, R., 1928. Particolarità formative in alcuni organi primitivi e sviluppo della coda nelle larve di Bufo vulgaris. Arch. Ital. Anat. e Embriol., **25**, 98–130.

Schechtman, A. M., 1938. Localization of the neural inductor and tail mesoderm in a frog egg (Hyla regilla). Proc. Soc. Exptl. Biol. and Med., **39**, 236–239.

Schubert, M., 1924. Untersuchungen über die Rückbildungserscheinungen der Epidermis des Froschlarvenschwanzes. Anat. Anz., **58**.

Vonwiller, P., 1919. Ueber die Reduktion der Schwanzmuskulatur bei der Metamorphose der Anuren. Vortr. Verhandl. Schweiz. Naturf. Ges., 1919.

SHOULDER GIRDLE AND FORELIMBS, PELVIC GIRDLE AND HINDLIMBS (all organ systems)

General publications

*Beetschen, J. C., 1952. Extension et limites du pouvoir régénérateur des membres après la métamorphose chez Xenopus laevis Daudin. Bull. Biol., **86**, 88–100.

*Bernasconi, A. F., 1951. Über den Ossifikationsmodus bei Xenopus laevis Daud. Thesis, Freiburg (Schweiz); Zürich.

Braus, H., 1904. Die Entwicklung der Form der Extremitäten und des Extremitätenskeletts. Hertwig, Handbuch der Vergleichende und experimentelle Entwicklung der Wirbeltiere, Jena, **3**, 167 ff.

————, 1909. Gliedmassenpfropfung und Grundfragen der Skelettbildung. i. Die Skelettanlage vor Auftreten des Vorknorpels und ihre Beziehung zu den späteren Differenzierungen. Exptl. Beitr. Morphol., **1**, 284–430.

BYRNES, E. F., 1898. On the regeneration of limbs in frogs after the extirpation of limb-rudiments. Anat. Anz., **15**.

*DALCQ, A., 1950. Les premiers stades de l'ossification des os longs chez Xenopus laevis. Compt. Rend. Ass. Anat., **61**, 85–89.

ERDMANN, K., 1933. Zur Entwicklung des knöchernen Skeletts von Triton und Rana unter besonderer Berücksichtigung der Zeitfolge der Ossifikationen. Zeitschr. Anat., **101**, 566–651.

*FOX, E. and J. T. IRVING, 1950. The ossification process in the long bones of Xenopus laevis. S. African Jour. Med. Sci., **15**, 5–10.

*GROBBELAAR, C. S., 1924. Beiträge zu einer anatomischen Monographie von Xenopus laevis (Daud.). Zeitschr. Anat. u. Entwicklungsgesch., **72**, 131–168.

————, 1935. The musculature of the Pipid genus Xenopus. S. African Jour. Sci., **32**, 395 ff (Abstr.).

GUARDABASSI, A., 1952. I sali di calcio del sacco endolinfatico ed i processi di calcificazione delle osse durante la metamorfosi normale e sperimentale nelle larve di Bufo vulgaris, Rana dalmatina e Rana esculenta. Boll. Soc. Ital. Biol. Sperim., **28**, 355–357.

HAMBURGER, V., 1925. Ueber den Einfluss des Nervensystems auf die Entwicklung der Extremitäten von Rana fusca. Arch. Entwicklungsmech., **105**, 149–201.

HOWES, G. B., 1893. Notes on the variation and development of the vertebral and limb-skeleton of the Amphibia. Proc. Zool. Soc. London, 1893, 268 ff.

*KÄLIN, J. and A. BERNASCONI, 1949. Ueber den Ossifikationsmodus bei Xenopus laevis Daud. Rev. Suisse Zool., **56**, 359–364.

NAKAMURA, O., 1936. Relation between limb development and tail resorption in Anura. Bot. and Zool. Jap., **6**.

POLEZAJEW, L. W., 1945. Limb regeneration in adult frog. Compt. Rend. Acad. Sci. U.R.S.S., **49**, 609–612.

————, 1946a. Morphological data on regenerative capacity in tadpole limbs as restored by chemical agents. Compt. Rend. Acad. Sci. U.R.S.S., **54**, 281–284.

————, 1946b. Further investigations on the regeneration of limbs in adult Anura. Compt. Rend. Acad. Sci. U.R.S.S., **54**, 461–464.

————, 1946c. Morphology of limb regenerates in adult Anura. Compt. Rend. Acad. Sci. U.R.S.S., **54**, 653–656.

*ROSTAND, J., 1932. De la faculté régénératrice chez Xenopus laevis. Compt. Rend. Soc. Biol., **111**, 451 ff.

TAYLOR, A. C., 1943. Development of the innervation pattern in the limb bud of the frog. Anat. Rec., **87**, 379–409.

————, 1944. Selectivity of nerve fibres from the dorsal and ventral roots in the development of the frog limb. Jour. Exptl. Zool., **96**, 159–185.

WEBER, A., 1933. Origine de l'ébauche des membres chez un Batracien anoure, Bombinator pachypus. Compt. Rend. Soc. Biol., **112**, 1269–1272.

YOUNGSTROM, K. A., 1937. Correlated anatomical and physiological studies of the developing legs of anuran tadpoles. Anat. Rec., **67**, 56 ff.

Shoulder girdle, forelimbs and opercular perforation

ALPHONSE, P. and G. BAUMANN, 1935. Contribution a l'étude de la métamorphose expérimentale des Amphibiens anoures sous l'action de la thyroxine. La perforation de l'opercule branchial. Arch. Anat. Histol. et Embryol., **19**.

*BERK, L., 1939. Studies in the reproduction of Xenopus laevis. III. The secondary sex characters of the male Xenopus laevis: the pads. S. Afr. Jour. Med. Sci., **4**, 47–60.

BLACHER, L. J., E. D. LIOSNER and M. A. WORONZOWA, 1934. Mechanismus der Perforation der operculären Membran der schwanzlosen Amphibien. Bull. Internatl. Acad. Polonaise Sci. et Lettres., Sér. B., 1934, 325 ff.

BRAUS, H., 1905. Ueber den Entbindungsmechanismus beim äuszerlichen Hervortreten der Vorderbeine der Unke und über künstliche Abrachie. Münchener med. Wochenschr., 1905 B.

————, 1906. Vordere Extremität und Operculum bei Bombinatorlarven. Ein Beitrag zur Kenntnis morphogener Correlation und Regulation. Morphol. Jahrbuch, **35**, 139–216.

————, 1907. Ueber Frühanlagen der Schultermuskeln bei Amphibien und ihre allgemeinere Bedeutung. Verhandl. Anat. Geselsch. Würzburg, **21**, 192–219.

BYRNES, E. F., 1904. Regeneration of the anterior limbs in the tadpoles of frogs. Arch. Entwicklungsmech., **18**, 171 ff.

EKMAN, G., 1912. Die Entstehung des Peribranchialraumes und seine Beziehungen zur Extremitätenanlage bei Bombinator. Vorläufige Mitteilungen über experimentelle Untersuchungen. Anat. Anz., **40**, 580–586.

FISCHER, E., 1928. (Carpus and tarsus of Anura.) Thesis, Bern.

*FUCHS, H., 1930. Beiträge zur Entwicklungsgeschichte und vergleichende Anatomie des Brustschultergürtels der Wirbeltiere. Gegenbaurs Jahrb., **64**, 1–132.

*GALLIEN, L., 1947. Caractère ambisexuel de la callosité chez Xenopus laevis. Bull. Soc. Zool. France, **72**, 192–193.

HELFF, O. M., 1926. Studies on amphibian metamorphosis. I. Formation of the opercular leg perforation in Anuran larvae during metamorphosis. Jour. Exptl. Zool., **45**, 1–67.

————, 1939. Studies on amphibian metamorphosis. XVI. The development of the forelimb opercular perforations in Rana temporaria and Bufo bufo. Jour. Exptl. Biol., **16**, 96–120.

*HOFFMAN, A. C., 1930. Opsomming van nuwe navorsinge oor die opbou en ontogenese van die zonaalskelet by Amphibie, vernaamlik van Cryptobranchus alleghaniensis, Necturus maculatus, Heleophryne en Xenopus laevis. S. African Jour. Sci., **27**, 446–450 (English summary).

*HOWES, G. B. and W. RIDEWOOD, 1888. On the carpus and tarsus of the Anura. Proc. Zool. Soc. London, **11**, 141–182.

HOWELL, A. B., 1935. Morphogenesis of the shoulder architecture. Part III. Amphibia. Quart. Rev. Biol., **10**, 397–431.

JANES, E. I., 1933. Observations on the pectoral musculature of Amphibia Salientia. A. The variability of the musculature in relation to the reduction of the clavicle, procoracoid, and episternum. Ann. Mag. Nat. Hist., **12**, 403–420.

*JONES, T. R., 1938. The hand and foot musculature of three South African Anurans, Rana fuscigula, Pyxicephalus adspersus, and Xenopus laevis, and its evolutionary significance. Thesis, Witwatersrand.

JORDAN, P., 1888. Die Entwicklung der vorderen Extremität der anuren Batrachier. Thesis, Leipzig.

*JUNGERSEN, H. F. E., 1891. Remarks on the structure of the hand in Pipa and Xenopus. Ann. Mag. Nat. Hist., **8**, 193–206.

LIGNITZ, W., 1897. Die Entwicklung des Schultergürtels beim Frosch. Thesis, Leipzig.

LIOSNER, L. D. and M. A. WORONZOWA, 1937. Ueber den Mechanismus der Operculum-Perforation bei thyreoïdbedingter Metamorphose. Bull. Biol. Med. Exptl. U.R.S.S., **4**, 303 ff.

———— and ————, 1940. On the mechanism of opercular membrane perforation during metamorphosis of Anura. Compt. Rend. Acad. Sci. U.R.S.S., N. S., **26,** 819 ff.

NAUCK, E. TH., 1928. Beiträge zur Kenntnis des Skeletts der paarigen Gliedmassen der Wirbeltiere. V. Die Entwicklung des ventralen Schultergürtelabschnittes bei Alytes obstetricans. Gegenbaur's Morphol. Jahrb., **60**, 61–77.

*NEWTH, D. R., 1948. The early development of the fore-limbs in Xenopus laevis. Proc. Zool. Soc. Lond., **118**, 559 ff.

*———, 1949. A contribution to the study of forelimb eruption in metamorphosing Anura. Proc. Zool. Soc. Lond., **119**, 643–659.

PLETZEN, R. VAN, 1953. Ontogenesis and morphogenesis of the breast-shoulder apparatus of Xenopus laevis. Ann. Univ. Stellenbosch, 29A, 137–184.

*PROCTER, J. B., 1921. On the variation of the scapula in the Batrachian groups Aglossa and Arcifera. Proc. Zool. Soc. London, 1921, 197–214.

ROMEIS, B., 1924. Histologische Untersuchungen zur Analyse der Schilddrüsenfütterung auf Froschlarven. II. Die Beeinflussung der Entwicklung der vorderen Extremitäten und des Brustschulterapparates. Arch. Entwicklungsmech., **101**.

SCHMALHAUSEN, J. J., 1907. Die Entwicklung des Skelettes der vorderen Extremität der anuren Amphibien. Anat. Anz., **31**, 177 ff.

VILLIERS, C. G. S. DE, 1922. Neue Beobachtungen über den Bau und die Entwicklung des Brustschulterapparates bei den Anuren, insbesondere bei Bombinator. Acta Zool. (Stockholm), **3**, 153 ff.

*———, 1924. On the anatomy of the breast-shoulder apparatus of Xenopus. Ann. Transvaal Mus., **10**, 197–211.

*———, 1929. The comparative anatomy of the breast-shoulder apparatus of the three Aglossal Anuran genera: Xenopus, Pipa and Hymenochirus. Ann. Transvaal Mus., **13**, 37–69; Trans. Roy. Soc. S. Africa, **17**, 23 ff.

WEBER, A., 1932. Pénétration des fibres nerveuses dans les bourgeons des membres antérieures chez un Batracien anoure, "Bombinator pachypus". Compt. Rend. Assoc. Anat., **27**, 610–612.

Pelvic girdle and hindlimbs

*BRETSCHER, A., 1947. Reduktion der Zehenzahl bei Xenopus-Larven nach lokaler Colchicinbehandlung. Rev. Suisse Zool., **54**, 273–279.

*———, 1949. Die Hinterbeinentwicklung von Xenopus laevis Daud. und ihre Beeinflussung durch Colchicin. Rev. Suisse Zool., **56**, 34–96.

FISCHER, E., 1928. (Carpus and tarsus of Anura.) Thesis, Bern.

*GAUPP, E., 1896. A. Ecker's und R. Wiedersheim's Anatomic des Frosches. I. (3rd ed.). Braunschweig.

*GREEN, T. L., 1931. On the pelvis of the Anura; a study in adaption and recapitulation. Proc. Zool. Soc. London, 1931, 1259–1290.

*GROBBELAAR, C. S., 1924. Beiträge zu einer anatomischen Monographie von Xenopus laevis (Daud.). Zeit. Anat. Entw.gesch., **72**, 131–168.

*HOFFMAN, A. C., 1930. Opsomming van nuwe navorsinge oor die opbou en ontogenese van die zonaalskelet by Amphibie, vernaamlik van Cryptobranchus alleghaniensis, Necturus maculatus, Heleophryne en Xenopus laevis. S. African Jour. Sci., **27**, 446–450 (English summary).

*HOWES, G. B. and W. RIDEWOOD, 1888. On the carpus and tarsus of the Anura. Proc. Zool. Soc. London, **11**, 141–182.

*JONES, T. R., 1938. The hand and foot musculature of three South African Anurans, Rana fuscigula, Pyxicephalus adspersus, and Xenopus laevis, and its evolutionary significance. Thesis, Witwatersrand.

SCHMALHAUSEN, J. J., 1908. Die Entwicklung des Skelettes der hinteren Extremität der anuren Amphibien. Anat. Anz., **33**, 337 ff.

TSCHERNOFF, N. D., 1907. Zur Embryonalentwicklung der hinteren Extremitäten des Frosches. Anat. Anz., **30**.

*TSCHUMI, P., 1954. Konkurrenzbedingte Rückbildungen der Hinterextremitäten von Xenopus nach Behandlung mit einem Chloraethylamin. Rev. Suisse Zool., **61**, 177–270.

*VILLIERS, C. G. S. DE, 1925. On the development of the "Epipubis" of Xenopus. Ann. Transvaal Mus., **11**, 61 ff.

HEART AND VASCULAR SYSTEM, LYMPHATIC SYSTEM AND SPLEEN

BAUMANN, A., 1933. Développement de l'innervation cardiaque primitive chez un Batracien anoure, Bombinator pachypus. Compt. Rend. Soc. Biol., **113**, 1381–1382.

———, 1934. Observations sur les premiers stades de l'innervation du cœur chez un Batracien anoure (Bombinator pachypus Bonap.). Rev. Suisse Zool., **41**, 235–261.

*BEDDARD, F. E., 1908. Some notes on the muscular and visceral anatomy of the batrachian genus Hemisus, with notes on the lymph-hearts of this and other genera. Proc. Zool. Soc. London, 1908, 894–934 (post. lymph-hearts and fat body, Xenopus).

CHORONSHITZKY, B., 1900. Die Entstehung der Milz, Leber, Gallenblase, Bauchspeicheldrüse und des Pfortadersystems bei den verschiedenen Abteilungen der Wirbeltiere. Anat. Hefte, I. Abt., **13**, H. 42/43, 369–620.

CLARK, E. R., 1909. Observations on living growing lymphatics in the tail of the frog larva. Anat. Rec., **3**.

EKMAN, G., 1929. Experimentelle Untersuchungen über die früheste Herzentwicklung bei Rana fusca. Arch. Entwicklungsmech., **116**, 327–347.

ENGELS, H. J., 1935. Ueber Umbildungsvorgänge im Kardinalvenensystem bei Bildung der Urniere. Morphol. Jahrb., **76**, 345–374.

FERNALD, R. L., 1943. The origin and development of the blood island of Hyla regilla. Univ. California Publ. Zool., **51**, 129–148.

FRIEDEL, H., 1933. Ueber das Inguinalorgan bei Anurenarten. Zeitschr. Anat. Entwicklungsgesch., **102**, 175–193.

GOLUBEW, A., 1869. Beiträge zur Kenntnis des Baues und der Entwicklungsgeschichte der Kapillargefässe des Frosches. Arch. Microsc. Anat., 1869.

*GROBBELAAR, C. S., 1924. On the venous and arterial systems of the Platanna (Xenopus laevis Daud.). S. African Jour. Sci., **21**, 392–398.

GRÖNBERG, G. and A. V. KLINCKOWSTRÖM, 1893. Zur Anatomie der Pipa americana III. Gefässsystem und subcutane Lymphsäcke. Zool. Jahrb., **7**, 647–666.

HOCHSTETTER, F., 1888. Beiträge zur vergleichenden Anatomie und Entwicklungsgeschichte des Venensystems der Amphibien und Fische. Morphol. Jahrb., **13**, 119–172.

HOYER, H., 1905. Ueber das Lymphgefässsystem der Froschlarven. Verhandl. deutsch. Anat. Gesellsch., **19**, 50 ff.

JOLLY, J., 1944. Recherches sur la formation du système vasculaire des Batraciens. Bull. Biol., **78**, 124–135.

——— and C. LIEURE, 1932. Sur les cœurs lymphatiques des larves d'Anoures. Compt. Rend. Soc. Biol., **110**. 12–14.

——— and ———, 1941. La culture du bourgeon caudal des Batraciens et la formation des vaisseaux de l'embryon. Compt. Rend. Soc. Biol., **135**, 702 ff.

——— and ———, 1942. La culture de la plaque médullaire des Batraciens et la formation des vaisseaux de l'embryon. Compt. Rend. Soc. Biol., **135**, 352 ff.

KAMPMEIER, O. F., 1915. On the origin of lymphatics in Bufo. Ann. Jour. Anat., **17**, 161–183.

———, 1920. The changes of the systemic venous plan during development and the relation of the lymph hearts to them in Anura. Anat. Rec., **19**, 83–96.

KNOWER, H. MCE., 1908. The origin and development of the anterior lymph hearts and the subcutaneous sacs in the frog. Anat. Rec., **2**, 59 ff.

KRAATZ, A., 1897. Zur Entstehung der Milz. Thesis. Marburg.

LILLIE, R. D., 1919. Early histogenesis of the blood in Bufo halophilus (Baird and Girard). Amer. Jour. Anat., **26**, 209–235.

MARCINOWSKI, K., 1906. Zur Entstehung der Gefässendothelien und des Blutes bei Amphibien. Jenaische Zeitschr. Naturwiss., **41**, 19–109.

MARSHALL, A. M. and E. J. BLES, 1890. The development of the blood vessels in the frog. Stud. Biol. Lab. Owens Coll., **2**, 81 pp.

MAURER, F. VON, 1888. Die Kiemen und ihre Gefässe bei anuren und urodelen Amphibien und die Umbildungen der beiden ersten Arterienbogen bei Teleostiern. Morphol. Jahrb., **14**, 175–222.

————, 1898. Die Vaskularisierung der Epidermis bei anuren Amphibien zur Zeit der Metamorphose. Morphol. Jahrb., **26**, 330–336.

MIETENS, H., 1909. Entstehung des Blutes bei Bufo vulgaris. Jenaische Zeitschr. Naturwiss., **45**, 299 ff.

MIJERS, M. A., 1928. A study of the tonsillar developments in the lingual region of Anurans. Jour. Morphol., **45**, 399–433.

*MILLARD, N., 1941. The vascular anatomy of Xenopus laevis (Daudin). Trans. Roy. Soc. S. Africa, **28**, 387–439.

*————, 1942. Abnormalities and variations in the vascular system of Xenopus laevis (Daudin). Trans. Roy. Soc. S. Africa, **29**, 9–28.

*————, 1945. The development of the arterial system of Xenopus laevis, including experiments on the destruction of the larval aortic arches. Trans. Roy. Soc. S. Africa, **30**, 217–234.

*————, 1949. The development of the venous system of "Xenopus laevis". Trans. Roy. Soc. S. Africa, **32**, 55–97.

MÖLLENDORF, W. VON, 1911. Ueber die Entwicklung der Darmarterien und des Vornierenglomerulus bei Bombinator. Ein Beitrag zur Kenntnis des visceralen Blutgefässsystems und seiner Genese bei den Wirbeltieren. Morphol. Jahrb., **43**, 579–646.

————, 1913. Ueber Anlage und Ausbildung des Kiemenbogenkreislaufs bei Anuren (Bombinator pachypus). Anat. Hefte, **47**, 249 ff.

*NIKITIN, B., 1925. Some particularities in the development of the vascular system o Xenopus. Bull. Soc. Nat. Moscou, Sec. Biol. N.S., **34**, 286–308 (English summary).

NUSSBAUM, J., 1894. Zur Entwicklungsgeschichte der Gefässendothelien und der Blutkörperchen bei den Anuren. Biol. Zentralbl., **13**, 93 ff.

OELLACHER, 1871. Ueber die erste Entwicklung des Herzens und der Pericardhöhle bei Bufo cinereus. Arch. Microsc. Anat., **7**.

*ORTLEPP, R. J., 1918. Note on the persistence of the right posterior cardinal vein in Xenopus laevis, and its significance. S. African Jour. Sci., **15**, 413–415.

*PATERSON, N. F., 1942. The anterior blood-vessels of Xenopus laevis. S. African Jour. Sci., **38**, 279–285.

PINNER, .G, 1950. Ueber die Entwicklung der Spindelzellen, sowie über die embryonale Blutbildung und die damit zusammenhängende Organe bei Rana esculenta. Oesterreich. zool. Zeitschr., **2**, 639–646.

RUFFINI, A., 1899. Sullo sviluppo della milza nella Rana esculenta. Monitore Zool. Ital., **10**, 91–92.

SCHMALHAUSEN, I. I., 1954. Die Entwicklung der Kiemen und ihrer Blutgefässe und Muskulatur bei den Amphibien. Zoologicheskii Zhurnal (U.S.S.R.), **33**, 848–868 (in Russian).

SHORE, T. W., 1902. The development of the renal-portals and fate of the posterior cardinal veins in the frog. Jour. Anat. Physiol., **36**, 20–53.

SOCHA, P., 1930. Die Entwicklung der Blutgefässe des Gehirns vom Grasfrosch. Bull. Acad. Polonaise Sci. et Lettres, Cl. Sci. Math. et Natl. Ser. B.I., 1930, 479–492.

SOLENSKY, W., 1895. Sur le développement du cœur chez les embryons de la grenouille. Compt. Rend. 3e Congr. Internat. Zool., Leyden.

WEBER, A., 1909. Etude sur la torsion de l'ébauche cardiacque chez Rana esculenta. Bibl. Anat., **18**, 136–141.

WEIDENREICH, F., 1933. Die Lymphgefässe der Anamnier. In: BOLK-GÖPPERT-KALLIUS-LUBOSCH, Handbuch der vergl. Anat. der Wirbeltiere, Berlin, **6**, 760–782.

WHITNEY, W. U., 1867. On the changes which accompany the metamorphosis of the tadpole, in reference especially to the respiratory and sanguiferous systems. Trans. Roy. Micr. Soc., **15**, 43.

NEPHRIC SYSTEM

CAMBAR, R., 1949. Données récentes sur le développement du système pronéphrétique chez les Amphibiens (anoures en particulier). Ann. Biol., **25**, 115–130.

FILATOW, D. P., 1904. Zur Entwicklungsgeschichte des Exkretionssystems der Amphibien. Anat. Anz., **25**, 33 ff.

GEERTRUYDEN, J. VAN, 1942. Quelques précisions sur le développement du pronéphros et de l'uretère primaire chez les Amphibiens anoures. Ann. Soc. Roy. Zool. Belgique, **73**, 180–195.

————, 1946. Recherches expérimentales sur la formation du mésonéphros chez les Amphibiens anoures. Arch. Biol., **57**, 145–181.

————, 1948. Les premiers stades de développement du mésonéphros chez les Amphibiens anoures. Acta Neerl. Morphol., **6**, 1–17.

GRAY, P., 1930. The development of the amphibian kidney. Pt. I. The development of the mesonephros of Rana temporaria. Quart. Jour. Microsc. Sci., **73**, 507–546.

————, 1936. The development of the amphibian kidney. Pt. III. The post-metamorphic development of the kidney, and the development of the vasa efferentia and seminal vesicles in Rana temporaria. Quart. Jour. Microsc. Sci., **78**, 445–473.

HALL, R. W., 1904. The development of the mesonephros and the Müllerian ducts in Amphibia. Bull. Mus. Comp. Zool. Harvard Univ., **45**, 32–123.

JAFFEE, O. CH., 1954. Morphogenesis of the pronephros of the Leopard frog (Rana pipiens). Jour. Morphol., **95**, 109–123.

*KUNST, J., 1936. Vergleichende Untersuchungen an Anurennieren insbesondere an der Niere von Xenopus laevis Daud. Zool. Jahrb. Abt. 2. Anat. u. Ontog. Tiere, **61**, 51–76.

MARSHALL, A. M. and E. J. BLES, 1890. The development of the kidneys and fatbodies in the frog. Stud. Biol. Lab. Owens Coll., **2**, 133–158.

MATSUKURA, Y., 1935. Ueber die Entwicklung des Vornieren-Kanälchensystems der Japan. Kröte (Bufo formosus). Folia Anat. Japonica, **13**, 417–441.

GONADAL SYSTEM AND ADRENALS

Gonads

ALLEN, B. M., 1907. An important period in the history of the sex-cells of Rana pipiens. Anat. Anz., **31**, 339–347.

BECCARI, N., 1924. Studii sulla prima origine delle cellule genitali nei Vertebrati. III: Ricerche nel Bufo viridis. Arch. Ital. Anat. e Embriol., **21**, 352–374.

————, 1925. Studii sulla prima origine delle cellule genitali nei Vertebrati. IV: Ovogenesi larvale, organo del Bidder e differenziamento dei sessi nel Bufo viridis. Arch. Ital. Anat. e Embriol., **22**, 483 ff.

BOUIN, M., 1900. Histogénèse de la glande génitale femelle chez Rana temporaria. Arch. Biol., **17**, 201–381.

BOUNOURE, L., 1929a. Sur l'existence des gonocytes primaires dans l'embryon de la Grenouille rousse à partir du début de la gastrulation; localisation et migration de ces gonocytes aux différents stades. Compt. Rend. Soc. Biol., **101**, 706–708.

————, 1929b. Sur un caractère cytologique essentiel des gonocytes primaires chez la Grenouille rousse. Compt. Rend. Soc. Biol., **101**.

————, 1935. Sur la possibilité de réaliser une castration dans l'œuf de la Grenouille rousse; résultats anatomiques. Compt. Rend. Soc. Biol., **120**, 1316–1319.

————, 1937a. La constitution des glandes génitales chez la Grenouille rousse après destruction étendue de la lignée germinale par l'action des rayons ultraviolets sur l'œuf. Compt. Rend. Acad. Sci., **204**, 1957–1959.

————, 1937b. Les suites de l'irradiation du déterminant germinal, chez la Grenouille rousse, par les rayons ultraviolets: résultats histologiques. Compt. Rend. Soc. Biol., **125**, 898–900.

————, 1939. L'origine des cellules reproductrices et le problème de la lignée germinale. Paris.

BUSETTO, I., 1943. Sull'origine interrenale del tessuto midollare della gonade nei Bufonidi e nei Discoglossidi (nota preliminare). Atti. R. Ist. Veneto Sc. Lett. Arti, Cl. Sc. Mat. Nat., **102**, 791 ff.

*CHANG, C. Y., 1953. Parabiosis and gonad transplantation in Xenopus laevis Daudin. Jour. Exptl. Zool., **123**, 1–28.

*———— and E. WITSCHI, 1955. Breeding of sex-reversed males of Xenopus laevis Daudin. Proc. Soc. Exptl. Biol. and Med. **89**, 150–152.

CHENG, T. H., 1932a. The germ-cell history of Rana cantabrigensis Baird. I. Germ-cell origin and gonad formation. Zeitschr. Zellforsch. Microsc. Anat., **16**, 497–541.

————, 1932b. The germ-cell history of Rana cantabrigensis Baird. II. Sex differentiation and development. Zeitschr. Zellforsch. microsc. Anat., **16**, 541–596.

CHRISTENSEN, K., 1930. Sex differentiation and development of oviducts in Rana pipiens. Amer. Jour. Anat., **45**, 159 ff.

DEAL, R. E., 1931. The development of sex characters in the tree frog. Anat. Rec., **48**, 27–45.

EGAWA, N., 1936. Ueber die postlarvale Entwicklung des Hodens japanischer Kröten (Bufo vulgaris japonicus). Kaibogaku Zasshi, **9**, 584 ff.

————, 1937. Ueber die larvale Entwicklung der indifferenten Keimdrüse der Rana nigromaculata. Osaka Ijishinshi, **8**, 532 ff.

*GALLIEN, L., 1947. Caractère ambisexuel de la callosité chez Xenopus laevis. Bull. Soc. Zool. France, **72**, 192–193.

————, 1951. Les recherches expérimentales sur l'action des hormones sexuelles dans la différenciation et l'organogénèse du sexe chez les Batraciens. Année Biol., **27**, 653–670.

*————, 1953. Inversion totale du sexe chez Xenopus laevis Daud., à la suite d'un traitement gynogène par le benzoate d'oestradiol, administré pendant la vie larvaire. Compt. Rend. Acad. Sci., **237**, 1565–1566.

*————, 1954. Action féminisante partielle du proprionate de testostérone chez Xenopus laevis Daud. Compt. Rend. Acad. Sci., **238**, 1539–1540.

GOTO, G., 1944. Studies on the sex-differentiation of a frog. II. On the male intersexuality of Rana nigromaculata. Kaibogaku Zasshi, **22**, 52 ff (in Japanese).

————, 1945. Studies on the sex-differentiation of a frog. III. On the male intersexuality of Rana limnocharis. Zool. Mag. (Dobutsugaku Zasshi), **56**, 3 ff (in Japanese).

HERTWIG, R., 1906. Weitere Untersuchungen über das Sexualitätsproblem. Verhandl. Deutsch. Zool. Gesellsch., 1906, 90–111.

HUMPHREY,. R R., 1933. The development and sex differentiation of the gonad in the wood frog (Rana sylvatica) following extirpation or orthotopic implantation of the intermediate segment and adjacent mesoderm. Jour. Exptl. Zool., **65**, 243–264.

IKEDA, J., 1896. On the development of Bidder's organ. Zool. Mag. (Dobutsugaku Zasshi), 1896, 8 pp.

IZADI, D., 1943. Développement de l'organe de Bidder du crapaud (Pseudo-ovocytes et ovogénèse vraie). Rev. Suisse Zool., **50**, 395–447.

KING, H. D., 1908. The structure and development of Bidder's organ in Bufo lentiginosus. Jour. Morphol., **19**, 439–465.

KRICHEL, W., 1931. Der Einfluss thyreoidaler Substanzen auf Larven von Bufo viridis und die Bedeutung dieser Stoffe für die Entwicklung der Keimdrüse bis zur Metamorphose. Zool. Jahrb. Abt. 3 Allg. Zool. u. Physiol. Tiere, **48**.

KUSCHAKEWITSCH, S., 1910. Die Entwicklungsgeschichte der Keimdrüsen von Rana esculenta. Festschr. R. Hertwig, Jena, **2**, 61–224.

PERLE, S., 1927. Origine de la première ébauche génitale chez Bufo vulgaris. Compt. Rend. Acad. Sci., **84**, 303–304.

SABBADIN, A., 1950. Comportamento delle cellule somatiche e germinali nello sviluppo e nel differenziamento delle gonadi in una razza differenziate di "Rana esculenta". Atti. Accad. Naz. Lincei. Rend. Cl. Sci. Fis. Mat. e Nat., **8**, 398–404.

————, 1951. Studio morfologico e quantitativo sullo sviluppo delle gonadi di una razza indifferenziata di Rana dalmatina. Considerazioni intorno al differenziamento sessuale negli Anfibi anuri. Arch. Zool. Ital., **36**, 167–216.

STEINHEIM, 1846. Die Entwicklung des Froschembryos, insbesondere des Muskel- und Genitalsystems. Abhandl. Nat. Wiss., Hamburg, 1846.

SWINGLE, W. W., 1921. The germ cells of Anurans. I. The male sexual cycle of Rana catesbeiana larvae. Jour. Exptl. Zool., **32**, 235–332.

————, 1925. The germ cells of Anurans. II. An embryological study of sex differentiation in Rana catesbeiana. Jour. Morphol. Physiol., **41**, 441–546.

TAKASHIMA, R., 1922. Ein Beitrag zur Abstammung der primordialen Keimzellen bei den anuren Amphibien. Folia Anat. Japonica, 1922 (in Japanese).

————, 1932. Untersuchungen über das Biddersche Organ. II Mitteilung. Ueber die Entwicklung des Bidderschen Organs bei unseren einheimischen Kröten. Folia Anat. Japonica, **10**, 291–314.

———— and N. EGAWA, 1937. Eine vergleichende-embryologische Untersuchung über die indifferente Keimdrüse bei den in Süd-Korea spezifischen Cacopoides tornieri (eine Art von Anura). Folia Anat. Japonica, **15**, 229–249.

VANNINI, E., 1938. Sviluppo delle gonadi e intersessualità transitoria in Rana agilis Thom. Arch. Zool. Ital., **25**, 41–83.

————, 1940. Nuove osservazioni sullo sviluppo delle gonadi e sul comportamento delle cellule germinative in una razza indifferenziata di "Rana agilis". Atti Reale Accad. Ital. Cl. Sci. Fis. Mat. e Nat., **12**, 790–799.

————, 1941. La participazione del tessuto interrenale nell'organogenesi della gonade e nei processi di intersessualità giovanile della "Rana agilis". Atti Reale Accad. Ital. Cl. Sci. Fis. Mat. e Nat., **9**, 777–785.

————, 1942a. Sull'origine interrenale dei "Cordoni della rete" e dei "corpi grassi" durante lo sviluppo delle gonadi e sulla partecipazione dell' interrenale ai processi di intersessualità giovanile nella "Rana agilis". Atti Reale Accad. Ital. Cl. Sci. Fis. Mat. e Nat., **13**, 731–790.

————, 1942b. Differenziamento sessuale, comportamento degli elementi germinativi e comparsa di cellule a sessualità intermedia in una razza indifferenziata di Rana agilis. Arch. Zool. Ital., **30**, 363–414.

————, 1943. Le varie fasi dello sviluppo delle gonadi negli Anfibi, in rapporto con il loro determinismo fisiologico e sperimentale. Atti Reale. Ist. Veneto Sci. Lett. Arti, Cl. Sc. Mat. e Nat., **102**, 7–42.

————, 1945. Experimenti sullo sviluppo normale della gonade in assenza del blastema mesonefrico nei girini di "Rana agilis." Atti Reale Ist. Veneto Sci. Lett. Arti, Cl. Sc. Mat. e Nat., **104**, 55–62.

————, 1946. Sviluppo e differenziamento delle gonadi nei girini di Bombina pachypus. Arch. Zool. Ital., **31**, 173–188.

———— and I. BUSETTO, 1946. Origine interrenale del tessuto midollare della gonade e sviluppo dell'organo di Bidder nel Bufo bufo (L) e nel Bufo viridis Laur. Atti Reale Ist. Veneto Sci. Lett. Arti, Cl. Sc. Mat. e Nat., **104**, 631–680.

WALTHER, M., 1926. Die Entwicklung des Bidderschen Organs von Bufo vulgaris L. Zeitschr. Anat. Entwicklungsgesch., **78**, 98–110.

WITSCHI, E., 1929a. Studies on sex differentiation and sex determination in Amphibians. I. Development and sexual differentiation of the gonads of Rana sylvatica. Jour. Exptl. Zool., **52**, 235–254.

————, 1929b. Studies on sex differentiation and sex determination in Amphibians. III. Rudimentary hermaphroditism and Y chromosome in Rana temporaria. Jour. Exptl. Zool., **54**, 157–214.

————, 1933. Studies on sex differentiation and sex determination in Amphibians. VI. The nature of Bidder's organ in the toad. Amer. Jour. Anat., **52**, 461 ff.

YOSHIKURA, M., 1953. The fate of testis-ova in Rana limnocharis. Jour. Sci. Hiroshima Univ. B 1, **14**, 165–172.

Genital tracts and fat bodies

*BEDDARD, F. E., 1908a. Some notes on the muscular and visceral anatomy of the batrachian genus Hemisus, with notes on the lymph-hearts of this and other genera. Proc. Zool. Soc. London, 1908, 894–934 (post. lymph-hearts and fat body, Xenopus).

BOONACKER, A. C., 1927. De ontwikkeling der afvoergangen van het urogenitaalstelsel bij Alytes obstetricans. Thesis, Leiden.

BOUIN, M., 1899. Origine des corps adipeux chez Rana temporaria. Bibl. Anat., **7**, 301 ff.

CHRISTENSEN, K., 1930. Sex differentiation and development of oviducts in Rana pipiens. Amer. Jour. Anat., **45**, 159 ff.

GIGLIO–TOS, E., 1895. Sull'origine dei corpi grassi negli Anfibi. Atti R. Acc. Sci. Torino, **31**.

GILES, A., 1889. Development of the fat-bodies in Rana temporaria ecc. Quart. Jour. Microsc. Sci., **29**, 133–142.

*GITLIN, G., 1941. Seasonal variations and sexual differences in the fat-bodies and other fat deposits of Xenopus laevis. S. African Jour. Med. Sci., **6**, 136–144.

GRAY, P., 1936. The development of the amphibian kidney. Part III. The post-metamorphic development of the vasa efferentia and seminal vesicles in Rana temporaria. Quart. Jour. Microsc. Sci., **78**, 445–473.

GRILLI, M., 1938. Sull'origine dei corpi grassi negli Anfibi (ricerche in Salamandrina perspicillata e Bufo viridis). Arch. Ital. Anat. e Embriol., **39**, 562–602.

MACBRIDE, E. W., 1892. The development of the oviduct in the frog. Quart. Jour. Microsc. Sci., **33**, 273–281.

NUSSBAUM, M., 1880. Ueber die Entwicklung der samenableitenden Wege der Anuren. Zool. Anz., **3**.

VANNINI, E., 1942. Sull'origine interrenale dei "Cordoni della rete" e dei "corpi grassi" durante lo sviluppo delle gonadi e sulla partecipazione dell'interrenale ai processi di intersessualità giovanile nella "Rana agilis". Atti Reale Accad. Ital. Cl. Sci. Fis. Mat. e Nat., **13**, 731–790.

Adrenals

BUSETTO, I., 1943. Sull'origine interrenale del tessuto midollare della gonade nei Bufonidi e nei Discoglossidi (nota preliminare.) Atti. R. Ist. Veneto Sc. Lett. Arti, Cl. Sc. Mat. Nat., **102**, 791 ff.

JONA, A., 1914a. Intorno alle origine e alla natura delle cellule acidofile nelle capsule surrenali di Rana. Arch. Ital. Anat. e Embriol., **12**, 295 ff.

————, 1914b. Sullo sviluppo del sistema interrenale e del sistema cromaffine negli Anfibi anuri. Arch. Ital. Anat. e Embriol., **12**, 311 ff.

*RIMER, G. E. G., 1931. Histology and morphological studies of the endocrine glands of Xenopus laevis. Thesis, Cape Town. 106 pp.

SRDINKO, G. V., 1900. Bau und Entwicklung der Nebenniere bei Anuren. Anat. Anz., **18**, 500–508.

STENGER, A. H. and H. A. CHARIPPER, 1946. A study of adrenal cortical tissue in Rana pipiens with special reference to metamorphosis. Jour. Morphol., **78**, 27–38.

VANNINI, E., 1941. La participazione del tessuto interrenale nell'organogenesi della gonade e nei processi di intersessualità giovanile della "Rana agilis". Atti Reale Accad. Ital. Cl. Sci. Fis. Mat. e Nat., **9**, 777–785.

————, 1942. Sull'origine interrenale dei "Cordoni della rete" e dei "corpi grassi" durante lo sviluppo delle gonadi e sulla partecipazione dell'interrenale ai processi di intersessualità giovanile nella "Rana agilis". Atti Reale Accad. Ital. Cl. Sci. Fis. Mat. e Nat., **13**, 731–790.

———— and I. BUSETTO, 1946. Origine interrenale del tessuto midollare della gonade e sviluppo dell'organo di Bidder nel Bufo bufo (L) e nel Bufo viridis Laur. Atti Reale Ist. Veneto Sci. Lett. Arti, Cl. Sc. Mat. e Nat., **104**, 631–680.

*ZWARENSTEIN, H. and I. SCHRIRE, 1932. The adrenal gland of Xenopus laevis. Proc. Roy. Soc. Edinburgh, **52**, 323–326.

ORO-PHARYNGEAL CAVITY; INCLUDING THYROID GLANDS, BRANCHIAL DERIVATIVES AND LUNGS

Oro-pharyngeal cavity and branchial derivatives (except middle ear)

BAMBEKE, CH. v., 1863. Recherches sur la structure de la bouche chez les têtards des batraciens anoures. Bull. Acad. Roy. Belgique, **16**.

BAUSENHARDT, D., 1939. Die Bildung der Choane und das Problem der Gesichtsfortsätze bei Amphibien. Zool. Anz., **128**, 24–35.

BROCK, G. T., 1929. The formation and fate of the operculum and gill chambers in the tadpole of Rana temporaria. Quart. Jour. Microsc. Sci., **73**, 335–343.

*COHN, L., 1905. Der Tentakelapparat von Dactylethra calcarata. Zeitschr. wiss. Zool., **78**, 620–644.

*————, 1906. Weitere Untersuchungen über den Tentakelapparat des Anurengenus Xenopus. Zool. Anz., **31**, 45–53.

FABRIZIO, M. and H. A. CHARIPPER, 1941. The morphogenesis of the thymus gland of Rana sylvatica as correlated with certain stages of metamorphosis. Jour. Morphol., **68**, 179–191.

*FAHRENHOLZ, C., 1937. Drüsen der Mundhöhle. In: BOLK-GÖPPERT-KALLIUS-LUBOSCH, Handbuch der vergl. Anat. der Wirbeltiere, Berlin, **3**, 115–200.

*HARMS, J. W., 1949. Der Thymus bei Xenopus laevis Daudin. Verhandl. Deutsch. Zool. Gesellsch., 1948, 232–242.

*————, 1952a. Experimentell-morphologische Untersuchungen über den Thymus von Xenopus laevis Daudin. Gegenbaurs Jahrb., 92, 256–338.

*————, 1952b. Extirpation mit anschliessender Transplantation des Thymus bei Xenopus laevis Daudin. Zeitschr. Naturforsch., 7b, 622–630.

HELFF, O. M., 1929. Studies on amphibian metamorphosis. IV. Growth and differentiation of anuran tongue during metamorphosis. Physiol. Zool., 2, 334 ff.

HÉRON-ROYER, A. F. and CH. VAN BAMBEKE, 1881. Sur les caractères fournis par la bouche des têtards des batraciens anoures d'Europe. Bull. Soc. Zool. France, 1881, 7 pp.

———— and ————, 1889. Le vestibule de la bouche chez les têtards des batraciens anoures d'Europe; sa structure, ses caractères chez les diverses espèces. Arch. Biol., 9, 185–302.

HINCKLEY, M. H., 1882. On some differences in the mouth structure of tadpoles of anurous Batrachia. Proc. Bost. Soc. Nat. Hist., 21 (Abstract in Amer. Nat., 17).

HORA, S. L., 1922. Some observations on the oral apparatus of the tadpoles of Megalophrys parva. Jour. and Proc. Asiatic Soc. Bengal. (N.S.), 18.

KEIFFER, H., 1889. Recherches sur la structure et le développement des dents et du bec cornés chez Alytes obstetricans. Arch. Biol., 9, 55–81.

LATASTE, F., 1877. Sur la position de la fente branchiale chez le têtard du Bombinator igneus. Jour. Zool., 6.

LIEBERT, J., 1894. Die Metamorphose des Froschmundes. Thesis, Leipzig.

MARCUS, E., 1930. Zur Entwicklungsgeschichte des Vorderdarmes der Amphibien. Zool. Jahrb. Abt. 2. Anat. u. Ontog. Tiere, 52, 405–486.

————, 1932. Weitere Versuche und Beobachtungen über die Vorderdarmentwicklung bei den Amphibien. Zool. Jahrb. Abt. 2. Anat. u. Ontog. Tiere, 55, 581–602.

*MUELLER, E., 1932. Untersuchungen über die Mundhöhlendrüsen der anuren Amphibien. Morphol. Jahrb., 70, 131–172.

OEDER, R., 1906. Die Entstehung der Munddrüsen und der Zahnleiste der Anuren. Zeitschr. Naturwiss. Jena, 41, 505–548.

*PATERSON, N. F., 1939. The olfactory organs and tentacles of Xenopus laevis. S. African Jour. Sci., 36, 390–404.

RIECK, W., 1932. Die Entwicklung des Mundhöhlenepithels der Anuren. Zool. Jahrb. Abt. 2. Anat. u. Ontog. Tiere, 55, 603–645.

SCHULZE, F. E., 1870. Die Geschmackorgane der Froschlarven. Arch. microsc. Anat., 6, 407–419.

*SHAPIRO, B. G., 1933. The topography and histology of the paratnyroid glandules in Xenopus laevis. Jour. Anat. London, 68, 39–44.

*STERBA, G., 1950. Ueber die morphologische und histogenetische Thymusprobleme bei Xenopus laevis D., nebst einige Bemerkungen über die Morphologie der Kaulquappen. Abhandl. Säch. Gesellsch. (Akad.) Wiss. Math.-Nat. Kl., 44.

*————, 1952a. Mitteilungen über die Altersinvolution des Amphibienthymus. I. Volumetrische Bestimmungen am Thymus des Krallenfrosches Xenopus laevis Daudin. Anat. Anz., 99, 106–114.

*————, 1952b. Regenerationsvorgänge am Thymus des Krallenfrosches Xenopus laevis Daudin. Gegenbaurs Jahrb., 92, 182–198.

STRICKER, S., 1858. Untersuchungen über die Papillen in der Mundhöhle der Froschlarven. Sitzungsber. Akad. Wiss. Wien, math.-nat. Kl. Abt. I, 26, 3–6.

VALLE, P. DELLA, 1914. L'apparato opercolare e la cavita peribranchiale nei Cordati. I. Lo sviluppo normale della regione nel Bufo vulgaris fino alla chiusura della cavita peribranchiale. Arch. Zool. Ital., 7, 115–230.

WEBER, A., 1925. Involution de la cavité péribranchiale après la métamorphose des Batraciens anoures. Compt. Rend. Soc. Biol., **93**, 410–411.

Thyroid glands

ALESCHIN, B. W., 1936. Die Schilddrüse in der Entwicklung und in der Metamorphose von Rana temporaria. Acta Zool., **17**, 1–54.

ALLEN, B. M., 1919. The development of the thyroid glands of Bufo and their normal relation to metamorphosis. Jour. Morphol., **32**, 489–506.

D'ANGELO, S. A., 1941. An analysis of the morphology of the pituitary and thyroid glands in amphibian metamorphosis. Amer. Jour. Anat., **69**, 407 ff.

———— and H. A. CHARIPPER, 1939. The morphology of the thyroid gland in metamorphosing Rana pipiens. Jour. Morphol., **64**, 355–372.

DEEN, R., 1953. Der Einfluss verschiedener Belichtungszeiten auf Hypophyse, Schilddrüse und Larvenentwicklung von Rana temporaria. Zool. Jahrb. Abt. 3 Allg. Zool. u. Physiol. Tiere, **63**, 477–500.

ETKIN, W., 1930. Growth of the thyroid gland of Rana pipiens in relation to metamorphosis. Biol. Bull., **59**, 285–292.

————, 1935. The mechanism of anuran metamorphosis. I. Thyroxine concentration and the metamorphic pattern. Jour. Exptl. Zool., **71**, 317–340.

————, 1936. The phenomena of anuran metamorphosis. III. The development of the thyroid gland. Jour. Morphol., **59**, 69–89.

*GASCHE, P., 1946. Zur Frage des Angriffspunktes des Thiouracils. Versuche an Xenopuslarven. I. Experientia, **2**, 6 pp.

*———— and J. DRUEY, 1946. Wirksamkeit schilddrüsenhemmender Stoffe auf die Xenopus-Metamorphose. II. Experientia, **2**, 4 pp.

GEIGY, R., 1941. Thyroxineinwirkung auf verschieden weit entwickelte Froschlarven. Verhandl. Schweiz. Naturf. Gesellsch. Basel, 4 pp.

HERINGA, G. C., 1924. Some notes on the thyroid-metamorphosis in tadpoles. Proc. Kon. Ned. Akad. Wet., **27**, 693–699.

HOSKINS, E. R. and M. M. HOSKINS, 1920. The inter-relation of the thyroid and hypophysis in the growth and development of frog larvae. Endocrinology, **4**, 1 ff.

JAMES, M. S., 1946. The role of the basibranchial cartilages in the early development of the thyroid of Hyla regilla. Univ. California Publ. Zool., **51**, 215–228.

KIYOTANI, K., 1935. Entwicklungsstudien über die Schilddrüsenanlage. II Mitteilung: Untersuchungen an den Anuren, besonders bei den Larven von Rhacophorus schlegelii. Okayama-Igakkai-Zasshi, **47**.

KRICHEL, W., 1931. Der Einfluss thyreoidaler Substanzen auf Larven von Bufo viridis und die Bedeutung dieser Stoffe für die Entwicklung der Keimdrüse bis zur Metamorphose. Zool. Jahrb. Abt. 3 Allg. Zool. u. Physiol. Tiere, **48**.

MIYAMOTO, K. and T. UETA, 1938. Ueber die erste Entwicklung der Schilddrüsenanlage beim Frosche (Rana nigromaculata). Mitteil. Med. Gesellsch. Osaka, **37**.

MIZOGUCHI, K., 1929. Ueber die Entwicklung der Schilddrüse bei den Anuren Amphibien Kaibogaku-Zasshi, **2**.

MORITA, S., 1932. Quantitative Untersuchungen über die Schilddrüse von Bufo in Beziehung zur Metamorphose. Jour. Sci. Hiroshima Univ., Ser. B. Div. 1, **2**.

MOSER, H., 1950. Ein Beitrag zur Analyse der Thyroxineinwirkung im Kaulquappenversuch und zur Frage nach dem Zustandekommen der Frühbereitschaft des Metamorphose-Reaktionssystems. Thesis, Basel; Rev. Suisse Zool., **57**, 1–144.

REISINGER, E., 1931. Zur Entstehung der Schilddrüse bei Amphibien. Verhandl. Deutsch. Zool. Gesellsch., Zool. Anz., **5** (suppl.).

*RIMER, G. E. G., 1931a. Histology and morphological studies of the endocrine glands of Xenopus laevis. Thesis, Cape Town, 106 pp.

*———, 1931b. The thyroid gland in Xenopus laevis. Trans. Roy. Soc. S. Africa, **19**, 331–339.

ROMEIS, B., 1924. Histologische Untersuchungen zur Analyse der Schilddrüsenfütterung auf Froschlarven. II. Die Beeinflussung der Entwicklung der vorderen Extremitäten und des Brustschulterapparates. Arch. Entwicklungsmech., **101**.

STEINMETZ, C. H., 1952. Thyroid function as related to the growth of tadpoles before metamorphosis. Endocrinology, **51**, 154–156.

TAKASIMA, R. and K. HASHIMOTO, 1932. Ueber die Entwicklung der Schilddrüse bei den japanischen Kröten. Kaibogaku-Zasshi, **4**.

———, K. ZAIMA and K. GOTO, 1938. Ergänzenden Mitteilung über die erste Anlage der Schilddrüse bei der Kröte (Bufo vulgaris japonicus). Folia Anat. Japonica, **16**, 357–386.

*WEISS, P. and F. ROSETTI, 1951. Growth responses of opposite sign among different neuron types exposed to thyroid hormone. Proc. Natl. Acad. Sci., **37**, 540–556.

Gills and lungs

GERHARDT, E., 1932. Die Kiemenentwicklung bei Anuren (Pelobates fuscus, Hyla arborea) und Urodelen (Triton vulgaris). Zool. Jahrb. Abt. 2. Anat. u. Ontog. Tiere, **55**, 137–314.

HEMPSTEAD, M., 1901. Development of the lungs in the frogs, Rana catesbiana, R. silvatica and R. virescens. Science, **12**.

HUSCHKE, E., 1826. Ueber die Umbildung des Darmkanales und der Kiemen der Froschquappen. Isis, **18**, 613 ff.

MAURER, F. VON, 1888. Die Kiemen und ihre Gefässe bei anuren und urodelen Amphibien und die Umbildungen der beiden ersten Arterienbogen bei Teleostiern. Morphol. Jahrb., **14**, 175–222.

NAUE, H., 1890. Ueber Bau und Entwicklung der Kiemen der Froschlarven. Zeitschr. Naturwiss., **63**, 129–174.

SCHMALHAUSEN, I. I., 1954. Die Entwicklung der Kiemen und ihrer Blutgefässe und Muskulatur bei den Amphibien. Zoologicheskii Zhurnal (U.S.S.R.), **33**, 848–868 (in Russian).

SCHULZE, F. L., 1888. Ueber die inneren Kiemen der Batrachierlarven. I Mitteilung: Ueber das Epithel der Lippen, der Mund-, Rachen- und Kiemenhöhle erwachsener Larven von Pelobates fuscus. Abhandl. Preuss. Akad. Wiss. Berlin, 1888, St. XXXII, 715–768.

———, 1892. Ueber die inneren Kiemen der Batrachierlarven. II Mitteilung: Skelett, Muskulatur, Blutgefässe, Filterapparat, respiratorische Anhänge und Atmungsbewegungen erwachsener Larven von Pelobates fuscus. Abhandl. Preuss. Akad. Wiss. Berlin, **13**, 205 ff.

SISIDO, S., 1936. Ueber die Entstehungsweise der ersten Lungenanlage bei R. nigromaculata. Osaka Ijisinsi, **7**.

TAKASIMA, R. and S. SISIDO, 1936. Ein Beitrag zur ersten Entwicklung der Krötenlunge (B. vulgaris jap.). Mitteil. Med. Gesellsch. Osaka, **35**.

———, S. WAKE and U. KONISIIKE, 1939. Ueber die erste Entwicklung der Lungenanlage bei Amphibien unter besonderer Berücksichtigung des Hynobius nebulosus. Folia Anat. Japonica, **17**, 575–591.

WEBER, A., 1931. Destinée des branchies internes et de la cavité qui les entoure lors de la métamorphose des Batraciens anoures. Compt. Rend. Soc. Biol., **107**, 437–438.

————, 1932. Les restes branchiaux d'un Batracien anoure, Bombinator pachypus. Compt. Rend. Soc. Biol., **111**, 952–954.

WHITNEY, W. U., 1867. On the changes which accompany the metamorphosis of the tadpole, in reference especially to the respiratory and sanguiferous systems. Trans. Roy. Micr. Soc., **15**, 43 ff.

YOKUSHIJI, T., 1932. Ueber die Morphogenese der Amphibienlunge. II Mitteilung: Untersuchungen an den Anuren, besonders bei den Larven von Bufo vulgaris japonicus. Okayama-Igakkei-Zasshi, **44**.

Middle ear

Fox, H., 1901. The development of the tympano-eustachian passage and associated structures in the common toad, Bufo lentiginosus. Proc. Acad. Nat. Sci. Philadelphia, **53**, 223 ff.

GAZAGNAIRE, CH., 1932. Origine de l'oreille moyenne chez Rana temporaria. Compt. Rend. Soc. Biol., **110**, 1076–1078.

HELFF, O. M., 1928. Studies on amphibian metamorphosis. III. The influence of the annular tympanic cartilage on the formation of the tympanic membrane. Physiol. Zool., **1**, 463 ff.

————, 1929. Formation of the yellow elastic fibrous region of the lamina propria of anuran tympanic membrane. Anat. Rec., **44**, 214 ff.

*SPANNHOF, L., 1954. Die Entwicklung des Mittelohres und des schallleitenden Apparates bei Xenopus laevis Daud. Zeitschr. wiss. Zool., **158**, 1–30.

SPEMANN, H., 1898. Ueber die erste Entwicklung der Tuba Eustachii und des Kopfskeletts von Rana temporaria. Zool. Jahrb. Abt. 2. Anat. u. Ontog. Tiere, **11**, 389–416.

VILLY, F., 1890. The development of the ear and the accessory organs in the common frog. Quart. Jour. Microsc. Sci., **30**, 523–550.

VIOLETTE, H. N., 1928. An experimental study on the formation of the middle ear in Rana. Proc. Soc. Exptl. Biol. and Med., **25**.

WITSCHI, E., 1947. Development and metamorphosis of the auxiliary apparatus of the ear of the frog. Anat. Rec., **99**, 568 (Abstr.).

————, 1949. The larval ear of the frog and its transformation during metamorphosis. Zeitschr. Naturforsch., **4**, 230–242.

INTESTINAL TRACT AND GLANDS

ARON, M., 1925. Dégénérescence du pancréas au cours de la métamorphose chez Rana esculenta. Compt. Rend. Soc. Biol., **93**.

BALINSKY, B. I., 1947. Kinematik des entodermalen Materials bei der Gestaltung der wichtigsten Teile des Darmkanals bei den Amphibien. Arch. Entwicklungsmech., **143**.

BEAUMONT, A., 1953a. Modifications histologiques du pancréas des larves de Batraciens anoures au cours de la métamorphose. Compt. Rend. Soc. Biol., **147**, 56–58.

————, 1953b. Le pancréas du têtard d'Alytes obstetricans (Laurenti). Arch. Anat. Microsc. et Morphol. Exptl., **42**, 32–40.

————, 1954. L'apparition du glycogène hépatique chez les larves de Batraciens anoures. Compt. Rend. Soc. Biol., **148**, 29–31.

BERGFELDT, A., 1896. Chordascheiden und Hypochorda bei Alytes obstetricans. Anat. Hefte, **7**.

BOWERS, M. A., 1909. Histogenesis and histolysis of the intestinal epithelium of Bufo lentiginosus. Amer. Jour. Anat., **9**, 263 ff.

BRAUN, W., 1906. Die Herkunft und Entwicklung des Pankreas bei Alytes obstetricans. Morphol. Jahrb., **36**, 27–51.

FAHRENHOLZ, C., 1923. Ueber eine ventrale Oeffnung der Leberanlage anurer Amphibien und deren morphologische Bedeutung. Zeitschr. Anat. Entwicklungsgesch., **69**, 344–381.

FARAGGIANA, R., 1935. Ricerche sull'istogenesi del pancreas di Anfibi anuri con particolare riguardo alle isole del Langerhans. Arch. Ital. Anat. e Embriol., **34**, 72–121.

CHORONSHITZKY, B., 1900. Die Entstehung der Milz, Leber, Gallenblase, Bauchspeicheldrüse und des Pfortadersystems bei den verschiedenen Abteilungen der Wirbeltiere. Anat. Hefte I. Abt., **13**, H. 42/43, 369–620.

*GEIGY, R. and F. ENGELMANN, 1954. Beitrag zur Entwicklung und Metamorphose des Darmes bei Xenopus laevis Daud. Rev. Suisse Zool., **61**, 335–347.

GIANELLI, L., 1903. Sulle prime fasi di sviluppo del pancreas negli Anfibi anuri (Rana esculenta). Monitore Zool. Ital., **14**, 33 ff.

GÖPPERT, E., 1891. Die Entwicklung und das spätere Verhalten des Pankreas der Amphibien. Morphol. Jahrb., **17**.

HUSCHKE, E., 1826. Ueber die Umbildung des Darmkanales und der Kiemen der Froschquappen. Isis, **18**, 613 ff.

JANES, R. G., 1934. Studies on the amphibian digestive system. I. Histological changes in the alimentary tract of anuran larvae during involution. Jour. Exptl. Zool., **67**, 73 ff.

————, 1937. Studies on the amphibian digestive system. II. Comparative histology of the pancreas, following early larval development, in certain species of Anura. Jour. Morphol., **61**, 581–611.

————, 1938. Studies on the amphibian digestive system. III. The origin and development of pancreatic islands in certain species of Anura. Jour. Morphol., **62**, 375–391.

KINDRED, J. E., 1924. Ciliogenesis in the oesophagal epithelium of the frog tadpole. Anat. Rec., **27**, 183 ff (Abstract).

KEMP, N. E., 1951. Development of intestinal coiling in anuran larvae. Jour. Exptl. Zool., **116**, 259–287.

KUNTZ, A., 1922. Metamorphic changes in the digestive system in Rana pipiens and Amblystoma tigrinum. Univ. Iowa Studies, **10**, 37 ff.

————, 1924. Anatomical and physiological changes in the digestive system during metamorphosis in R. pipiens and A. tigrinum. Jour. Morphol., **38**, 581–598.

MARCELIN, R. H., 1903. Histogénèse de l'épithélium intestinal chez la grenouille (Rana esculenta). Rev. Suisse Zool., **11**, 369 ff.

NAKAMURA, O. and Y. TAHARA, 1953. Formation of the stomach in Anura. Mem. Osaka Univ. Lib. Arts Educ. B. Nat. Sci., **2**, 1–8.

———— and ————, 1954. Formation of the intestine in Anura. Mem. Osaka Univ. Lib. Arts Educ. B. Nat. Sci., **3**, 77–89.

*PASTEELS, J., 1943. Fermeture du blastopore, anus et intestin caudal chez les Amphibiens anoures. Acta Neerl. Morphol., **5**, 11–25.

RATNER, G., 1891. Zur Metamorphose des Darmes bei der Froschlarve. Thesis, Dorpat.

REICHENOW, E., 1908. Die Rückbildungserscheinungen am Anurendarm während der Metamorphose und ihre Bedeutung für die Zellforschung. Arch. microsc. Anat., **72**.

REUTER, K., 1900a. Ueber die Entwicklung der Darmspirale bei Alytes obstetricans. Anat. Hefte, **13**, H. 42/43, 339–359.

————, 1900b. Ueber die Rückbildungserscheinungen am Darmkanal der Larve von Alytes obstetricans. Anat. Hefte, H. **45**, 433 ff.

*RIMER, G. E. G., 1931. Histology and morphological studies of the endocrine glands of Xenopus laevis. Thesis, Cape Town, 106 pp.

RUFFINI, A., 1899. Sullo sviluppo e sul tardivo contegno dello strato glandulare dello stomaco nella Rana esculenta. Monitore Zool. Ital., **10**, suppl., 63–68.

STÖHR, PH., 1895. Ueber die Entwicklung der Hypochorda und des dorsalen Pankreas bei Rana temporaria. Morphol. Jahrb., **23**, 123–140.

WEYSSE, A. W., 1895. Ueber die ersten Anlagen der Hauptanhangsorgane des Darm-kanals beim Frosch. Arch. Microsc. Anat., **46**.

YOUNG, E., 1904. De l'influence du régime alimentaire sur la longueur de l'intestin chez les larves de Rana esculenta. Compt. Rend. Acad. Sci. Paris, **139**, 749 ff.

NORMAL TABLES

Anura

ADLER, W., 1901. Die Entwicklung der äusseren Körperform und des Mesoderms be Bufo vulgaris. Internatl. Monatschr. Anat. Physiol., **18**.

*BRETSCHER, A., 1949. Die Hinterbeinentwicklung von Xenopus laevis Daud. und ihre Beeinflussung durch Colchicin. Rev. Suisse Zool., **56**, 34–96.

CAMBAR, R. and BR. MARROT, 1954. Table chronologique du développement de la grenouille agile (Rana Dalmatina Bon.). Bull. Biol. Franco-Belg., **88**, 168–177.

DEL CONTE, E. and J. L. SIRLIN, 1952. Pattern series of the first embryonary stages in Bufo arenarum. Anat. Rec., **112**, 125–135.

EAKIN, R. M., 1946. Determination and regulation of polarity in the retina of Hyla regilla. Univ. California Publ. Zool., **51**, 245 ff.

GALLIEN, L. and CH. HOUILLON, 1951. Table chronologique du développement chez Discoglossus pictus. Bull. Biol. Franco-Belg., **85**, 373–375.

*GASCHE, P., 1944. Beginn und Verlauf der Metamorphose bei Xenopus laevis Daud. Festlegung von Umwandlungsstadien. Helvetica Physiol. et Pharmacol. Acta, **2**, 607–626.

HAMBURGER, V., 1947. A manual of experimental embryology (2nd impr.). Chicago (contains the figures of the Normal Tables of SHUMWAY, 1940 and POLLISTER and MOORE, 1937 and figures after HARRISON's unpublished series by S. E. SCHWEICH).

KOPSCH, FR., 1952. Die Entwicklung des braunen Grasfrosches Rana fusca Roesel (dargestellt in der Art der Normentafeln zur Entwicklungsgeschichte der Wirbel-thiere). Stuttgart, 1952.

MILLER, D. C., 1940. Normal table of Rana pipiens. Proc. Indiana Acad. Sci., **49**.

MOSER, H., 1950. Ein Beitrag zur Analyse der Thyroxineinwirkung im Kaulquappen-versuch und zur Frage nach dem Zustandekommen der Frühbereitschaft des Metamorphose-Reaktionssystems. Thesis, Basel; Rev. Suisse Zool., **57** (Suppl. 2), 1–144.

*PETER, K., 1931. The development of the external features of Xenopus laevis, based on material collected by the late E. J. BLES. Jour. Linn. Soc. Zool., **37**, 515–523.

POLLISTER, A. W. and J. A. MOORE, 1937. Tables for the normal development of Rana sylvatica. Anat. Rec., **68**, 489 ff.

RUGH, R., 1948. Experimental embryology, a manual of techniques and procedures (revised ed.). Minneapolis (contains the figures of the Normal Tables of SHUMWAY, 1940, MUELLER, TAYLOR and KOLLROS, 1946, POLLISTER and MOORE, 1937, WEISZ, 1945, EAKIN, 1947, ANDERSON, 1943 and TWITTY and BODENSTEIN, and figures after HARRISON's unpublished series by N. LEAVITT).

————, 1951. The frog, its reproduction and development (Rana pipiens). Philadel-phia, 1951.

SHUMWAY, W., 1940. Stages in the normal development of Rana pipiens. I. External forms. Anat. Rec., **78**, 139–147.

————, 1942. Stages in the normal development of Rana pipiens. II. Identification of the stages from sectioned material. Anat. Rec., **83**, 309–315.

TAYLOR, A. C. and J. J. KOLLROS, 1946. Stages in the normal development of Rana pipiens larvae. Anat. Rec., **94**, 7–23.

*WEISZ, P. B., 1945. The normal stages in the development of the South African clawed toad, Xenopus laevis. Anat. Rec., **93**, 161–169.

Urodela

ANDERSON, P. L., 1943. The normal development of Triturus pyrrhogaster. Anat. Rec., **86**, 58 ff.

EYCLESHYMER, A. C. and J. M. WILSON, 1910. Normal plates of the development of Necturus maculosus. F. Keibel's Normentafeln zur Entwicklungsgeschichte der Wirbelthiere, H. 11.

FANKHAUSER, G., Stages in normal development of Triturus viridescens. Unpublished.

GLÄSNER, L., 1925. Normentafel zur Entwicklung des gemeinen Wassermolchs. F. Keibel's Normentafeln zur Entwicklungsgeschichte der Wirbelthiere, H. 14.

GLÜCKSOHN, S., 1931. Aüssere Entwicklung der Extremitäten und Stadieneinteilung der Larvenperiode von Triton taeniatus Leyd. und Triton cristatus Laur. Arch. Entwicklungsmech., **125**, 341–405.

HAMBURGER, V., 1947. A manual of experimental embryology (2nd impr.). Chicago (contains the figures of the Normal Tables of SHUMWAY, 1940 and POLLISTER and MOORE, 1937, and figures after HARRISON's unpublished series by S. E. SCHWEICH).

HARRISON, R. G. Normal stages of Amblystoma punctatum. Unpublished.

KNIGHT, F. C. E., 1938. Die Entwicklung von Triton alpestris bei verschiedenen Temperaturen, mit Normentafel. Arch. Entwicklungsmech., **137**, 461–473.

KUDÔ, T., 1938. Normentafel zur Entwicklungsgeschichte des Japanischen Riesensalamanders (Megalobatrachus japonicus Temminck). F. Keibel's Normentafeln zur Entwicklungsgeschichte der Wirbelthiere, H. 16.

OKADA, Y. K. and M. ICHIKAWA, 1947. A new Normal Table of the development of Triturus pyrrhogaster (Boie). Exptl. Morphol. (Tokyo), **3**, 1–6 (in Japanese; English translation with figures present at Hubrecht Laboratory).

OYAMA, J., 1930. A normal table of the development of the Japanese newt, Diemictylus (Triturus) pyrrhogaster. Zool. Mag. (Tokyo), **42**, 465–473 (Japanese).

ROTMANN, E., 1940. Die Bedeutung der Zellgrösse für die Entwicklung der Amphibienlinse. Arch. Entwicklungsmech., **140**, 124–156 (contains a comparative table of several Urodelan Normal Tables).

RUGH, R., 1948. Experimental embryology, a manual of techniques and procedures (revised ed.). Minneapolis (contains the figures of the Normal Tables of SHUMWAY, 1940, MUELLER, TAYLOR and KOLLROS, 1946, POLLISTER and MOORE, 1937, WEISZ, 1945, EAKIN, 1947, ANDERSON, 1943 and TWITTY and BODENSTEIN, and figures after HARRISON's unpublished series by N. LEAVITT).

TWITTY, V. C. and D. BODENSTEIN. Normal stages of Triturus torosus. (Quoted from Rugh, Experimental Embryology).

WUNDERER, H., 1910. Die Entwicklung der äusseren Körperform des Alpensalamanders (Salamandra atra Laur.). Zool. Jahrb. Abt. 2. Anat. u. Ontog. Tiere, **29**.

* * *

APPENDIX TO THE BIBLIOGRAPHY

This appendix lists titles on anuran development that have come to our attention since the publication of the first edition.

Titles relating to electron microscopy, nuclear transplantation, and biochemistry have as a rule been omitted. Titles on regeneration have been listed only in so far as they deal with the morphology of limb, tail, and lens regeneration in *Xenopus laevis*.

The regeneration of the central nervous system in *Xenopus laevis* and the influence of the nervous system and some other factors on *Xenopus* tail regeneration are discussed in several articles, not listed here, which have appeared since 1954 in the Polish journal "Folia Biologica (Kraków)" (authors: Jurand, Kosciuszko, Kwiatkowski, Maron, Olekiewicz, Roguski, Rzehak, Skowron, Srebro).

The appendix is in three sections. The first deals with *Xenopus laevis*, and includes several recent articles on reproduction, breeding, and rearing of this species. The second section deals with the developmental morphology of other anuran forms, while the third section is a supplementary list of anuran and urodelan Normal Tables.

XENOPUS LAEVIS

Reproduction, breeding and rearing

BALINSKY, B. I., 1957. South African amphibia as material for biological research. S. Afr. J. Sci., **53**, 383–391.

BROCAS, J. and F. VERZÁR, 1961. The aging of Xenopus laevis, a South African frog. Gerontologia, **5**, 228–240.

HENRIQUES, U., 1964. Breeding of Xenopus laevis Daudin. Acta Endocr. Copenh., Suppl. **90**, 89–98.

KALK, M., 1960. Climate and breeding in Xenopus laevis, S. Afr. J. Sci., **56**, 271–276.

KUCIAS, J., 1961. Einfluss der Futterart auf die Entwicklungsgeschwindigkeit der Larven von Xenopus laevis Daud. Biol. Zbl., **80**, 1, 93–95.

SCHNEIDER, L., 1956. Eine neue Methode für die Aufzucht von Larven des Krallenfrosches Xenopus laevis Daudin. Zool. Anz., **156**, 70.

UEHLINGER, V., 1966. Facteurs influençant la fertilité des mâles de Xenopus laevis D. (Batracien anoure). Experientia, **22**, 556.

ZIELENIEWSKI, J., 1964. A simplified method of breeding Xenopus laevis frogs. Pol. Endocr., **15**, 118–119.

Development and regeneration

BAGNARA, J. T., 1960. Tail melanophores of Xenopus in normal development and regeneration. Biol. Bull., **118**, 1–8.

BALINSKY, B. I., 1955. Histogenetic and organogenetic processes in the development of specific characters in some South African tadpoles. J. Embryol. exp. Morph., **3**, 93–120.

239

BALINSKY, B. I., 1957. On the factors determining the size of the lens rudiment in amphibian embryos. J. exp. Zool., **135**, 255–300.

BIJTEL, J. H., 1958. De ontwikkeling van de ductus nasolacrimalis bij Xenopus laevis. Ned. Tijdschr. Geneesk., **102** (abstract, in Dutch).

BLACKLER, A. W., 1958. Contribution to the study of germ-cells in the Anura. J. Embryol. exp. Morph., **6**, 491–503.

————, 1962. Transfer of primordial germ-cells between two subspecies of Xenopus laevis. J. Embryol. exp. Morph., **10**, 641–651.

————, 1965. Germ-cell transfer and sex ratio in Xenopus laevis. J. Embryol. exp. Morph., **13**, 51–61.

———— and M. FISCHBERG, 1961. Transfer of primordial germ-cells in Xenopus laevis. J. Embryol. exp. Morph., **9**, 634–641.

————, ———— and D. R. NEWTH, 1965. Hybridization of two subspecies of Xenopus laevis (Daudin). Revue suisse Zool., **72**, 841–857.

BOSQUE, P. G., 1956. Die Topographie der vegetativen Ganglien vor und nach der Metamorphose bei Xenopus laevis Daudin. Morph. Jb., **97**, 28–44.

BRAHMA, S. K., 1959. Studies on the process of lens induction in Xenopus laevis (Daudin). Wilhelm Roux Arch. EntwMech. Org., **151**, 181–187.

CHANG, C. Y. and E. WITSCHI, 1955. Breeding of sex-reversed males of Xenopus laevis Daudin. Proc. Soc. exp. Biol. Med., **89**, 150–152.

———— and ————, 1956. Genic control and hormonal reversal of sex differentiation in Xenopus. Proc. Soc. exp. Biol. Med., **93**, 140–144.

CORNER, M. A., 1963. Development of the brain in Xenopus laevis after removal of parts of the neural plate. J. exp. Zool., **153**, 301–312.

————, 1964. Localization of capacities for functional development in the neural plate of Xenopus laevis. J. comp. Neurol., **123**, 243–256.

————, 1966. Morphogenetic field properties of the forebrain area of the neural plate in an anuran. Experientia, **22**, 188.

CURTIS, A. S. G., 1960. Cortical grafting in Xenopus laevis. J. Embryol. exp. Morph., **8**, 163–173.

DENT, J. N., 1962. Limb regeneration in larvae and metamorphosing individuals of the South African clawed toad. J. Morph., **110**, 61–78.

ELSDALE, T. R., J. B. GURDON and M. FISCHBERG, 1960. A description of the technique for nuclear transplantation in Xenopus laevis. J. Embryol. exp. Morph., **8**, 437–444.

FOX, H. and L. HAMILTON, 1964. Origin of the pronephric duct in Xenopus laevis. Arch. Biol. (Liège), **75**, 245–251.

FREEMAN, G., 1963. Lens regeneration from the cornea in Xenopus laevis. J. exp. Zool., **154**, 39–66.

GALLIEN, L., 1956. Inversion expérimentale du sexe chez un anoure inférieur Xenopus laevis Daudin. Analyse des conséquences génétiques. Bull. biol. Fr. Belg., **90**, 163–183.

————, 1957. Inversion du phénotype sexuel, altérations de la gonadogenèse et déterminisme de la différenciation des gonies chez les amphibiens. J. Fac. Sci. Hokkaido Univ. (Series 6, Zoology), **13**, 373–378.

GAZE, R. M. and A. PETERS, 1961. The development, structure and composition of the optic nerve of Xenopus laevis (Daudin). Q. Jl exp. Physiol., **46**, 299–309.

HAMILTON, L., 1963. An experimental analysis of the development of the haploid syndrome in embryos of Xenopus laevis. J. Embryol. exp. Morph., **11**, 267–278.

HAUSER, R., 1965. Autonome Regenerationsleistungen des larvalen Schwanzes von Xenopus laevis und ihre Abhängigkeit vom Zentralnervensystem. Wilhelm Roux Arch. EntwMech. Org., **156**, 404–448.

HUGHES, A., 1957. The development of the primary sensory system in Xenopus laevis (Daudin). J. Anat. **91,** 323–338.

———, 1959. Studies in embryonic and larval development in amphibia. II. The spinal motor-root. J. Embryol. exp. Morph., **7,** 128–145.

———, 1961. Cell degeneration in the larval ventral horn of Xenopus laevis (Daudin). J. Embryol. exp. Morph., **9,** 269–284.

——— and P. A. TSCHUMI, 1958. The factors controlling the development of the dorsal root ganglia and ventral horn in Xenopus laevis (Daud.). J. Anat., **92,** 498–527.

KERR, T., 1966. The development of the pituitary in Xenopus laevis Daudin. Gen. comp. Endocr., **6,** 303–311.

KOMALA, Z., 1957. Comparative investigations on the course of ontogenesis and regeneration of the limbs in Xenopus laevis tadpoles in various stages of development. Folia biol. (Kraków), **5,** 1–51 (Polish, with English summary).

KOMNICK, H., 1961. Über Herkunft, Bedeutung und Schicksal der Melanocyten im Cerebralliquor von Krallenfroschlarven (Xenopus laevis). Wilhelm Roux Arch. EntwMech. Org., **153,** 14–31.

KREINER, J., 1956. The roof of the third ventricle of the brain in platanna (Xenopus laevis Daud.). Folia biol. (Kraków), **4,** 237–290 (Polish, with English summary).

LEWIS, P. R. and A. F. W. HUGHES, 1960. Patterns of myo-neural junctions and cholinesterase activity in the muscles of tadpoles of Xenopus laevis. Q. Jl microsc. Sci., **101,** 55–67.

McMURRAY, V. M., 1954. The development of the optic lobes in Xenopus laevis. The effect of repeated crushing of the optic nerve. J. exp. Zool., **125,** 247–264.

MIKAMO, K., 1961. Overripeness of the egg in Xenopus laevis Daudin. Thesis, Hokkaido Univ.

——— and E. WITSCHI, 1963. Functional sex-reversal in genetic females of Xenopus laevis, induced by implanted testes. Genetics, **48,** 1411–1421.

——— and ———, 1964. Masculinization and breeding of the WW Xenopus. Experientia, **20,** 622.

MIODOŃSKI, A., 1957. The ontogenesis of the ganglion of the eighth nerve. Folia biol. (Kraków), **5,** 151–174 (Polish, with English summary).

NIEUWKOOP, P. D., 1956. Are there direct relationships between the cortical layer of the fertilized egg and the development of the future axial system in Xenopus laevis embryos? Pubbl. Staz. Zool. Napoli, **28,** 241–249.

——— and E. H. SUMINSKI, 1959. Does the so called "germinal cytoplasm" play an important role in the development of the primordial germ cells. Archs Anat. microsc. Morph. exp., **48** bis, 189–198.

OLSSON, R., 1955. Structure and development of Reissner's fibre in the caudal end of Amphioxus and some lower vertebrates. Acta zool. (Stockh.), **36,** 167–198.

———, 1956. The development of Reissner's fibre in the brain of the salmon. Acta zool. (Stockh.), **37,** 235–250.

ORTOLANI, G. and F. VANDERHAEGHE, 1965. L'activation de l'oeuf de Xenopus laevis laevis. Revue suisse Zool., **72,** 652–658.

OVERTON, J., 1963. Patterns of limb regeneration in Xenopus laevis. J. exp. Zool., **154,** 153–161.

PRESTIGE, M. C., 1965. Cell turnover in the spinal ganglia of Xenopus laevis tadpoles. J. Embryol. exp. Morph., **13,** 63–72.

RAPOLA, J., 1962. Development of the amphibian adrenal cortex. Morphological and physiological studies on Xenopus laevis Daudin. Annls Acad. scient. fenn. (Series A, IV. Biologica), **64,** 81 pp.

REYNAUD, J. and V. UEHLINGER, 1965. Une mutation létale récessive "yr" (yolky rectum) chez Xenopus laevis Daudin. Revue suisse Zool., **72,** 675–680.

SAXÉN, L., E. SAXÉN, S. TOIVONEN and K. SALIMÄKI, 1957. The anterior pituitary and the thyroid function during normal and abnormal development of the frog. Ann. Zool. Soc. "Vanamo", **18,** no. 4, 1–44.

———— and S. TOIVONEN, 1955. The development of the ultimobranchial body in Xenopus laevis Daudin and its relation to the thyroid gland and epithelial bodies. J. Embryol. exp. Morph., **3,** 376–384.

SEDRA, S. N. and M. I. MICHAEL, 1957. The development of the skull, visceral arches, larynx and visceral muscles of the South African clawed toad, Xenopus laevis (Daudin) during the process of metamorphosis (from stage 55 to stage 66). Verh. K. ned. Akad. Wet. (Afd. Natuurk., Ser. 2), **51,** 1–80.

SHAFFER, B. M., 1963. The isolated Xenopus laevis tail: a preparation for studying the central nervous system and metamorphosis in culture. J. Embryol. exp. Morph., **11,** 77–90.

SIMS, R. T., 1961. The blood-vessels of the developing spinal cord of Xenopus laevis. J. Embryol. exp. Morph., **9,** 32–41.

————, 1964. The relationship between size and vascularity in the spinal cord of developing Xenopus laevis. J. Embryol. exp. Morph., **12,** 491–499.

————, 1966. Experiments on the pattern of the blood vessels in the central nervous system of Xenopus laevis. J. Anat., **100,** 91–98.

SMIT, A. L., 1953. The ontogenesis of the vertebral column of Xenopus laevis (Daudin) with special reference to the segmentation of the metotic region of the skull. Ann. Univ. Stellenbosch, **29** (Section A), 79–136.

TSCHUMI, P. A., 1955. Versuche über die Wachstumsweise von Hinterbeinknospen von Xenopus laevis Daud. und die Bedeutung der Epidermis. Revue suisse Zool., **62,** 281–288.

————, 1956. Die Bedeutung der Epidermisleiste für die Entwicklung der Beine von Xenopus laevis Daud. Revue suisse Zool., **63,** 707–716.

————, 1957. The growth of the hindlimb bud of Xenopus laevis and its dependence upon the epidermis. J. Anat., **91,** 149–173.

UEHLINGER, V. and J. REYNAUD, 1965. Une anomalie héréditaire "kt" (kinky tailtip) chez Xenopus laevis D. Revue suisse Zool., **72,** 680–685.

VANABLE, J. W., 1964. Granular gland development during Xenopus laevis meta-morphosis. Dev. Biol., **10,** 331–357.

WALLACE, H., 1960. The development of anucleolate embryos of Xenopus laevis. J. Embryol. exp. Morph., **8,** 405–413.

WITSCHI, E., 1950. The bronchial diverticula of Xenopus laevis Daudin. Anat. Rec., **108,** 590.

OTHER ANURA

Developmental morphology

ADAMSON, L., R. G. HARRISON and I. BAYLEY, 1960. The development of the whistling frog Eleutherodactylus martinicensis of Barbados. Proc. Zool. Soc. (London), **133,** 453–469.

ALBERT, J., 1961. Recherches descriptives et expérimentales sur le développement de l'appareil digestif chez la larve de la Grenouille agile (Rana dalmatina Bon.). Thesis, Bordeaux.

ALBERT, J., 1963. Contribution à l'étude de la morphogenèse de l'appareil digestif chez la larve de la Grenouille agile (Rana dalmatina Bonaparte). Bull. biol. Fr. Belg., **97,** 483–513.

BAFFONI, G. M., 1959. L'andamento dell'attività mitotica nel prosencefalo e nel mesencefalo durante lo sviluppo di un anfibio anuro. Rend. Accad. naz. Lincei (Ser. 8), **26,** 598–603.

————, 1959. Osservazioni sulla morfogenesi ed istogenesi cerebellare in un anfibio anuro (Bufo bufo L.). Riv. Neurobiol., **5,** 33–73 (English summary).

————, 1960. Contributo allo studio della metamorfosi negli anfibî. Riv. Biol., **53** (N.S. **13**), 293–340 (English summary).

———— and R. PINACCI, 1958. L'andamento dell'attività mitotica nel midollo spinale di anfibî anuri durante lo sviluppo. Rend. Accad. naz. Lincei (Ser. 8), **25,** 128–134.

BEAUMONT, A. and B. GAUDIN, 1962. Développement, structure et régression de la columelle bronchique de Rana temporaria L. C. r. Séanc. Soc. Biol., **156,** 1962.

BOSCHWITZ, D., 1960. The ultimobranchial body of the anura of Israel. Herpetologica, **16,** 91–I00.

BYTINSKI-SALZ, H., 1960. Chromatophore studies. VI. The behaviour of the melanophores in the developing limb buds of Discoglossus pictus. Bull. Res. Coun. Israel, **9B,** 24–34.

CAMBAR, R., 1957. Recherches descriptives et expérimentales sur l'involution du système pronéphrétique pendant la métamorphose des amphibiens anoures. Act. Soc. linn. Bordeaux, **97,** 33 pp.

CHACKO, T., 1965. The development and metamorphosis of the hyobranchial skeleton in Rana tigrina, the Indian bull frog. Acta zool. (Stockh.), **46,** 311–328.

CHIBON, P., 1960. Développement au laboratoire d'Eleutherodactylus martinicensis Tschudi, batracien anoure à développement direct. Bull. Soc. zool. Fr., **85,** 412–418.

————, 1962. Différenciation sexuelle de Eleutherodactylus martinicensis Tschudi, batracien anoure à développement direct. Bull. Soc. zool. Fr., **87,** 509–515.

CLAIRAMBAULT, P., 1963. Le tèlencéphale de Discoglossus pictus (Oth.). Etude anatomique chez le têtard et chez l'adulte. J. Hirnforsch., **6,** 87–121.

————, 1965. Le tèlencéphale du jeune têtard de Discoglossus pictus (Oth.). J. Hirnforsch., **7,** 499–512.

ETKIN, W., 1965. The phenomena of amphibian metamorphosis. IV. The development of the median eminence. J. Morph., **116,** 371–378.

FOX, H., 1962. Growth and degeneration of the pronephric system of Rana temporaria. J. Embryol. exp. Morph., **10,** 103–114.

————, 1963. The amphibian pronephros. Q. Rev. Biol., **38,** 1–25.

————, 1966. Thyroid growth and its relationship to metamorphosis in Rana temporaria. J. Embryol. exp. Morph., **16,** 487–496.

GAUDIN, A. J., 1965. Larval development of the tree frogs Hyla regilla and Hyla californiae. Herpetologica, **21,** 117–130.

GILLOIS, M. and A. BEAUMONT, 1964. Etude quantitative du pancréas d'Alytes obstetricans Laur. pendant la vie larvaire et la métamorphose. C. r. Séanc. Soc. Biol., **158,** 8.

GIPOULOUX, J.-D., 1962. Mise en évidence du "cytoplasme germinal" dans l'oeuf et l'embryon du discoglosse: Discoglossus pictus Otth. (Amphibien anoure). C. r. Séanc. Acad. Sci. (Paris), **254,** 2433–2435.

GRIFFITHS, I., 1959. The embryonic origin of the intrinsic limb musculature in Amphibia, Salientia. Experientia, **15,** 150.

HAMPTON, S. H. and E. P. VOLPE, 1963. Development and interpopulation variability of the mouthparts of Scaphiopus holbrooki. Am. Midl. Nat., **70,** 319–328.

HOSHINO, T., 1961. Histogenesis of the intestinal epithelium of anuran larvae (Bufo vulgaris japonicus) with reference to tissue iron. Okajimas Folia anat. jap., **37**, 105–121.

HUGHES, A., 1959. Studies in embryonic and larval development in Amphibia. I. The embryology of Eleutherodactylus ricordii, with special reference to the spinal cord. J. Embryol. exp. Morph., **7**, 22–38.

IWASAWA, H., 1965. Histological studies on the relation between a short larval period and thyroidal activity in the frog, Rana tagoi. Acta anat. nippon., **40**, 214–219 (Japanese, with English summary).

————, 1966. Comparative morphological studies of the thyroid glands in metamorphosing anuran larvae. Sci. Rep. Niigata Univ. (Ser. D), **3**, 9–17.

IZOARD, F., 1963. Formation et développement normaux du muscle gastrocnémien chez les amphibiens anoures. Bull. Soc. zool. Fr., **88**, 359–363.

KOBAYASHI, H., 1954. Hatching mechanism in the toad, Bufo vulgaris formosus. 1. Observations and some experiments on perforation of gelatinous envelope and hatching J. Fac. Sci. Tokyo Univ., **7**, 79–87.

KOLLROS, J. J. and V. M. McMURRAY, 1955. The mesencephalic V nucleus in anurans. I. Normal development in Rana pipiens. J. comp. Neurol., **102**, 47–64.

LAMOTTE, M. and various collaborators, ± 1954 and later. Contribution à l'étude des Batraciens de l'Ouest-africain. (A series of close to 20 papers, containing descriptions, mostly of larval and metamorphic stages, of a number of little-known African anuran species). Bull. Inst. fr. Afr. noire (IFAN) (sér. A).

LANOT, R., 1962. Evolution des arcs artériels postérieurs au cours de la métamorphose chez la Grenouille rousse (Rana temporaria L.). Bull. biol. Fr. Belg., **96**, 703–721.

LØVTRUP, S., 1965. Morphogenesis in the amphibian embryo. Gastrulation and neurulation. Acta Universitatis Gothoburgensis. Zoologica Gothoburgensia I, 1–139.

————, 1966. Morphogenesis in the amphibian embryo. Cell type distribution, germ layers, and fate maps. Acta zool. (Stockh.), **47**, 209–276.

LYNN, W. G. and A. M. PEADON, 1955. The role of the thyroid gland in direct development in the anuran, Eleutherodactylus martinicensis. Growth, **19**, 263–286.

MARINI, M., 1962. L'organo sottocommessurale durante lo sviluppo di un anfibio anuro (Hyla arborea). Rend. Accad. naz. Lincei (Ser. 8), **33**, 170–175.

MARTIN, S., 1959. Recherches descriptives et expérimentales sur les modalités du développement et étude de l'évolution du système pronéphrétique chez l'embryon et la larve du Crapaud accoucheur (Alytes obstetricans Laur.). Thesis, Bordeaux.

MATSUMOTO, T., 1959. Embryologische Studien über das Sinnesorgan der Seitenlinie bei den Amphibien. J. Kansai med. School, **11**, 392–418 (Japanese, with German summary).

MEDDA, J. and A. BOSE, 1965. Morphogenesis of the caudal skeletal elements of Anura with reference to localizations of alkaline phosphatase. Folia biol. (Kraków), **13**, 281–288.

MULHERKAR, L., K. V. RAO and G. V. SURYAVANSHI, 1964. A study of the development of lens in Microhyla ornata. J. Anim. Morph. Physiol., **11**, 45–50.

NAKAMURA, O., 1961. Presumptive rudiments of the endodermal organs on the surface of the anuran gastrula. Embryologia, **6**, 99–109.

OHARA, Y., 1956. The metamorphosis of the eye muscles in the common Japanese toad (Bufo vulgaris japonicus). Acta anat. nippon, **31**, 453–471 (Japanese, with English summary).

OLEDZKA-SŁOTWIŃSKA, H., 1963. Contribution à l'étude du développement de la muqueuse olfactive chez la salamandre (Salamandra salamandra L.) et la grenouille (Rana esculenta L.). Acta biol. cracov. (Ser. Zoologia), **6**, 79–81.

Padoa, E., 1959. L'origine dell'interrenale e della medulla dei corpi genitali nella "Rana esculenta". Monitore zool. ital., **66**, 21 pp. (German summary).

Pehlemann, F.-W., 1962. Experimentelle Untersuchungen zur Determination und Differenzierung der Hypophyse bei Anuren (Pelobates fuscus, Rana esculenta). Wilhelm Roux Arch. EntwMech. Org., **153**, 551–602.

Pinacci, R., 1960. L'andamento dell'attività mitotica nel mielencefalo di un Anfibio anuro durante lo sviluppo. Rend. Accad. naz. Lincei (Ser. 8), **29**, 150–153.

Robertson, D. R. and G. E. Swartz, 1964. The development of the ultimobranchial body in the frog Pseudacris nigrita triseriata. Trans. Am. microsc. Soc., **83**, 330–337.

Rossi, A., 1965. Sul valore prospettico degli strati esterno ed interno della placca neurale degli anfibi anuri (Rana esculenta, Bufo bufo e Bufo viridis). Boll. Zool., **32**, 991–1017.

Rostand, J., 1962. Sur l'ordre de sortie des membres antérieurs chez les amphibiens anoures. Archs Anat. Histol. Embryol. norm. expl., **44**, Suppl., 211–213.

Saxén, L., 1956. The initial formation and subsequent development of the double visual cells in Amphibia. J. Embryol. exp. Morph., **4**, 57–65.

Schreckenberg, M. G., 1956. The embryonic development of the thyroid gland in the frog, Hyla brunnea. Growth, **20**, 295–313.   \

Sedra, S. N. and M. I. Michael, 1958. The metamorphosis and growth of the hyobranchial apparatus of the Egyptian toad, Bufo regularis Reuss. J. Morph., **103**, 1–30.

———— and ————, 1959. The ontogenesis of the sound conducting apparatus of the Egyptian toad, Bufo regularis Reuss, with a review of this apparatus in Salientia. J. Morph., **104**, 359–376.

———— and A. A. Moursi, 1958. The ontogenesis of the vertebral column of Bufo regularis Reuss. Ceskoslovenská Morf., **6**, 7–32.

Shimozawa, A., 1959. Histological studies of corium of Rana catesbiana at metamorphosis and adult stage. Zool. Mag. (Tokyo), **68**, 297–299 (Japanese, with English summary).

————, 1961. Embryological studies on the corium of the larva of Rana catesbiana Zool. Mag. (Tokyo), **70**, 348–352 (Japanese, with English summary).

Smithberg, M., 1954. The origin and development of the tail in the frog, Rana pipiens. J. exp. Zool., **127**, 397–425.

Stokely, P. S. and J. C. List, 1955. Observations on the development of the anuran urostyle. Trans. Am. microsc. Soc., **74**, 112–115.

Tahara, Y. and O. Nakamura, 1961. Topography of the presumptive rudiments in the endoderm of the anuran neurula. J. Embryol. exp. Morph., **9**, 138–158.

Toney, M. E., 1954. Histology of the anterior pituitary gland of the frog during normal development. Growth, **18**, 215–225.

Triplett, E. L., 1958. The development of the sympathetic ganglia, sheath cells, and meninges in amphibians. J. exp. Zool., **138**, 283–312.

Vannini, E., 1956. Impostazione di nuove ricerche sullo sviluppo degli organi di Bidder e delle gonadi in girini di Bufo. Boll. Zool., **23**, 525–532.

———— and A. Sabbadin, 1954. The relation of the interrenal blastema to the origin of the somatic tissues of the gonad in frog tadpoles. J. Embryol. exp. Morph., **2**, 275 ff.

Volpe, E. P., 1957. The early development of Rana capito sevosa. Tulane Stud. Zool., **5**, 207–225.

———— and J. L. Dobie, 1959. The larva of the oak toad, Bufo quercicus Holbrook. Tulane Stud. Zool., **7**, 145–152.

VOLPE, E. P., and S. M. HARVEY, 1958. Hybridization and larval development in Rana palmipes Spix. Copeia, **1958,** 197–207.

————, M. A. WILKENS and J. L. DOBIE, 1961. Embryonic and larval development of Hyla avivoca. Copeia, **1961,** 340–349.

WITSCHI, E., 1955. The bronchial columella of the ear of larval Ranidae. J. Morph., **96,** 497–512.

WU HAO-LING and JU-CHI LI, 1964. The study of the development of Kaloula borealis (Barbour). III. The origin and changes of some of the main blood vessels. Acta zool. sin., **16,** 520–533 (Chinese, with English summary).

YANAI, T., 1958. Notes on the hatching glands of the frog, Polypedates buergeri. Annotnes zool. jap., **31,** 222–224.

———— and H. TAKAYANAGI, 1958. Notes on the hatching glands of the frog, Hyla arborea japonica. Annotnes zool. jap., **31,** 39–42.

NORMAL TABLES[1])

General

WITSCHI, E., 1953. Proposals for an international agreement on normal stages in vertebrate embryology. XIV Intern. Zool. Congr. Copenhagen, 1 p.

————, 1972, Equivalent numerical designations for staging systems: amphibians and fishes. In: "Biological Data Book" (ed. by P. L. Altman and D. S. Dittmer), 2nd ed., vol. **1,** Washington, D. C., Fed. of Am. Soc. for Exp. Biol., 173.

Anura

CAMBAR, R. and J.-D. GIPOULOUX, 1956. Table chronologique du développement embryonnaire et larvaire du crapaud commun: Bufo bufo L. Bull. biol. Fr. Belg., **90,** 198–217.

———— and S. MARTIN, 1959. Table chronologique du développement embryonnaire et larvaire du crapaud accoucheur (Alytes obstetricans Laur.). Act. Soc. linn. Bordeaux, **98,** 1–20.

GOSNER, K. L., 1960. A simplified table for staging anuran embryos and larvae with notes on identification. Herpetologica, **16,** 183–190.

HING, L. K., 1959. The breeding habits and development of Rana chalconata (Schleg.) (Amphibia). Treubia, **25,** 89.

JORQUERA, B. and L. IZQUIERDO, 1964. Tabla de desarrollo normal de Calyptocephalella gayi (Rana chilena). Biologica (Santiago), **36,** 43–53.

————, E. PUGIN and O. GOICOECHEA, 1972. Tabla de desarrollo normal de Rhinoderma darwini. Archos Med. vet., **4,** 1–15.

KHAN, M. S., 1965. A normal table of Bufo melanostictus Schneider. Biologia (Lahore), **11,** 1–39.

LIMBAUGH, B. A. and E. P. VOLPE, 1957. Early development of the Gulf coast toad, Bufo valliceps Wiegmann. Am. Mus. Novit., **1842,** 1–32.

MANELLI, H. and F. MARGARITORA, 1961. Tavole cronologiche dello sviluppo di Rana esculenta. Rend. Accad. naz. 40 (Ser. 4), **12,** 183–195.

MICHNIEWSKA-PREDYGIER, Z. and A. PIGON, 1957. Early developmental stages of Rana temporaria L., R. terrestris Andrz., R. esculenta L. and Bufo bufo (L.). Studia Soc. Sci. Torun., **3,** 147–157 (Polish, with brief English summary).

[1]) updated until 1974.

RAMASWAMI, L. S. and A. B. LAKSHMAN, 1959. The skipper-frog as a suitable embryological animal and an account of the action of mammalian hormones on spawning the same. Proc. nat. Inst. Sci. India, **25**, B, 68–79 (Contains a Normal Table of Rana cyanophlyctis).

ROSSI, A., 1959. Tavole cronologiche dello sviluppo embrionale e larvale del Bufo bufo (L.). Monitore zool. ital., **66**, 17 pp.

SCHREIBER, G., 1937. La definizione degli stadi della metamorfosi del Bufo. Rend. accad. naz. Lincei (Ser. 6), **25**, 342–348.

SEDRA, S. N. and M. I. MICHAEL, 1961. Normal table of the egyptian toad, Bufo regularis Reuss, with an addendum on the standardization of the stages considered in previous publications. Ceskoslovenská Morf., **9**, 333–351.

SIBOULET, R., 1970. Table chronologique du développement embryonnaire et larvaire du crapaud de Maurétanie Bufo mauretanicus Schlegel, 1841 à différentes températures. Vie Milieu, Sér. C., **21**, 179–198.

TAHARA, Y., 1959. Table of the normal developmental stages of the frog, Rana japonica.

I. Early development (stages 1–25). Jap. J. exp. Morph., **13**, 49–60 (Japanese, with English summary).

WITSCHI, E., 1956. Standard stages of frog development (Rana pipiens). In: "Development of Vertebrates" (by E. Witschi), Philadelphia, W. B. Saunders, 79–81.

Urodela

FANKHAUSER, G., 1967. Examples of corresponding stages in normal development of five species of urodeles. In: "Methods in Developmental Biology" (ed. by F. H. Wilt and N. K. Wessells), New York, Crowell, 89.

GALLIEN, L. and O. BIDAUD, 1959. Table chronologique du développement chez "Triturus helveticus" Razoumowsky. Bull. Soc. zool. Fr., **84**, 22–32.

———— and M. DUROCHER, 1957. Table chronologique du développement chez Pleurodeles waltlii Michah. Bull. biol. Fr. Belg., **91**, 97–114.

HARRISON, R. G., 1969. Harrison stages and description of the normal development of the spotted salamander, Amblystoma punctatum (Linn.). In: "Organization and Development of the Embryo" (by R. G. Harrison; ed. by S. Wilens), New Haven, etc., Yale University Press, 44–66.

IWAMA, H., 1968. Normal table of Megalobatrachus japonicus. (ed. by T. Kajishima and K. Takata), Nagoya, R. Iwana, 28 pp. (Japanese).

SAWANO, J., 1947. Normal table of Hynobius lichenatus Boulenger. Sapporo, Suguru-Bunko, 24 pp. (Japanese).

USUI, M. and M. HAMASAKI, 1939. Tafeln zur Entwicklungsgeschichte von Hynobius nigrescens Stejneger. Zool. Mag. (Tokyo), **51**, 195–206 (Japanese).

VANDEL, A., J. DURAND and M. BOUILLON, 1966. Contribution à l'étude du développement de Proteus anguinus Laurenti (Batraciens, Urodèles). Annls Spéléol., **21**, 609–619.

CHAPTER X

INDEX

In this chapter a *general alphabetical index* is given as well as a separate list of *anatomical descriptions* of *organ systems* (in systematical order) and a separate *systematical index* to the *development during metamorphosis*.

ALPHABETICAL INDEX

ANATOMICAL DESCRIPTION OF ORGAN SYSTEMS

(systematical order)

DEVELOPMENT DURING METAMORPHOSIS

(systematical order)

* *
*

NOTE TO READERS – In the original book, the following pages were fold-out pages allowing the reader to view the image page while reading the relevant text in the book. It was not possible to maintain this format when the book was made Print on Demand. As a result, we recommend that you make a copy of the following pages so that you will be able to view the images alongside the text as originally intended. We appreciate your understanding.

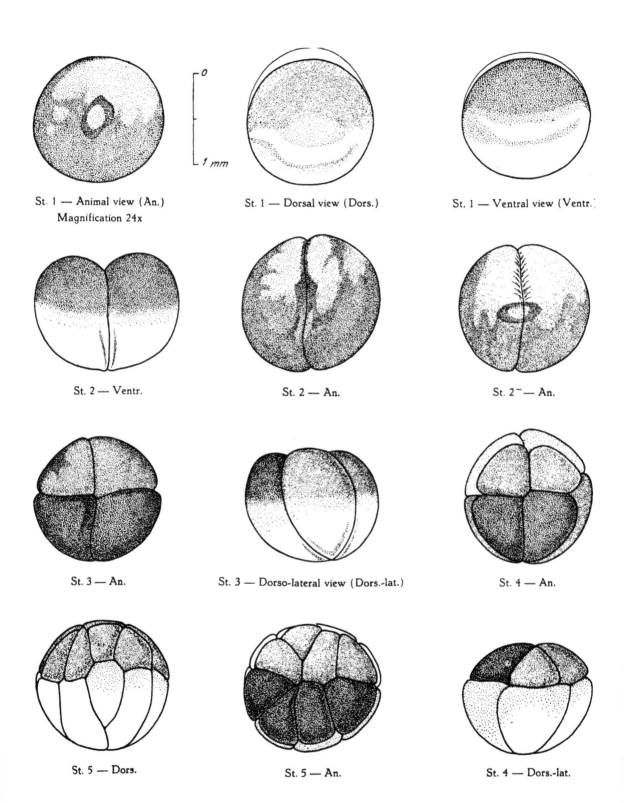

St. 1 — Animal view (An.)
Magnification 24x

St. 1 — Dorsal view (Dors.)

St. 1 — Ventral view (Ventr.)

St. 2 — Ventr.

St. 2 — An.

St. 2⁻— An.

St. 3 — An.

St. 3 — Dorso-lateral view (Dors.-lat.)

St. 4 — An.

St. 5 — Dors.

St. 5 — An.

St. 4 — Dors.-lat.

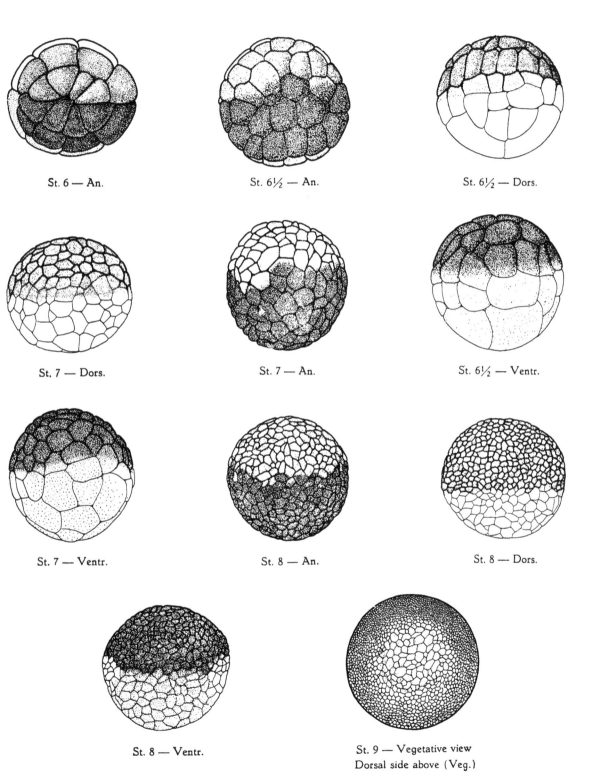

St. 6 — An.

St. 6½ — An.

St. 6½ — Dors.

St. 7 — Dors.

St. 7 — An.

St. 6½ — Ventr.

St. 7 — Ventr.

St. 8 — An.

St. 8 — Dors.

St. 8 — Ventr.

St. 9 — Vegetative view
Dorsal side above (Veg.)

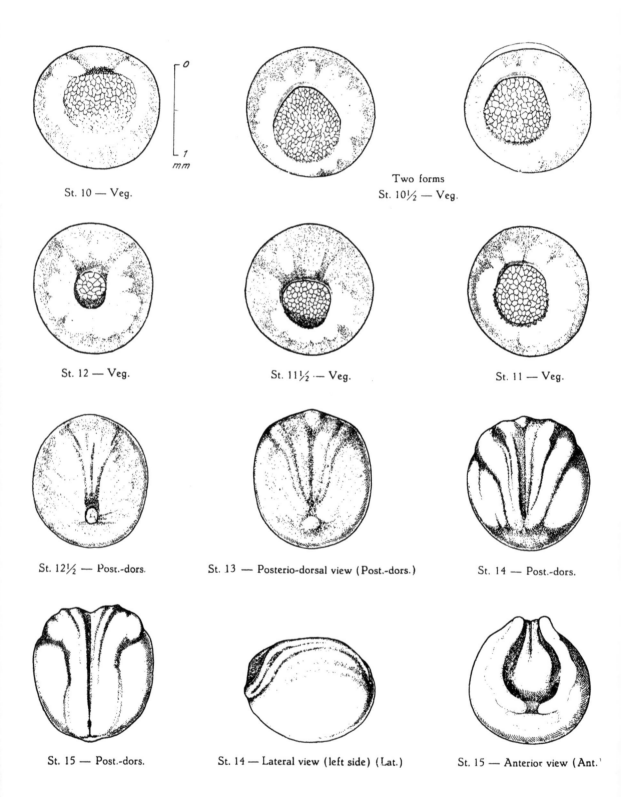

St. 10 — Veg.

Two forms
St. 10½ — Veg.

St. 12 — Veg.

St. 11½ — Veg.

St. 11 — Veg.

St. 12½ — Post.-dors.

St. 13 — Posterio-dorsal view (Post.-dors.)

St. 14 — Post.-dors.

St. 15 — Post.-dors.

St. 14 — Lateral view (left side) (Lat.)

St. 15 — Anterior view (Ant.

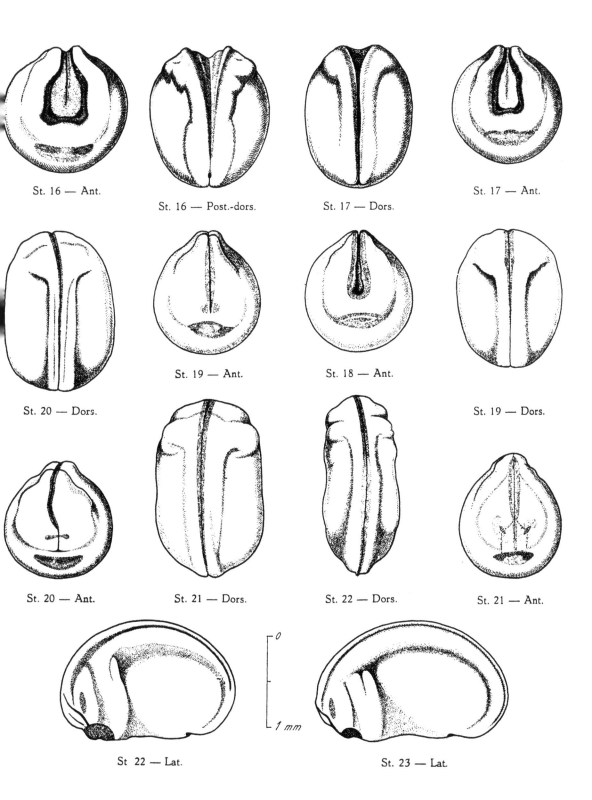

St. 16 — Ant.

St. 16 — Post.-dors.

St. 17 — Dors.

St. 17 — Ant.

St. 20 — Dors.

St. 19 — Ant.

St. 18 — Ant.

St. 19 — Dors.

St. 20 — Ant.

St. 21 — Dors.

St. 22 — Dors.

St. 21 — Ant.

0

1 mm

St 22 — Lat.

St. 23 — Lat.

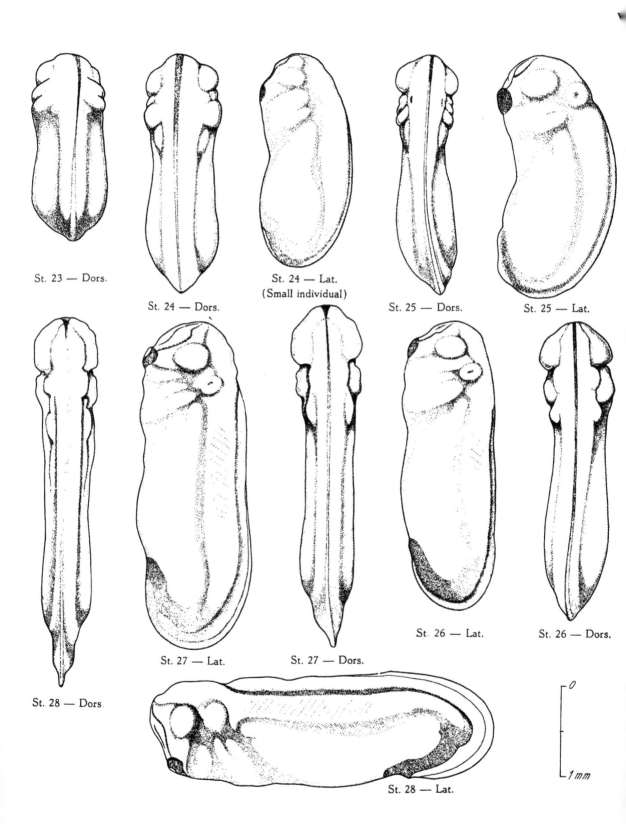

St. 23 — Dors.

St. 24 — Dors.

St. 24 — Lat.
(Small individual)

St. 25 — Dors.

St. 25 — Lat.

St. 28 — Dors.

St. 27 — Lat.

St. 27 — Dors.

St. 26 — Lat.

St. 26 — Dors.

St. 28 — Lat.

1 mm

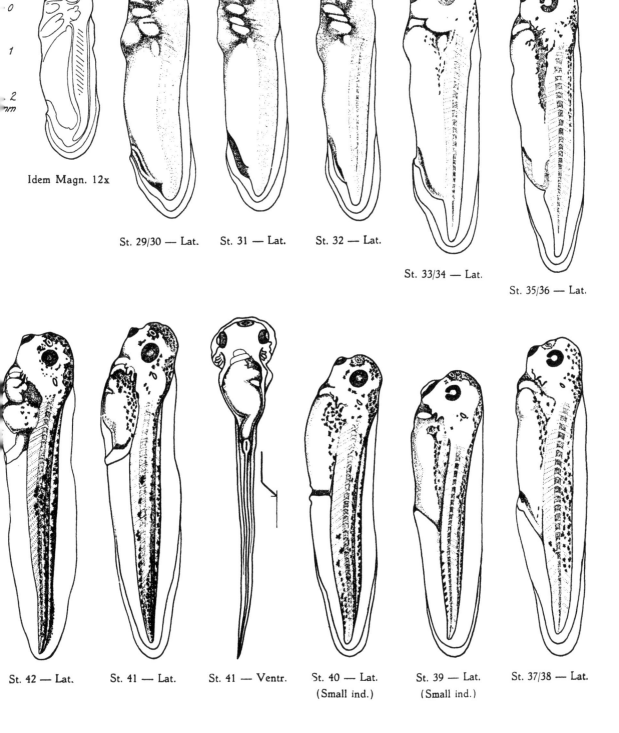

St. 29/30 — Lat. St. 31 — Lat. St. 32 — Lat.

St. 33/34 — Lat.

St. 35/36 — Lat.

Idem Magn. 12x

St. 42 — Lat. St. 41 — Lat. St. 41 — Ventr. St. 40 — Lat. St. 39 — Lat. St. 37/38 — Lat.
(Small ind.) (Small ind.)

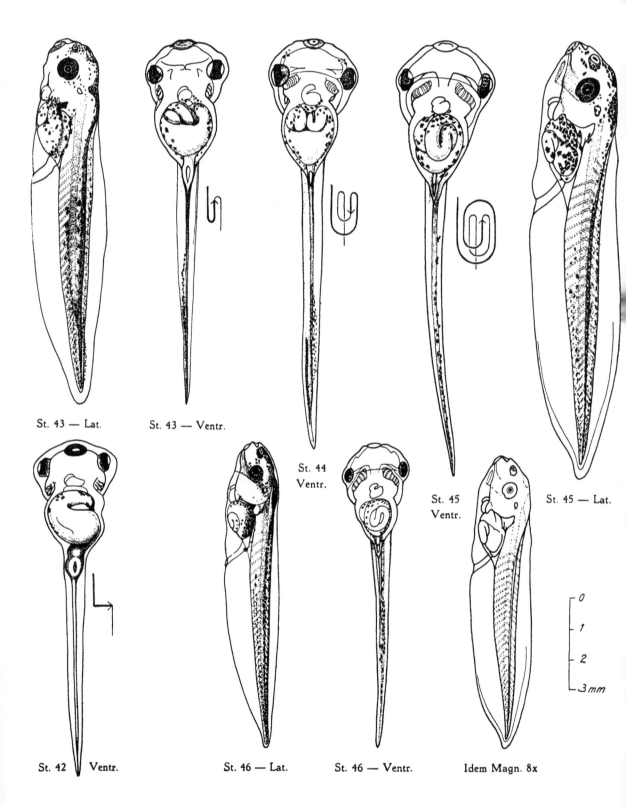

St. 43 — Lat.

St. 43 — Ventr.

St. 44
Ventr.

St. 45
Ventr.

St. 45 — Lat.

St. 42   Ventr.

St. 46 — Lat.

St. 46 — Ventr.

Idem Magn. 8x

0
1
2
3mm

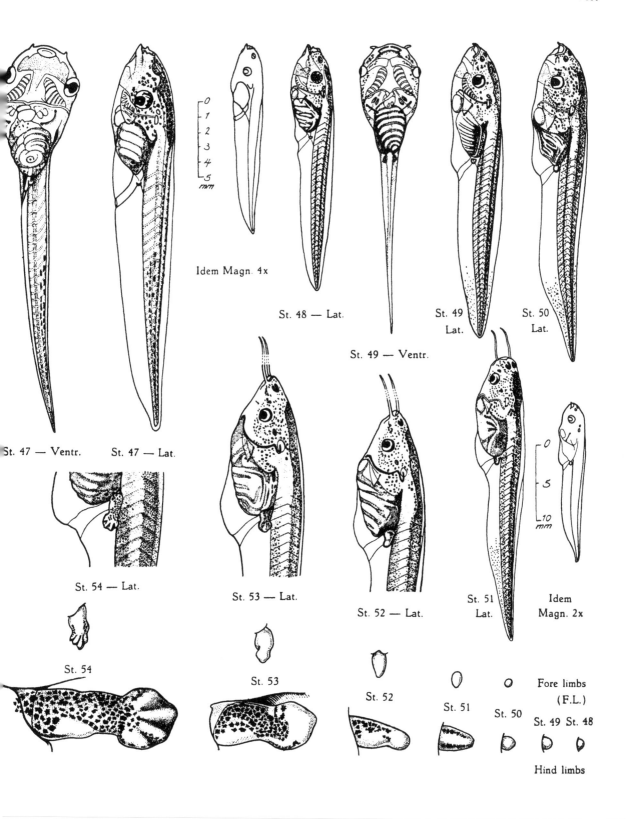

Idem Magn. 4x

St. 48 — Lat.

St. 49 — Ventr.

St. 49
Lat.

St. 50
Lat.

St. 47 — Ventr.   St. 47 — Lat.

St. 54 — Lat.

St. 53 — Lat.

St. 52 — Lat.

St. 51
Lat.

Idem
Magn. 2x

St. 54

St. 53

St. 52

St. 51

St. 50

Fore limbs
(F.L.)

St. 49   St. 48

Hind limbs

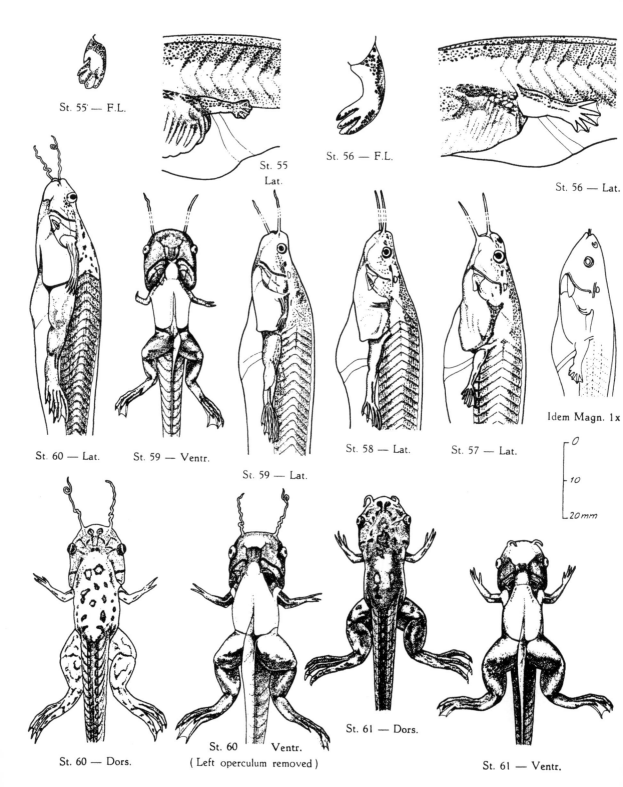

St. 55 — F.L.

St. 55
Lat.

St. 56 — F.L.

St. 56 — Lat.

St. 60 — Lat.

St. 59 — Ventr.

St. 59 — Lat.

St. 58 — Lat.

St. 57 — Lat.

Idem Magn. 1x

$$\begin{array}{l} 0 \\ 10 \\ 20\,mm \end{array}$$

St. 60 — Dors.

St. 60   Ventr.
(Left operculum removed)

St. 61 — Dors.

St. 61 — Ventr.

X

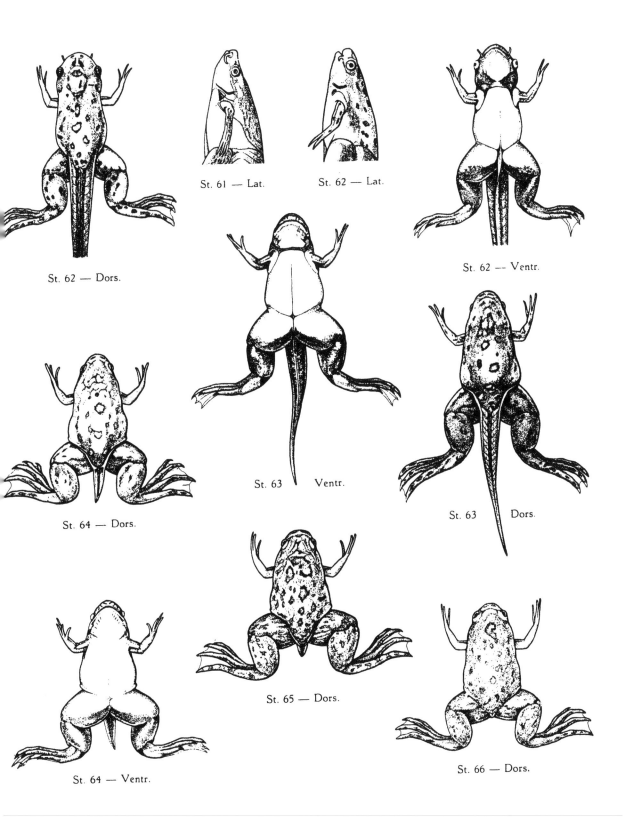

St. 61 — Lat.

St. 62 — Lat.

St. 62 — Dors.

St. 62 — Ventr.

St. 63 — Ventr.

St. 64 — Dors.

St. 63 — Dors.

St. 65 — Dors.

St. 64 — Ventr.

St. 66 — Dors.

For Product Safety Concerns and Information please contact our EU
representative GPSR@taylorandfrancis.com
Taylor & Francis Verlag GmbH, Kaufingerstraße 24, 80331 München, Germany

www.ingramcontent.com/pod-product-compliance
Ingram Content Group UK Ltd.
Pitfield, Milton Keynes, MK11 3LW, UK
UKHW051834180425
457613UK00022B/1244